MW00777703

MATHEMATICS FOR ECONOMISTS MADE SIMPLE

Viatcheslav V. Vinogradov

CHARLES UNIVERSITY IN PRAGUE
KAROLINUM PRESS
2010

Scientific reviewers:
Prof. Jan Kmenta, PhD. (University of Michigan and CERGE–EI, Prague)
Prof. Elena Kustova, PhD. (Saint Petersburg State University)

*This publication was kindly supported by the foundation
Nadace CERGE–EI, Prague, grant nr. NIF2004/04.*

ISBN 978-80-246-1657-5

TO MY TEACHERS

Contents

He liked those literary cooks
Who skim the cream of others' books
And ruin half an author's graces
By plucking bon-mots from their places.

Hannah More, *Florio (1786)*

Introduction

This textbook is based on an extended collection of handouts I distributed since 1999 to the graduate students in economics attending my summer mathematics class at the Center for Economic Research and Graduate Education (CERGE) at Charles University in Prague. A pilot version of the project entitled "A Cook-Book of Mathematic" opened CERGE Lecture Notes series in 1999.

Two considerations motivated me to write this book. First, I wanted to write a short textbook, which could be covered in the course of two months and which, in turn, covers the most significant issues of mathematical economics. I have attempted to maintain a balance between being overly detailed and overly schematic. Therefore this text should resemble (in the 'ideological' sense) a "hybrid" of Chiang's classic textbook *Fundamental Methods of Mathematical Economics* and the comprehensive reference manual by Berck, Strom and Sydsæter (exact references appear at the end of this section).

My second objective in writing this text was to provide my students with simple "cook-book" recipes for solving problems they might face in their studies of economics. Since the target audience was supposed to have some mathematical background (admittance to the program requires at least BA level mathematics), my main goal was to refresh students' knowledge of mathematics rather than teach them math 'from scratch'. Students were expected to be familiar with the basics of set theory, the real-number system, the concept of a function, polynomial, rational, exponential and logarithmic functions, inequalities and absolute values. In general, the text anticipates the development of micro- and macroeconomics curricula at CERGE and includes topics varying from matrix algebra and calculus to static optimization to differential and difference equations, and optimization in continuous time. Optimization in discrete time (dynamic programming) is a central pillar of the first year of graduate macroeconomic curriculum and as such is largely left out of this book.

Bearing in mind the applied nature of the course, I usually refrained from presenting complete proofs of theoretical statements. Instead, I chose to allocate more time and space to examples of problems and their solutions and economic applications. I strongly believe that for students in economics -- for whom this text is meant -- the *application* of mathematics in their studies takes precedence over *das Glasperlenspiel* of abstract theoretical constructions.

Mathematics is an ancient science and, therefore, it is little wonder that these notes may remind the reader of other textbooks which have already been written and published. To be candid, I did not intend to be entirely original, since that would be impossible. On the contrary, I tried to benefit from books already in existence and adapted some interesting examples, problems, and worthy pieces of theory presented there. If the reader requires further proofs or more detailed discussion, I have included a useful, but hardly exhaustive, reference guide at the end of each section.

With very few exceptions, the analysis is limited to the case of real numbers, the theory of complex numbers being beyond the scope of these notes.

Finally, I would like to express my deep gratitude to my wife Natalia Churikova for her advice, to Professor Jan Kmenta for his valuable comments and suggestions, to my colleagues for their interest and stimulating discussion, and in particular to Professor Sergey Slobodyan for sharing his excellent lecture notes on macroeconomics with me, to my teaching and research assistants for their help in preparing the Cookbook to print, to CERGE and the Economics Institute of the Czech Academy of Sciences (EI) management for publication support, to the CERGE-EI English Department for editorial assistance, and, last but not least, to all students who inspired me to write this book and provided very helpful feedback on earlier versions of the text.

All remaining mistakes and misprints are solely mine.

I wish you success in your mathematical kitchen! *Bon Appetit!*

Supplementary Reading (General)

- Arrow, K. and Intriligator, M. (Eds.). (1987). *Handbook of Mathematical Economics* (Vols. 1–4). New York: North Holland.

- Berck P., Strom, A., and Sydsæter, K. (2005). *Economist's Mathematical Manual* (4th ed.). Berlin: Springer.

- Chiang, A. (2005). *Fundamental Methods of Mathematical Economics* (4th ed.). New York, NY: McGraw-Hill.

- Debreu, G. (1989). *Mathematical Economics: Twenty Papers by Gerard Debreu.* London: Cambridge University Press.

- Dixit, A. K. (1990). *Optimization in Economic Theory* (2nd ed.). Oxford: Oxford University Press.

- Leonard D. and Ngo, V. L. (1992). *Optimal Control Theory and Static Optimization in Economics.* Cambridge: Cambridge University Press.

- Ostaszewski, I. (1993). *Mathematics in Economics: Models and Methods.* Oxford: Blackwell.

- Samuelson, P. (1947). *Foundations of Economic Analysis.* Cambridge: Harvard University Press.

- Schofield, N. (2004). *Mathematical Methods in Economics and Social Choice.* Springer.

- Sydsæter, K. (1981). *Topics in Mathematical Analysis for Economists.* London: Academic Press.

- Sydsæter, K., Hammond, P., Seierstad, A., and Strom, A. (2005). *Further Mathematics for Economic Analysis* (2nd ed.). Verlag: Prentice Hall.

- Silberberg, E. (2000). *The Structure of Economics: A Mathematical Analysis* (3rd ed.). New York, NY: McGraw-Hill.

- Simon C. and Blume, L. (1994). *Mathematics for Economists.* Norton.

- Solow D. (2010). *How to Read and Do Proofs: An Introduction to Mathematical Thought Process* (5th ed.). New York, NY: John Wiley.

- Takayama, A. (1985). *Mathematical Economics* (2nd ed.). Cambridge: Cambridge University Press.

- Takayama, A. (1994). *Analytical Methods in Economics.* Ann Arbor: University of Michigan Press.

- Weintraub, E. R. (2002) *How Economics Became a Mathematical Science* Durham, NC: Duke University Press.

- Yamane, T. (1968). *Mathematics for Economists* (2nd ed.). Englewood Cliffs, N.J. : Prentice-Hall.

Preface

Professor Vinogradov has tragically died before this book was put into its final form. He has left hand-written notes which I attempted to incorporate as much as possible while typesetting the project for printing. Otherwise, I did not add any new material and left the book in a somewhat unfinished state in which I discovered it. In particular, there is no index.

July 2010   Sergey Slobodyan, CERGE–EI

Chapter 0

Preliminaries

0.1 Basic Mathematical Notation

Statements: $A, B, C, \ldots$

True/False: all statements are either true or false.

Negation: $\neg A$ 'not A'

Conjunction: $A \wedge B$ 'A and B'

Disjunction: $A \vee B$ 'A or B'

Implication: $A \Rightarrow B$ 'A implies B'

(A is a sufficient condition for B; B is a necessary condition for A.)

Equivalence: $A \Leftrightarrow B$ 'A if and only if B' (A iff B, for short)

(A is necessary and sufficient for B; B is necessary and sufficient for A.)

Example 1 $(\neg A) \wedge A \Leftrightarrow FALSE.$

$(\neg(A \vee B)) \Leftrightarrow ((\neg A) \wedge (\neg B))$ *(De Morgan rule)*.

Quantifiers:

Existential: $\exists$ 'There exists' or 'There is'

Universal: $\forall$ 'For all' or 'For every'

Uniqueness: ∃!   'There exists a unique ...' or
                               'There is a unique...'

The colon : and the vertical line | are widely used as abbreviations for 'such that'

$a \in \mathcal{S}$ means 'a is an element of (belongs to) set $\mathcal{S}$'

Example 2 (Definition of continuity)

$$f \text{ is continuous at } x \text{ if}$$

$$((\forall \epsilon > 0)(\exists \delta > 0) : (\forall y \in |y - x| < \delta \Rightarrow |f(y) - f(x)| < \epsilon)$$

Set Operations:

∪   Union of sets

∩   Intersection of sets

\   Difference of sets

∅   Empty set

$\mathcal{A} \subseteq \mathcal{B}$ or $\mathcal{A} \subset \mathcal{B}$ Set $\mathcal{A}$ is a subset of set $\mathcal{B}$ (some authors use $\subseteq$ sign to stress that $\mathcal{A}$ can be equal to $\mathcal{B}$)

$2^{\mathcal{A}}$ The set of all subsets of $\mathcal{A}$

$\bar{\mathcal{A}}$ Closure of set $\mathcal{A}$

In this text, optional information which might be helpful is typeset in footnotesize; sans serif font is used for emphasis in definitions and recipes.

The symbol ⚠ is used to draw the reader's attention to potential pitfalls.

0.2   Methods of Mathematical Proof

Among all, the most common methods of proof are as follows:

- Working forward

- Working backward

- Proof by contradiction

- Proof by contrapositive

- Direct proof

- Proof by definition

- Proof by induction

The examples below illustrate the usage of these methods.

Example 3 *Show that for any non-negative real numbers a and b,
$(a + b)/2 \geq \sqrt{ab}$*

Proof (working forward):
$(\sqrt{a} - \sqrt{b})^2 \geq 0$ *is a true statement* $\Rightarrow a - 2\sqrt{a}\sqrt{b} + b \geq 0 \Rightarrow$
$a + b \geq 2\sqrt{a}\sqrt{b} \Rightarrow (a + b)/2 \geq \sqrt{ab}$ *Q.E.D.*

Proof (working backwards):
Statement that has to be proven $(a + b)/2 \geq \sqrt{ab} \Leftrightarrow (a + b)^2 \geq 4ab$
$\Leftrightarrow a^2 + 2ab + b^2 \geq 4ab \Leftrightarrow a^2 - 2ab + b^2 \geq 0 \Leftrightarrow (a - b)^2 \geq 0$, *that is true.*

Example 4 *If a real number r is such that $r^2 = 2$ then r is not rational.*

Proof by contradiction:
Assume $r^2 = 2$ and r is rational $\Rightarrow r = n/m$, *where m and n have
no common divisors* $\Rightarrow \frac{n^2}{m^2} = 2 \Rightarrow n^2 = 2m^2$, *so n is even.* $\Rightarrow n = 2k$
$\Rightarrow 4k^2 = 2m^2 \Rightarrow m^2 = 2k^2$, *so m is even* $\Rightarrow m = 2l \Rightarrow m$ *and n have
2 as a common divisor, a contradiction to the assumption.*

Example 5 *Prove that $2^n - 1$ is prime $\Rightarrow n$ is prime.*

Proof by contrapositive:
We need to show that n is not prime $\Rightarrow 2^n - 1$ *is not prime.*
*Indeed, if n is not prime then $n = km$ for $1 < k$ and $1 < m \Rightarrow$
$2^{km} - 1 = (2^k)^m - 1^m \Rightarrow$ using a well-known expansion $a^m - b^m =$*

$(a-b)(a^{m-1}+a^{m-2}b+\cdots+ab^{m-2}+b^{m-1})$, $2^{km}-1=(2^k-1)(2^{k(m-1)}+2^{k(m-2)}+\cdots+2^k+1)$ $\Rightarrow$ $2^n-1=k'm'$ where $1<k'$ and $1<m'$ $\Rightarrow$ 2^n-1 is not prime.

Example 6 $m,\ n$ even implies that $m+n$ is even.

Direct proof:

m and n are even $\Rightarrow$ $m=2m'$ and $n=2n'$ $\Rightarrow$ $m+n=2m'+2n'=2(m'+n')$, thus $m=n$ is even.

Example 7 Show that $f(x)=x^2$ is a convex function.

Proof by definition:

Recall the definition of convexity: f is a convex function if and only if

$$(\forall x)(\forall y)(\forall t\in[0,1]):f(tx+(1-t)y)\le tf(x)+(1-t)f(y)$$

Since $f(tx+(1-t)y)=(tx+(1-t)y)^2=t^2x^2+2t(1-t)xy+(1-t)^2y^2$ and $tf(x)+(1-t)f(y)=tx^2+(1-t)y^2$, we evaluate $f(tx+(1-t)y)-(tf(x)+(1-t)f(y))==-t(1-t)x^2+2t(1-t)xy-(1-t)ty^2=-t(1-t)(x^2-2xy+y^2)=-t(1-t)(x-y)^2\le 0$ that completes the proof.

Method of **induction** can be refined as follows:

Given a sequence of statements $\{A_n\}$, prove that they are all true in two steps:

Step 1: Prove that A_1 is true.

Step 2: Prove that $(\forall n:A_n\Rightarrow A_{n+1})$ is true.

Example 8 Show that $n!\le n^n$ for all $n\ge 1$.

Proof by induction:

Step 1: A_1 is true since $1!=1\le 1=1^1$

Step 2: If $A_n:n!\le n^n$ is true then $(n+1)(n!)\le(n+1)n^n\le(n+1)(n+1)^n$ $\Rightarrow$ $(n+1)!\le(n+1)^{n+1}$, indicating that A_{n+1} is also true.

0.3   Powers, Exponents, Logarithms and Complex Numbers

- Powers: for any real parameters p and q and positive real variables x and y,

  1. $x^p \cdot x^q = x^{p+q}, \quad \frac{x^p}{x^q} = x^{p-q}$

  2. $(x^p)^q = x^{pq}$

  3. $(xy)^p = x^p x^p, \quad \left(\frac{x}{y}\right)^p = \frac{x^p}{y^p}$

  4. $x^{-p} = \frac{1}{x^p}, \quad x^{\frac{1}{p}} = \sqrt[p]{x}$

- Exponents: similarly for powers, for any positive real parameters a and b and real variables x and y,

  1. $a^x \cdot a^y = a^{x+y}, \quad \frac{a^x}{a^y} = a^{x-y}$

  2. $(a^x)^y = a^{xy}$

  3. $(ab)^x = a^x b^x, \quad \left(\frac{a}{b}\right)^x = \frac{a^x}{b^x}$

  4. $a^{-x} = \frac{1}{a^x}$

- Logarithms

  1. The natural logarithm function $\ln(x)$, that is the logarithm to the base e, is defined through the identity $e^{\ln(x)} = x$, where the number e is $e = \lim_{n \to \infty} \left(1 + \frac{1}{n}\right)^n = 2.7182818284590452\ldots$

  2. $\ln(xy) = \ln(x) + \ln(y)$,
     $\ln(\frac{x}{y}) = \ln(x) - \ln(y)$,
     $\ln(x^p) = p\ln(x)$,
     $\ln(\frac{1}{x}) = -\ln(x)$

  3. The logarithm to the base a ($a > 0$, $a \neq 1$) is the function $a^{\log_a(x)} = x$

  4. For any a, $b > 0$, $a \neq 1$, $b \neq 1$, $\log_a(b) \cdot \log_b(a) = 1$

  5. $\log_a(x) = \frac{\ln(x)}{\ln(a)}, \quad \log_a(x) = \log_a(e) \cdot \ln(x)$

- Complex numbers

1. A complex number z is a linear combination $z = a + ib$, where a and b are real numbers, and i is such that $i^2 = -1$, or $i = \sqrt{-1}$ (i is often called the *imaginary unit*). The real and imaginary parts of z are denoted $Re(z)$ and $Im(z)$, respectively and defined as $Re(z) = a$, $Im(z) = b$.

   A complex number can be thought of as a point with Cartesian coordinates (a, b) in the Real–Imaginary plane.

   Alternatively, a complex number can be presented in the *trigonometric* or *polar* form: $z = a+ib = \rho(\cos(\theta)+i\sin(\theta)) = \rho e^{i\theta}$, where $\rho = \sqrt{a^2 + b^2} = |z|$, $\cos(\theta) = a/\rho$, $\sin(\theta) = b/\rho$.

   The pair (ρ, θ) can be understood as polar coordinates of point z in the Real–Imaginary plane.

2. A *conjugate* of a complex number $z = a+ib$ is $\bar{z} = a-ib$. One useful property of conjugates is that for any two complex numbers z_1 and z_2, $\overline{z_1 + z_2} = \bar{z}_1 + \bar{z}_2$.

3. The *modulus* of a complex number $z = a + ib$ is $|z| = \sqrt{a^2 + b^2}$. Modulus has the following properties: $|z| = |\bar{z}|$, $z \cdot \bar{z} = |z|^2$, and for any two complex numbers z_1 and z_2, $|z_1 \cdot z_2| = |z_1| \cdot |z_2|$. Furthermore, the triangle inequality holds: $|z_1 + z_2| \le |z_1| + |z_2|$.

4. Algebraic operations with complex numbers are defined in a straightforward way (note that complex numbers form a *field*): Given two complex numbers $z_1 = a_1 + ib_1$ and $z_2 = a_2 + ib_2$,

$$z_1 + z_2 = (a_1 + a_2) + i(b_1 + b_2)$$

$$z_1 - z_2 = (a_1 - a_2) - i(b_1 + b_2)$$

$$z_1 \cdot z_2 = (a_1 a_2 - b_1 b_2) + i(a_1 b_2 + b_1 a_2)$$

$$\frac{z_1}{z_2} = \frac{z_1 \cdot \bar{z}_2}{z_2 \cdot \bar{z}_2} = \frac{z_1 \bar{z}_2}{|z_2|^2} = \frac{(a_1 a_2 + b_1 b_2) = i(b_1 a_2 - a_1 b_2)}{a_2^2 + b_2^2}$$

In the trigonometric form, with $z_1 = \rho_1(\cos(\theta_1)+i\sin(\theta_1)) = \rho_1 e^{i\theta_1}$ and $z_2 = \rho_2(\cos(\theta_2) + i\sin(\theta_2)) = \rho_1 e^{i\theta_1}$, the multiplication and division formulas become

$$
\begin{aligned}
z_1 \cdot z_2 &= \rho_1 e^{i\theta_1} \cdot \rho_2 e^{i\theta_2} = \rho_1 \rho_2 e^{i(\theta_1 + \theta_2)} \\
&= \rho_1 \rho_2 (\cos(\theta_1 + \theta + 2) + i\sin(\theta_1 + \theta_2))
\end{aligned}
$$

$$
\frac{z_1}{z_2} = \frac{\rho_1 e^{i\theta_1}}{\rho_2 e^{i\theta_2}} = \frac{\rho_1}{\rho_2}(\cos(\theta_1 - \theta_2) + i\sin(\theta_1 - \theta_2))
$$

It is also convenient to use the trigonometric representation of a complex number $z = \rho(\cos(\theta) + i\sin(\theta)) = \rho e^{i\theta}$ to define roots and powers:

$$
\begin{aligned}
z^n &= \rho^n (\cos(\theta) + i\sin(\theta))^n \\
&= \rho^n e^{ni\theta} = \rho^n (\cos(n\theta) + i\sin(n\theta))
\end{aligned}
$$

In particular, if $\rho = 1$ then for any integer n,

$$
(\cos(\theta) + i\sin(\theta))^n = \cos(n\theta) + i\sin(n\theta)
$$

(De Moivre's formula)

Further, the algebraic equation $z_n = c$ (where $c = r(\cos(t) + i\sin(t))$ is a given complex constant) has exactly n roots over the field of complex numbers, that are

$$
z_k = \sqrt[n]{r}\left(\cos\left(\frac{t + 2k\pi}{n}\right) + i\sin\left(\frac{t + 2k\pi}{n}\right)\right)
$$

5. The complex exponential function e^ζ of a complex argument $\zeta = \alpha + i\beta$ is defined as $e^\zeta = e^{\alpha + i\beta} = e^\alpha \cdot e^{i\beta} = e^\alpha(\cos(\beta) + i\sin(\beta))$. This definition yields a number of useful corollaries:

$$
e^{ik\pi} = (-1)^k, \qquad \text{in particular, } e^{i\pi} = -1
$$

$$
e^{\bar{\zeta}} = \overline{e^\zeta}, \qquad e^{\zeta + i2\pi} = e^\zeta
$$

$$
e^{\zeta_1 + \zeta_2} = e^{\zeta_1} \cdot e^{\zeta_2}, \qquad e^{\zeta_1 - \zeta_2} = \frac{e^{\zeta_1}}{e^{\zeta_2}}
$$

Finally, for any real γ,

$$
\cos\gamma = \frac{e^{i\gamma} + e^{-i\gamma}}{2}, \qquad \sin(\gamma) = \frac{e^{i\gamma} - e^{-i\gamma}}{2i}
$$

(Euler's formulas)

Further Reading

- Cupillari, A. (2005). *The Nuts and Bolts of Proofs* (3rd ed.). Boston: Elsevier Academic Press.

- Korn, G. A., and Korn, T. M. (2000). *Mathematical Handbook for Scientists and Engineers* (2nd ed.). Mineola, NY: Dover Publications.

- Solow, D. (2005). *How to Read and Do Proofs: An Introduction to Mathematical Thought Processes* (4th ed.). Hoboken, N.J.: John Wiley.

Chapter 1

Linear Algebra

1.1 Matrix Algebra

Definition 1 *An $m \times n$ matrix is a rectangular array of real numbers with m rows and n columns.*

$$A = \begin{pmatrix} a_{11} & a_{12} & \cdots & a_{1n} \\ a_{21} & a_{22} & \cdots & a_{2n} \\ \vdots & \vdots & & \vdots \\ a_{m1} & a_{m2} & \cdots & a_{mn} \end{pmatrix},$$

$m \times n$ is called the *dimension* or *order* of A. If $m = n$, the matrix is the *square* of order n.

A subscripted element of a matrix is always read as $a_{\text{row,column}}$ ⚠

A shorthand notation is $A = (a_{ij})$, $i = 1, 2, \ldots, m$ and $j = 1, 2, \ldots, n$, or $A = (a_{ij})_{[m \times n]}$.

A vector is a special case of a matrix, when either $m = 1$ (row vectors $v = (v_1, v_2, \ldots, v_n)$) or $n = 1$ – column vectors

$$u = \begin{pmatrix} u_1 \\ u_2 \\ \vdots \\ u_m \end{pmatrix}.$$

Matrix Operations

- Addition and subtraction: given $A = (a_{ij})_{[m \times n]}$ and $B = (b_{ij})_{[m \times n]}$

$$A \pm B = (a_{ij} \pm b_{ij})_{[m \times n]},$$

i.e. we simply add or subtract corresponding elements.

Note that these operations are defined only if A and B are of the same dimension.

- Scalar multiplication:

$$\lambda A = (\lambda a_{ij}), \qquad \text{where } \lambda \in \mathbf{R},$$

i.e. each element of A is multiplied by the same scalar λ.

- Matrix multiplication: if $A = (a_{ij})_{[m\times n]}$ and $B = (a_{ij})_{[n\times k]}$ then

$$A \cdot B = C = (c_{ij})_{[m\times k]}, \qquad \text{where } c_{ij} = \sum_{l=1}^{n} a_{il}b_{lj}.$$

Note the dimensions of matrices!

Recipe 1 – How to Multiply Two Matrices:
In order to get the element c_{ij} of matrix C you need to multiply the ith row of matrix A by the jth column of matrix B.

Example 9
$$\begin{pmatrix} 3 & 2 & 1 \\ 4 & 5 & 6 \end{pmatrix} \cdot \begin{pmatrix} 9 & 10 \\ 12 & 11 \\ 14 & 13 \end{pmatrix} =$$

$$= \begin{pmatrix} 3\cdot 9 + 2\cdot 12 + 1\cdot 14 & 3\cdot 10 + 2\cdot 11 + 1\cdot 13 \\ 4\cdot 9 + 5\cdot 12 + 6\cdot 14 & 4\cdot 10 + 5\cdot 11 + 6\cdot 13 \end{pmatrix}.$$

Example 10 *A system of m linear equations for n unknowns*

$$\begin{cases} a_{11}x_1 + a_{12}x_2 + \ldots + a_{1n}x_n &= b_1, \\ \qquad\qquad \ldots & \vdots \\ a_{m1}x_1 + a_{m2}x_2 + \ldots + a_{mn}x_n &= b_m, \end{cases}$$

can be written as $Ax = b$, where $A = (a_{ij})_{[m\times n]}$, and

$$x = \begin{pmatrix} x_1 \\ \vdots \\ x_n \end{pmatrix}, \qquad b = \begin{pmatrix} b_1 \\ \vdots \\ b_m \end{pmatrix}.$$

Laws of Matrix Operations

- Commutative law of addition: $A + B = B + A$.

- Associative law of addition: $(A + B) + C = A + (B + C)$.

- Associative law of multiplication: $A(BC) = (AB)C$.

- Distributive law:
  $A(B + C) = AB + AC$ (premultiplication by A),
  $(B + C)A = BA + CA$ (postmultiplication by A).

The commutative law of multiplication is not applicable in the matrix case, $AB \neq BA$!!! ⚠

Example 11   *Let* $\quad A = \begin{pmatrix} 2 & 0 \\ 3 & 8 \end{pmatrix}, \quad B = \begin{pmatrix} 7 & 2 \\ 6 & 3 \end{pmatrix}.$

$$\text{Then} \quad AB = \begin{pmatrix} 14 & 4 \\ 69 & 30 \end{pmatrix} \neq BA = \begin{pmatrix} 20 & 16 \\ 21 & 24 \end{pmatrix}.$$

Example 12   *Let* $\quad v = (v_1, v_2, \ldots, v_n), \quad u = \begin{pmatrix} u_1 \\ u_2 \\ \vdots \\ u_n \end{pmatrix}.$

Then $vu = v_1 u_1 + v_2 u_2 + \ldots + v_n u_n$ *is a scalar (a so-called* scalar product *of two vectors), and* $uv = C = (c_{ij})_{[n \times n]}$ *(often called a* vector product *of two vectors) is a n by n matrix, with* $c_{ij} = u_i v_j$, $i, j = 1, \ldots, n$.

We introduce the *identity* or *unit* matrix of dimension n I_n as

$$I_n = \begin{pmatrix} 1 & 0 & \ldots & 0 \\ 0 & 1 & \ldots & 0 \\ \vdots & \vdots & \ddots & \vdots \\ 0 & 0 & \ldots & 1 \end{pmatrix}.$$

Note that I_n is always a square $[n \times n]$ matrix (further on the subscript n will be omitted). I_n has the following properties:

a) $AI = IA = A$,

b) $AIB = AB$ for all A, B.

In this sense the identity matrix corresponds to 1 in the case of scalars.
    The *null* matrix is a matrix of any dimension for which all elements are zero:

$$\mathbf{0} = \begin{pmatrix} 0 & \cdots & 0 \\ \vdots & \ddots & \vdots \\ 0 & \cdots & 0 \end{pmatrix}.$$

Properties of a null matrix:

a) $A + \mathbf{0} = A$,

b) $A + (-A) = \mathbf{0}$.

⚠ Note that $AB = \mathbf{0} \nRightarrow A = \mathbf{0}$ or $B = \mathbf{0}$; $\quad AB = AC \nRightarrow B = C$.

Definition 2 *A* diagonal matrix *is a square matrix whose only non-zero elements appear on the principle (or main) diagonal.*

Definition 3 *A* triangular matrix *is a square matrix which has only zero elements above or below the principle diagonal.*

Inverses and Transposes

Definition 4 *We say that* $B = (b_{ij})_{[n \times m]}$ *is the transpose of* $A = (a_{ij})_{[m \times n]}$ *if* $a_{ji} = b_{ij}$ *for all* $i = 1, \ldots, n$ *and* $j = 1, \ldots, m$.

Usually transposes are denoted as A' (or as A^T).

Recipe 2 – How to Find the Transpose of a Matrix:
    The transpose A' *of* A *is obtained by making the columns of* A *into the rows of* A'.

Example 13 $\qquad A = \begin{pmatrix} 1 & 2 & 3 \\ 0 & a & b \end{pmatrix}, \qquad A' = \begin{pmatrix} 1 & 0 \\ 2 & a \\ 3 & b \end{pmatrix}.$

Definition 5 *If $A' = A$, A is called symmetric.*
   If $A' = -A$, A is called anti-symmetric (or skew-symmetric).
   If $A'A = I$, A is called orthogonal.
   If $A = A'$ and $AA = A$, A is called idempotent.

Properties of transposition:

a) $(A')' = A$

b) $(A + B)' = A' + B'$

c) $(AB)' = B'A'$

   Note the order of transposed matrices!   ⚠

d) $(\alpha A)' = \alpha A'$, where α is a real number.

Definition 6 *The inverse matrix A^{-1} is defined as $A^{-1}A = AA^{-1} = I$.*

Note that A as well as A^{-1} are square matrices of the same dimension
(it follows from the necessity to have the preceding line defined).   ⚠

Example 14   *If* $A = \begin{pmatrix} 1 & 2 \\ 3 & 4 \end{pmatrix}$ *then the inverse of A is*

$$A^{-1} = \begin{pmatrix} -2 & 1 \\ \frac{3}{2} & -\frac{1}{2} \end{pmatrix}.$$

We can easily check that

$$\begin{pmatrix} 1 & 2 \\ 3 & 4 \end{pmatrix} \begin{pmatrix} -2 & 1 \\ \frac{3}{2} & -\frac{1}{2} \end{pmatrix} = \begin{pmatrix} -2 & 1 \\ \frac{3}{2} & -\frac{1}{2} \end{pmatrix} \begin{pmatrix} 1 & 2 \\ 3 & 4 \end{pmatrix} = \begin{pmatrix} 1 & 0 \\ 0 & 1 \end{pmatrix}.$$

Other important characteristics of inverse matrices and inversion:

- Not all square matrices have their inverses. If a square matrix
  has its inverse, it is called *regular* or *non-singular*. Otherwise it
  is called *singular* matrix.

- If A^{-1} exists, it is unique.

- $A^{-1}A = I$ is equivalent to $AA^{-1} = I$.

Properties of inversion:

a) $(A^{-1})^{-1} = A$

b) $(\alpha A)^{-1} = \frac{1}{\alpha}A^{-1}$, where α is a real number, $\alpha \neq 0$.

c) $(AB)^{-1} = B^{-1}A^{-1}$

 Note the order of matrices!   ⚠

d) $(A')^{-1} = (A^{-1})'$

Determinants and a Test for Non-Singularity

The formal definition of the determinant is as follows: given $n \times n$ matrix $A = (a_{ij})$,

$$\det(A) = \sum_{(\alpha_1,...,\alpha_n)} (-1)^{I(\alpha_1,...,\alpha_n)} a_{1\alpha_1} \cdot a_{2\alpha_2} \cdot .... \cdot a_{n\alpha_n}$$

where $(\alpha_1, \ldots, \alpha_n)$ are all different permutations of $(1, 2, \ldots, n)$, and $I(\alpha_1, \ldots, \alpha_n)$ is the number of inversions.

Usually we denote the determinant of A as $\det(A)$ or $|A|$.

For practical purposes, we can give an alternative recursive definition of the determinant. Given the fact that the determinant of a scalar is a scalar itself, we arrive at the following

Definition 7 (Laplace expansion formula)

$$\det(A) = \sum_{k=1}^{n} (-1)^{l+k} a_{lk} \cdot \det(M_{lk}) \quad \text{for some integer } l, \ 1 \leq l \leq n.$$

Here M_{lk} is the *minor* of element a_{lk} of matrix A, which is obtained by deleting the lth row and the kth column of A. $(-1)^{l+k} \det(M_{lk})$ is called the *cofactor* of the element a_{lk}.

Example 15 *Given matrix*

$$A = \begin{pmatrix} 2 & 3 & 0 \\ 1 & 4 & 5 \\ 7 & 2 & 1 \end{pmatrix},$$

the minor of the element a_{23} is $\quad M_{23} = \begin{pmatrix} 2 & 3 \\ 7 & 2 \end{pmatrix}.$

Note that in the above expansion formula we expanded the determinant by elements of the lth row. Alternatively, we can expand it by elements of the lth column. Thus the Laplace expansion formula can be re-written as

$$\det(A) = \sum_{k=1}^{n}(-1)^{k+l}a_{kl} \cdot \det(M_{kl}) \quad \text{for some integer } l, \ 1 \leq l \leq n.$$

Example 16 *The determinant of 2×2 matrix:*

$$\det \begin{pmatrix} a_{11} & a_{12} \\ a_{21} & a_{22} \end{pmatrix} = a_{11}a_{22} - a_{12}a_{21}.$$

Example 17 *The determinant of 3×3 matrix:*

$$\det \begin{pmatrix} a_{11} & a_{12} & a_{13} \\ a_{21} & a_{22} & a_{23} \\ a_{31} & a_{32} & a_{33} \end{pmatrix} =$$

$$= a_{11} \cdot \det \begin{pmatrix} a_{22} & a_{23} \\ a_{32} & a_{33} \end{pmatrix} -$$

$$-a_{12} \cdot \det \begin{pmatrix} a_{21} & a_{23} \\ a_{31} & a_{33} \end{pmatrix} +$$

$$+a_{13} \cdot \det \begin{pmatrix} a_{21} & a_{22} \\ a_{31} & a_{32} \end{pmatrix} =$$

$$= a_{11}a_{22}a_{33} + a_{12}a_{23}a_{31} + a_{13}a_{21}a_{32} -$$

$$-a_{13}a_{22}a_{31} - a_{12}a_{21}a_{33} - a_{11}a_{23}a_{32}.$$

Properties of the determinant:

a) $\det(A \cdot B) = \det(A) \cdot \det(B)$.

b) In general, $\det(A + B) \neq \det(A) + \det(B)$. ⚠

Recipe 3 – How to Calculate the Determinant:
We can apply the following useful rules:

1. *The multiplication of any one row (or column) by a scalar k will change the value of the determinant k-fold.*

2. *The interchange of any two rows (columns) will alter the sign but not the numerical value of the determinant.*

3. *If a multiple of any row is added to (or subtracted from) any other row it will not change the value or the sign of the determinant. The same holds true for columns. (I.e. the determinant is not affected by linear operations with rows (or columns)).*

4. *If two rows (or columns) are identical, the determinant will vanish.*

5. *The determinant of a triangular matrix is a product of its principal diagonal elements.*

Using these rules, we can simplify the matrix (e.g. obtain as many zero elements as possible) and then apply Laplace expansion.

Example 18   *Let* $A = \begin{pmatrix} 4 & -2 & 6 & 2 \\ 0 & -1 & 5 & -3 \\ 2 & -1 & 8 & -2 \\ 0 & 0 & 0 & 2 \end{pmatrix}.$

Subtracting the first row, divided by 2, from the third row, we get

$$\det A = \begin{pmatrix} 4 & -2 & 6 & 2 \\ 0 & -1 & 5 & -3 \\ 2 & -1 & 8 & -2 \\ 0 & 0 & 0 & 2 \end{pmatrix} = 2 \cdot \det \begin{pmatrix} 2 & -1 & 3 & 1 \\ 0 & -1 & 5 & -3 \\ 2 & -1 & 8 & -2 \\ 0 & 0 & 0 & 2 \end{pmatrix} =$$

$$2 \cdot \det \begin{pmatrix} 2 & -1 & 3 & 1 \\ 0 & -1 & 5 & -3 \\ 0 & 0 & 5 & -3 \\ 0 & 0 & 0 & 2 \end{pmatrix} = 2 \cdot 2 \cdot (-1) \cdot 5 \cdot 2 = -40$$

If we have a block-diagonal matrix, i.e. a partitioned matrix of the form $P = \begin{pmatrix} P_{11} & P_{12} \\ 0 & P_{22} \end{pmatrix}$, where P_{11} and P_{22} are square matrices, then

$$\det(P) = \det(P_{11}) \cdot \det(P_{22}).$$

If we have a partitioned matrix $P = \begin{pmatrix} P_{11} & P_{12} \\ P_{21} & P_{22} \end{pmatrix}$, where P_{11}, P_{22} are square matrices, then

$$\begin{aligned}
\det(P) &= \det(P_{22}) \cdot \det(P_{11} - P_{12}P_{22}^{-1}P_{21}) \\
&= \det(P_{11}) \cdot \det(P_{22} - P_{21}P_{11}^{-1}P_{12}).
\end{aligned}$$

Proposition 1 (The Determinant Test for Non-Singularity)
Matrix A is non-singular $\Leftrightarrow \det(A) \neq 0$.

As a corollary, we get

Proposition 2 *A^{-1} exists $\Leftrightarrow \det(A) \neq 0$.*

Recipe 4 – How to Find an Inverse Matrix:
There are two ways of finding inverses.
Assume that matrix A is invertible, i.e. $\det(A) \neq 0$.

1. Method of adjoint matrix. *For the computation of an inverse matrix A^{-1} we use the following algorithm: $A^{-1} = (d_{ij})$, where*

$$d_{ij} = \frac{1}{\det(A)}(-1)^{i+j} \det(M_{ji}).$$

Note the order of indices at M_{ji}! ⚠

> *This method is called "method of adjoint" because we have to compute the so-called adjoint of matrix A, which is defined as matrix $\mathrm{adj}A = C' = (|C_{ji}|)$, where $|C_{ij}|$ is the cofactor of the element a_{ij}.*

2. Gauss elimination method or pivotal method. *An identity matrix is placed along side matrix A that is to be inverted. Then, the same elementary row operations are performed on* both *matrices until A has been reduced to an identity matrix. The identity matrix upon which the elementary row operations have been performed will then become the inverse matrix we seek.*

Example 19 (method of adjoint)

$$A = \begin{pmatrix} a & b \\ c & d \end{pmatrix} \implies A^{-1} = \frac{1}{ad - bc} \begin{pmatrix} d & -b \\ -c & a \end{pmatrix}.$$

Example 20 (Gauss elimination method)

Let $A = \begin{pmatrix} 2 & 3 \\ 4 & 2 \end{pmatrix}$. *Then*

$$\begin{pmatrix} 2 & 3 \\ 4 & 2 \end{pmatrix} \begin{pmatrix} 1 & 0 \\ 0 & 1 \end{pmatrix} \sim \begin{pmatrix} 1 & 3/2 \\ 4 & 2 \end{pmatrix} \begin{pmatrix} 1/2 & 0 \\ 0 & 1 \end{pmatrix} \sim$$

$$\sim \begin{pmatrix} 1 & 3/2 \\ 0 & -4 \end{pmatrix} \begin{pmatrix} 1/2 & 0 \\ -2 & 1 \end{pmatrix} \sim \begin{pmatrix} 1 & 3/2 \\ 0 & 1 \end{pmatrix} \begin{pmatrix} 1/2 & 0 \\ 1/2 & -1/4 \end{pmatrix} \sim$$

$$\sim \begin{pmatrix} 1 & 0 \\ 0 & 1 \end{pmatrix} \begin{pmatrix} -1/4 & 3/8 \\ 1/2 & -1/4 \end{pmatrix}.$$

Therefore, the inverse is

$$A^{-1} = \begin{pmatrix} -1/4 & 3/8 \\ 1/2 & -1/4 \end{pmatrix}.$$

Rank of a Matrix

A linear combination of vectors $a^1, a^2, \ldots, a^k$ is a sum

$$q_1 a^1 + q_2 a^2 + \ldots + q_k a^k,$$

where $q_1, q_2, \ldots, q_k$ are real numbers.

Definition 8 *Vectors* $a^1, a^2, \ldots, a^k$ *are linearly dependent if and only if there exist numbers* $c_1, c_2, \ldots, c_k$ *not all zero, such that*

$$c_1 a^1 + c_2 a^2 + \ldots + c_k a^k = \mathbf{0}.$$

Example 21 *Vectors* $a^1 = (2, 4)$ *and* $a^2 = (3, 6)$ *are linearly dependent: if, say,* $c_1 = 3$ *and* $c_2 = -2$ *then* $c_1 a^1 + c_2 a^2 = (6, 12) + (-6, -12) = \mathbf{0}$.

Recall that if we have n linearly independent vectors $e^1, e^2, \ldots, e^n$, they are said to *span* an n-dimensional vector space or to *constitute a basis* in an n-dimensional vector space. For more details see the section "Vector Spaces".

Definition 9 *The rank of matrix A* rank(A) *can be defined as*
– the maximum number of linearly independent rows;
– or the maximum number of linearly independent columns;
– or the order of the largest non-zero minor of A.

Example 22
$$\text{rank} \begin{pmatrix} 1 & 3 & 2 \\ 2 & 6 & 4 \\ -5 & 7 & 1 \end{pmatrix} = 2.$$

The first two rows are linearly dependent, therefore the maximum number of linearly independent rows is equal to 2.

Properties of the rank:

- The column rank and the row rank of a matrix are equal.

- $\text{rank}(AB) \leq \min(\text{rank}(A), \text{rank}(B))$.

- $\text{rank}(A) = \text{rank}(AA') = \text{rank}(A'A)$.

Using the notion of rank, we can re-formulate the condition for non-singularity:

Proposition 3 *If A is a square matrix of order n, then*
$\text{rank}(A) = n \Leftrightarrow \det(A) \neq 0.$

1.2 Systems of Linear Equations

Consider a system of n linear equations for n unknowns $Ax = b$.

Recipe 5 – How to Solve a Linear System $Ax = b$ (general rules):

$b = 0$ (homogeneous case)
If $\det(A) \neq 0$ *then the system has a unique trivial (zero) solution.*
If $\det(A) = 0$ *then the system has an infinite number of solutions.*

$b \neq 0$ (non-homogeneous case)
If $\det(A) \neq 0$ *then the system has a unique solution.*
If $\det(A) = 0$ *then*

a) $\text{rank}(A) = \text{rank}(\tilde{A}) \Rightarrow$ *the system has an infinite number of solutions.*

b) $\text{rank}(A) \neq \text{rank}(\tilde{A}) \Rightarrow$ *the system is inconsistent.*

Here $\tilde{A}$ is a so-called *augmented matrix*,

$$\tilde{A} = \begin{pmatrix} a_{11} & \cdots & a_{1n} & b_1 \\ \vdots & & \vdots & \vdots \\ a_{n1} & \cdots & a_{nn} & b_n \end{pmatrix}.$$

Recipe 6 – How to Solve the System of Linear Equations, if $b \neq 0$ and $\det(A) \neq 0$:

1. The inverse matrix method:

   Since A^{-1} exists, the solution x can be found as $x = A^{-1}b$.

2. Gauss method:

   We perform the same elementary row operations on matrix A and vector b until A has been reduced to an identity matrix. The vector b upon which the elementary row operations have been performed will then become the solution.

3. Cramer's rule:

   We can consequently find all elements $x_1, x_2, \ldots, x_n$ of vector x using the following formula:

   $$x_j = \frac{\det(A_j)}{\det(A)},$$

   where

   $$A_j = \begin{pmatrix} a_{11} & \cdots & a_{1j-1} & b_1 & a_{1j+1} & \cdots & a_{1n} \\ \vdots & & \vdots & \vdots & \vdots & & \vdots \\ a_{n1} & \cdots & a_{nj-1} & b_n & a_{nj+1} & \cdots & a_{nn} \end{pmatrix}.$$

Another method of solving linear systems is called the *LU* decomposition. It provides an efficient way to solve a system of linear equations $A\mathbf{x}=\mathbf{b}$ for different values of $\mathbf{b}$, requires fewer arithmetic operations compared to matrix inversion, and even works for non-square matrix A. The core idea of this method is to factor matrix A as a product of a lower triangular matrix L and upper triangular matrix U, so that $A = LU$. Using the *LU* decomposition, the original system of linear equations can be re-written as $LU\mathbf{x}=\mathbf{b}$. Letting $U\mathbf{x}=\mathbf{z}$, we first solve the system of linear equations $L\mathbf{z}=\mathbf{b}$ for $\mathbf{z}$. Once $\mathbf{z}$ is known, we are able to solve the system $U\mathbf{x}=\mathbf{z}$ for $\mathbf{x}$. Since both systems are triangular, only backward substitution is required to find the solution. Note, however, that it might be problematic to find matrices L and U.

Example 23 *Let us solve*

$$\begin{pmatrix} 2 & 3 \\ 4 & -1 \end{pmatrix} \begin{pmatrix} x_1 \\ x_2 \end{pmatrix} = \begin{pmatrix} 12 \\ 10 \end{pmatrix}$$

for x_1, x_2 using Cramer's rule:

$$\det(A) = 2 \cdot (-1) - 3 \cdot 3 = -14,$$

$$\det(A_1) = \det \begin{pmatrix} 12 & 3 \\ 10 & -1 \end{pmatrix} = 12 \cdot (-1) - 3 \cdot 10 = -42,$$

$$\det(A_2) = \det \begin{pmatrix} 2 & 12 \\ 4 & 10 \end{pmatrix} = 2 \cdot 10 - 12 \cdot 4 = -28,$$

therefore

$$x_1 = \frac{-42}{-14} = 3, \qquad x_2 = \frac{-28}{-14} = 2.$$

Economics Application 1 (General Market Equilibrium)
    Consider a market for three goods. Demand and supply for each good are given by:

$$\begin{cases} D_1 = 5 - 2P_1 + P_2 + P_3 \\ S_1 = -4 + 3P_1 + 2P_2 \end{cases}$$

$$\begin{cases} D_2 = 6 + 2P_1 - 3P_2 + P_3 \\ S_2 = 3 + 2P_2 \end{cases}$$

$$\begin{cases} D_3 = 20 + P_1 + 2P_2 - 4P_3 \\ S_3 = 3 + P_2 + 3P_3 \end{cases}$$

where P_i is the price of good i, $i = 1, 2, 3$.
    The equilibrium conditions are: $D_i = S_i$, $i = 1, 2, 3$, that is

$$\begin{cases} 5P_1 & +P_2 & -P_3 & = 9 \\ -2P_1 & +5P_2 & -P_3 & = 3 \\ -P_1 & -P_2 & +7P_3 & = 17 \end{cases}$$

This system of linear equations can be solved in at least two ways.

a) Using Cramer's rule:

$$A_1 = \begin{vmatrix} 9 & 1 & -1 \\ 3 & 5 & -1 \\ 17 & -1 & 7 \end{vmatrix}, \qquad A = \begin{vmatrix} 5 & 1 & -1 \\ -2 & 5 & -1 \\ -1 & -1 & 7 \end{vmatrix}$$

$$P_1^* = \frac{A_1}{A} = \frac{356}{178} = 2.$$

Similarly $P_2^* = 2$ and $P_3^* = 3$. The vector of (P_1^*, P_2^*, P_3^*) describes the general market equilibrium.

b) Using the inverse matrix rule:

Let us denote

$$A = \begin{pmatrix} 5 & 1 & -1 \\ -2 & 5 & -1 \\ -1 & -1 & 7 \end{pmatrix}, \qquad P = \begin{pmatrix} P_1 \\ P_2 \\ P_3 \end{pmatrix}, \qquad B = \begin{pmatrix} 9 \\ 3 \\ 17 \end{pmatrix}.$$

The matrix form of the system is:

$AP = B$, which implies $P = A^{-1}B$.

$$A^{-1} = \frac{1}{|A|} \begin{pmatrix} 34 & -6 & 4 \\ 15 & 34 & 7 \\ 7 & 4 & 27 \end{pmatrix},$$

$$P = \frac{1}{178} \begin{pmatrix} 34 & -6 & 4 \\ 15 & 34 & 7 \\ 7 & 4 & 27 \end{pmatrix} \begin{pmatrix} 9 \\ 3 \\ 17 \end{pmatrix} = \begin{pmatrix} 2 \\ 2 \\ 3 \end{pmatrix}.$$

Again, $P_1^* = 2$, $P_2^* = 2$ and $P_3^* = 3$.

Economics Application 2 (Leontief Input-Output Model)

This model addresses the following planning problem: Assume that n industries produce n goods (each industry produces only one good) and the output good of each industry is used as an input in the other $n - 1$ industries. In addition, each good is demanded for 'non-input' consumption. What are the efficient amounts of output each of the n industries should produce? ('Efficient' means that there will be no shortage and no surplus in producing each good).

The model is based on an input matrix:

$$A = \begin{pmatrix} a_{11} & a_{12} & \cdots & a_{1n} \\ a_{21} & a_{22} & \cdots & a_{2n} \\ & & \ddots & \\ a_{n1} & a_{n2} & \cdots & a_{nn} \end{pmatrix},$$

where a_{ij} denotes the amount of good i used to produce one unit of good j.

To simplify the model, let us set the price of each good equal to $1. Then the value of inputs should not exceed the value of output:

$$\sum_{i=1}^{n} a_{ij} \le 1, \, j = 1, \ldots n.$$

If we denote an additional (non-input) demand for good i by b_i, then the optimality condition reads as follows: the demand for each input should equal the supply, that is

$$x_i = \sum_{j=1}^{n} a_{ij} x_j + b_i, \, i = 1, \ldots n,$$

or

$$x = Ax + b,$$

or

$$(I - A)x = b.$$

The system $(I - A)x = b$ can be solved using either Cramer's rule or the inverse matrix.

Example 24 (Numerical Illustration)

$$\text{Let} \quad A = \begin{pmatrix} 0.2 & 0.2 & 0.1 \\ 0.3 & 0.3 & 0.1 \\ 0.1 & 0.2 & 0.4 \end{pmatrix}, \quad b = \begin{pmatrix} 8 \\ 4 \\ 5 \end{pmatrix}$$

Thus the system $(I - A)x = b$ becomes

$$\begin{cases} 0.8x_1 & -0.2x_2 & -0.1x_3 & = 8 \\ -0.3x_1 & +0.7x_2 & -0.1x_3 & = 4 \\ -0.1x_1 & -0.2x_2 & +0.6x_3 & = 5 \end{cases}$$

Solving it for x_1, x_2, x_3 we find the solution $\left(\frac{4210}{269}, \frac{3950}{269}, \frac{4260}{269}\right)$.

1.3   Quadratic Forms

Generally speaking, a *form* is a polynomial expression, in which each term has a uniform degree (e.g. $L = ax + by + cz$ is an example of a *linear* form in three variables x, y, z, where a, b, c are arbitrary real constants).

Definition 10 *A quadratic form Q in n variables $x_1, x_2, \ldots, x_n$ is a polynomial expression in which each component term has a degree two (i.e. each term is a product of x_i and x_j, where $i, j = 1, 2, \ldots, n$):*

$$Q = \sum_{i=1}^{n} \sum_{j=1}^{n} a_{ij} x_i x_j,$$

where a_{ij} are real numbers. For convenience, we assume that $a_{ij} = a_{ji}$. In matrix notation, $Q = x'Ax$, where $A = (a_{ij})$ is a symmetric matrix, and

$$x = \begin{pmatrix} x_1 \\ x_2 \\ \vdots \\ x_n \end{pmatrix}.$$

Example 25 *A quadratic form in two variables:*

$$Q = a_{11} x_1^2 + 2 a_{12} x_1 x_2 + a_{22} x_2^2.$$

Definition 11 *A quadratic form Q is said to be*

$$\left.\begin{array}{l} \textit{positive definite (PD)} \\ \textit{negative definite (ND)} \\ \textit{positive semi-definite (PSD)} \\ \textit{negative semi-definite (NSD)} \end{array}\right\} \textit{if } Q = x'Ax \textit{ is} \left\{\begin{array}{ll} > 0 & \forall x \neq \mathbf{0} \\ < 0 & \forall x \neq \mathbf{0} \\ \geq 0 & \forall x \\ \leq 0 & \forall x \end{array}\right.$$

otherwise Q is called indefinite (ID).

Example 26

$$Q = x^2 + y^2 \textit{ is PD,}$$

$$Q = (x + y)^2 \textit{ is PSD,}$$

$Q = x^2 - y^2$ *is ID.*

Leading principal minors D_k, $k = 1, 2, \ldots, n$ of a matrix $A = (a_{ij})_{[n \times n]}$ are defined as

$$D_k = \det \begin{pmatrix} a_{11} & \cdots & a_{1k} \\ \vdots & & \vdots \\ a_{k1} & \cdots & a_{kk} \end{pmatrix}.$$

Proposition 4

1. *A quadratic form* Q *is PD* $\Leftrightarrow$ $D_k > 0$ *for all* $k = 1, 2, \ldots, n$.

2. *A quadratic form* Q *is ND* $\Leftrightarrow$ $(-1)^k D_k > 0$ *for all* $k = 1, 2, \ldots, n$.

Note that if we replace $>$ by $\geq$ in the above statement, it does NOT give us the criteria for the semi-definite case! $\triangle$

Proposition 5 *A quadratic form* Q *is PSD* $\Leftrightarrow$ *all the principal minors of* A *are* ≥ 0.

By definition, the principal minor

$$A \begin{pmatrix} i_1 \ldots i_p \\ i_1 \ldots i_p \end{pmatrix} = \det \begin{pmatrix} a_{i_1 i_1} & \cdots & a_{i_1 i_p} \\ \vdots & & \vdots \\ a_{i_p i_1} & \cdots & a_{i_p i_p} \end{pmatrix},$$

where $1 \leq i_1 < i_2 < \ldots < i_p \leq n$, $p \leq n$.

Proposition 6 *A quadratic form* Q *is NSD* $\Leftrightarrow$ $-Q$ *is PSD.*

Example 27 *Consider the quadratic form* $Q(x, y, z) = 3x^2 + 3y^2 + 5z^2 - 2xy$. *The corresponding matrix has the form*

$$A = \begin{pmatrix} 3 & -1 & 0 \\ -1 & 3 & 0 \\ 0 & 0 & 5 \end{pmatrix}.$$

Leading principal minors of A *are*

$$D_1 = 3 > 0,$$

$$D_2 = \det \begin{pmatrix} 3 & -1 \\ -1 & 3 \end{pmatrix} = 8 > 0,$$

$$D_3 = \det \begin{pmatrix} 3 & -1 & 0 \\ -1 & 3 & 0 \\ 0 & 0 & 5 \end{pmatrix} = 40 > 0,$$

therefore, the quadratic form is positive definite.

1.4   Eigenvalues and Eigenvectors

Definition 12 *Let A be a square matrix. Any number λ such that the equation*

$$Ax = \lambda x \tag{1.1}$$

has a non-zero vector-solution x is called an eigenvalue (or a characteristic root) of the equation (1.1)

Definition 13 *Any non-zero vector x satisfying (1.1) is called an eigenvector (or characteristic vector) of A for the eigenvalue λ.*

Recipe 7 – How to Calculate Eigenvalues:
   $Ax - \lambda x = 0 \Rightarrow (A - \lambda I)x = 0$. Since x is non-zero, the determinant of $(A - \lambda I)$ should vanish. Therefore all eigenvalues can be calculated as roots of the equation (which is often called the characteristic equation or the characteristic polynomial of A)

$$\det(A - \lambda I) = 0.$$

Example 28 *Let us consider the quadratic form from Example 27.*

$$\det(A - \lambda I) = \det \begin{pmatrix} 3 - \lambda & -1 & 0 \\ -1 & 3 - \lambda & 0 \\ 0 & 0 & 5 - \lambda \end{pmatrix} =$$

$$= (5 - \lambda)(\lambda^2 - 6\lambda + 8) = (5 - \lambda)(\lambda - 2)(\lambda - 4) = 0,$$

therefore the eigenvalues are $\lambda = 2$, $\lambda = 4$ and $\lambda = 5$.

In many instances, eigenvalues play a very important role:

Proposition 7 (Characteristic Root Test for Sign Definiteness)
   A quadratic form Q with the corresponding eigenvalues $\lambda_1, \lambda_2, \ldots, \lambda_n$ is

$$\left. \begin{array}{l} \text{positive definite} \\ \text{negative definite} \\ \text{positive semi-definite} \\ \text{negative semi-definite} \end{array} \right\} \Leftrightarrow \left\{ \begin{array}{l} \lambda_i > 0 \text{ for all } i = 1, 2, \ldots, n \\ \lambda_i < 0 \text{ for all } i = 1, 2, \ldots, n \\ \lambda_i \geq 0 \text{ for all } i = 1, 2, \ldots, n \\ \lambda_i \leq 0 \text{ for all } i = 1, 2, \ldots, n \end{array} \right.$$

A form is indefinite if at least one positive and one negative eigenvalues exist.

Some useful results:

- $\det(A) = \lambda_1 \cdot \ldots \cdot \lambda_n$.

- if $\lambda_1, \ldots, \lambda_n$ are eigenvalues of A then $1/\lambda_1, \ldots, 1/\lambda_n$ are eigen-values of A^{-1}.

- if $\lambda_1, \ldots, \lambda_n$ are eigenvalues of A then $f(\lambda_1), \ldots, f(\lambda_n)$ are eigen-values of $f(A)$, where $f(\cdot)$ is a polynomial.

- the rank of a symmetric matrix is the number of non-zero eigen-values it contains.

- the rank of any matrix A is equal to the number of non-zero eigenvalues of $A'A$.

- if we define the *trace* of a square matrix of order n as the sum of the n elements on its principal diagonal $\mathrm{tr}(A) = \sum_{i=1}^{n} a_{ii}$, then $\mathrm{tr}(A) = \lambda_1 + \ldots + \lambda_n$.

Properties of the trace:

a) if A and B are of the same order, $\mathrm{tr}(A + B) = \mathrm{tr}(A) + \mathrm{tr}(B)$;

b) if λ is a scalar, $\mathrm{tr}(\lambda A) = \lambda \mathrm{tr}(A)$;

c) $\mathrm{tr}(AB) = \mathrm{tr}(BA)$, whenever AB is square;

d) $\mathrm{tr}(A') = \mathrm{tr}(A)$;

e) $\mathrm{tr}(A'A) = \sum_{i=1}^{n} \sum_{j=1}^{n} a_{ij}^2$.

1.5 Diagonalization and Spectral Theorems

Let us start with the definitions.

Definition 14 *A square real matrix A is* normal *if $A'A = AA'$. In other words, A commutes with its transpose.*

Definition 15 *A square real matrix U is called* orthogonal *if $U'U = I$.*

Definition 16 *Two square matrices A and B are called* similar *if there exists a non-singular matrix S such that $B = S^{-1}AS$.*

Definition 17 *A square matrix A is called* diagonalizable *if it is similar to a diagonal matrix.*

In other words,

Definition 18 *Matrix A is diagonalizable $\Leftrightarrow P^{-1}AP = D$ for a non-singular matrix P and a diagonal matrix D.*

NOTE: As a conventional notation, $\text{diag}(\lambda_1, \ldots, \lambda_n)$ denotes a diagonal matrix having elements $\lambda_1 \ldots, \lambda_n$ on its principal diagonal.

Definition 19 *A square matrix A is called* orthogonally diagonalizable *if $U'AU = \Lambda$, where U is orthogonal and Λ is a diagonal matrix.*

It can be easily shown that the following statements are equivalent:

- Matrix U is orthogonal;

- U is non-singular and $U' = U^{-1}$;

- $UU' = I$;

- U' is orthogonal.

The two theorems below shed light on diagonalizability conditions:

Proposition 8 *A square matrix A of order n is diagonalizable if and only if there exists a system of n linearly independent vectors, each being an eigenvector of A.*

Proposition 9 *If a square matrix of order n has n distinct eigenvalues then it is diagonalizable.*

The following statement is often called *the Spectral Theorem for Normal Matrices* (real case):

Proposition 10 *Let $A = \{a_{ij}\}$ be a square real matrix of order n with eigenvalues $\lambda_1, \ldots, \lambda_n$. The following statements are equivalent:*

1. *A is a normal matrix;*

2. *A is orthogonally diagonalizable;*

3. $\sum_{i,j}^{n} |a_{ij}|^2 = \sum_{i=1}^{n} |\lambda_i|^2$;

4. *There exist a set $V_1, \ldots, V_n$ of eigenvectors of A such that $V_i'V_j = 0$ for $i \neq j$ (i.e., V_i and V_j are orthogonal), and $V_i'V_i = 1$ (that is, V_i has unit length) for all $i = 1, \ldots, n$.*

Finally, note that if a matrix A is symmetric, $A' = A$, then A is normal, and Proposition 10 implies the so-called Spectral Theorem for Symmetric Matrices:

Proposition 11 (Spectral Theorem for Symmetric Matrices)
  If A is a symmetric matrix of order n and $\lambda_1, \ldots, \lambda_n$ are its eigenvalues, there exists an orthogonal matrix U such that

$$U^{-1}AU = \begin{pmatrix} \lambda_1 & & 0 \\ & \ddots & \\ 0 & & \lambda_n \end{pmatrix}.$$

Usually, U is the normalized matrix formed by eigenvectors. It has the property that $U'U = I$ (i.e. U is an orthogonal matrix; $U' = U^{-1}$). "Normalized" means that for any column u of the matrix U, $u'u = 1$.

It is essential that A be symmetric!   ⚠

Example 29 *Diagonalize the matrix*

$$A = \begin{pmatrix} 1 & 2 \\ 2 & 4 \end{pmatrix}.$$

First, we need to find the eigenvalues:

$$\det \begin{pmatrix} 1-\lambda & 2 \\ 2 & 4-\lambda \end{pmatrix} = (1-\lambda)(4-\lambda) - 4 = \lambda^2 - 5\lambda = \lambda(\lambda - 5),$$

i.e. $\lambda = 0$ and $\lambda = 5$.
For $\lambda = 0$ we solve

$$\begin{pmatrix} 1-0 & 2 \\ 2 & 4-0 \end{pmatrix} \begin{pmatrix} x_1 \\ x_2 \end{pmatrix} = \begin{pmatrix} 0 \\ 0 \end{pmatrix}$$

or

$$x_1 + 2x_2 = 0,$$
$$2x_1 + 4x_2 = 0.$$

The second equation is redundant and the eigenvector, corresponding to $\lambda = 0$, is $v_1 = C_1 \cdot (2, -1)'$, where C_1 is an arbitrary real constant. For $\lambda = 5$ we solve

$$\begin{pmatrix} 1-5 & 2 \\ 2 & 4-5 \end{pmatrix} \begin{pmatrix} x_1 \\ x_2 \end{pmatrix} = \begin{pmatrix} 0 \\ 0 \end{pmatrix}$$

or

$$-4x_1 + 2x_2 = 0,$$
$$2x_1 - x_2 = 0.$$

Thus the general expression for the second eigenvector is $v_2 = C_2 \cdot (1, 2)'$.

Let us normalize the eigenvectors, i.e. let us pick constants C such that $v_1' v_1 = 1$ and $v_2' v_2 = 1$. After normalization we get $v_1 = (2/\sqrt{5}, -1/\sqrt{5})'$, $v_2 = (1/\sqrt{5}, 2/\sqrt{5})'$. Thus the diagonalization matrix U is

$$U = \begin{pmatrix} \frac{2}{\sqrt{5}} & \frac{1}{\sqrt{5}} \\ -\frac{1}{\sqrt{5}} & \frac{2}{\sqrt{5}} \end{pmatrix}.$$

We can easily check that

$$U^{-1} A U = \begin{pmatrix} 0 & 0 \\ 0 & 5 \end{pmatrix}.$$

More rigorously, Proposition 11 can be presented as follows:

Proposition 12 (Spectral Theorem for Symmetric Matrices)
If a real square matrix A of order n is symmetric then

1. *All eigenvalues $\lambda_1, \ldots, \lambda_n$ are real;*

2. *A is orthogonally diagonalizable, i.e. $U' A U = diag(\lambda_1, \ldots, \lambda_n)$, where $U = (V_1, \ldots, V_n)$ is an orthogonal matrix formed by orthogonal eigenvectors $V_1, \ldots, V_n$ of unit length corresponding to eigenvalues $\lambda_1, \ldots, \lambda_n$.*

The second part of Proposition 12 deserves comment. We construct U by taking normalized eigenvectors of A as its columns. By definition of orthogonal matrix, for any $i \neq j$, $V_i'V_j = 0$, i.e. each two different columns of U should be orthogonal. The latter means that pairwise orthogonality of all eigenvectors of A is essential. It can be easily shown that eigenvectors corresponding to different eigenvalues are orthogonal. If an eigenvalue λ_i is a repeated eigenvalue of order k, then it generates k linearly independent eigenvectors, and these eigenvectors should be chosen **orthogonal to each other** (otherwise $U'AU$ may *not* be diagonal). Note that orthogonal vectors are always linearly independent, but the inverse statement is not necessarily true.

Example 30 *Find all eigenvalues of the matrix* $A = \begin{pmatrix} 2 & 2 & -2 \\ 2 & 5 & -4 \\ -2 & -4 & 5 \end{pmatrix}$

and find matrix U such that $U'U = I$ and $U'AU$.

Solution: To construct U, we first find the eigenvalues of A by solving the characteristic equation:

$$
\begin{aligned}
\det(A - \lambda I) &= \det \begin{pmatrix} 2-\lambda & 2 & -2 \\ 2 & 5-\lambda & -4 \\ -2 & -4 & 5-\lambda \end{pmatrix} \\
&= \det \begin{pmatrix} 2-\lambda & 2 & -2 \\ 0 & 1-\lambda & 1-\lambda \\ -2 & -4 & 5-\lambda \end{pmatrix} \\
&= (1-\lambda) \det \begin{pmatrix} 2-\lambda & 2 & -2 \\ 0 & 1 & 1 \\ -2 & -4 & 5-\lambda \end{pmatrix} \\
&= (1-\lambda)(\lambda^2 - 11\lambda + 10) \\
&= (1-\lambda)(\lambda - 1)(\lambda - 10) = 0.
\end{aligned}
$$

Thus A has one simple eigenvalue $\lambda_1 = 10$, and two repeated eigenvalues $\lambda_2 = \lambda_3 = 1$.

A is a real symmetric matrix, therefore under Proposition 12 it is diagonalizable, and U can be found as a normalized matrix formed by orthogonal eigenvectors of A.

One can easily check that the eigenvector corresponding to $\lambda = 10$ should have the form $V_1 = \begin{pmatrix} C \\ 2C \\ -2C \end{pmatrix}$, where C is an arbitrary non-zero constant. For instance, let us take $C = 1$, then $V_1 = \begin{pmatrix} 1 \\ 2 \\ -2 \end{pmatrix}$.

The eigenvalue $\lambda = 1$ is a repeated eigenvalue of order two, and we need to find two orthogonal eigenvectors corresponding to this repeated eigenvalue, V_2 and V_3, such that $V_2'V_3 = 0$.

Let $V_2 = (x_1, x_2, x_3)'$. The components of V can be evaluated from the equation $(A - I)V_2 = 0$, that reduces to a linear equation in three unknowns: $x_1 + 2x_2 - 2x_3 = 0$. If we arbitrarily assign numerical values to x_1 and x_2 —— for instance, set $x_1 = 0$ and $x_2 = 1$, —— then $x_3 = x_2 = 1$, and V_2 is found as $V_2 = \begin{pmatrix} 0 \\ 1 \\ 1 \end{pmatrix}$.

While computing $V_3 = (y_1, y_2, y_3)'$, the same basic equation $(A - I)V_3 = 0$, or $y_1 + 2y_2 - 2y_3 = 0$, still applies. Let us assign to y_1 a numerical value different from x_1, say, $y_1 = 1$. Note, however, that while searching for V_3, along with the equation $1 + 2y_2 - 2y_3 = 0$ we also need to impose the orthogonality condition of V_2 and V_3, that reads as $x_1y_1 + x_2y_2 + x_3y_3 = 0$. Taking these two linear equations in y_2 and y_3 into consideration (recall that $x_1y_1 = 0$ by construction) we find V_3 as $V_3 = \begin{pmatrix} 1 \\ -1/4 \\ 1/4 \end{pmatrix}$.

Thus, after the normalization of eigenvalues, the orthogonal matrix U becomes

$$U = \begin{pmatrix} 1/3 & 0 & 4/3\sqrt{2} \\ 2/3 & 1/\sqrt{2} & -1/3\sqrt{2} \\ -2/3 & 1/\sqrt{2} & 1/3\sqrt{2} \end{pmatrix}.$$

One can check oneself that indeed

$$U'AU = \begin{pmatrix} 10 & 0 & 0 \\ 0 & 1 & 0 \\ 0 & 0 & 1 \end{pmatrix}.$$

1.6 Appendix: Vector Spaces

Basic Concepts

Definition 20 *A* (real) vector space *is a nonempty set V of objects together with an additive operation $+ : V \times V \to V$, $+(u,v) = u + v$ and a scalar multiplicative operation $\cdot : R \times V \to V$, $\cdot(a,u) = au$ which satisfies the following axioms for any $u, v, w \in V$ and any $a, b \in R$ (R is the set of all real numbers):*

A1) $(u+v)+w=u+(v+w)$
A2) $u+v=v+u$
A3) $0+u=u$
A4) $u+(-u)=0$
S1) $a(u+v)=au+av$
S2) $(a+b)u=au+bu$
S3) $a(bu)=(ab)u$
S4) $1u=u$.

Definition 21 *The objects of a vector space V are called* vectors, *the operations $+$ and $\cdot$ are called* vector addition *and* scalar multiplication, *respectively. The element $0 \in V$ is the* zero vector *and $-v$ is the additive inverse of V.*

Example 31 (n-Dimensional Vector Space R^n)
Define $R^n = \{(u_1, u_2, \ldots, u_n)' | u_i \in R, i = 1, \ldots, n\}$ (the apostrophe denotes the transpose). Consider $u, v \in R^n$, $u = (u_1, u_2, \ldots, u_n)'$, $v = (v_1, v_2, \ldots, v_n)'$ and $a \in R$.
Define the additive operation and the scalar multiplication as follows:

$$
\begin{aligned}
u + v &= (u_1 + v_1, \ldots u_n + v_n)', \\
au &= (au_1, \ldots au_n)'.
\end{aligned}
$$

It is not difficult to verify that R^n together with these operations is a vector space.

Definition 22 *Let V be a vector space. An* inner product *or* scalar product *in V is a function $s : V \times V \to R$, $s(u,v) = u \cdot v$ which satisfies the following properties:*

$$u \cdot v = v \cdot u,$$
$$u \cdot (v + w) = u \cdot v + u \cdot w,$$
$$a(u \cdot v) = (au) \cdot v = u \cdot (av),$$
$$u \cdot u \geq 0 \text{ and } u \cdot u = 0 \text{ iff } u = 0.$$

Example 32 *Let* $u, v \in R^n$, $u = (u_1, u_2, \ldots, u_n)'$, $v = (v_1, v_2, \ldots, v_n)'$. *Define* $u \cdot v = u_1 v_1 + \ldots + u_n v_n$. *Then this rule is an inner product in* R^n.

Definition 23 *Let* V *be a vector space and* $\cdot : V \times V \to R$ *an inner product in* V. *The* norm *of* magnitude *is a function* $\| \cdot \| : V \to R$ *defined as* $\|v\| = \sqrt{v \cdot v}$.

Proposition 13 *If* V *is a vector space, then for any* $v \in V$ *and* $a \in R$
  i) $\|au\| = |a| \|u\|$;
  ii)   *(Triangle inequality)* $\|u + v\| \leq \|u\| + \|v\|$;
  iii) *(Schwarz inequality)* $|u \cdot v| \leq \|u\| \|v\|$.

Example 33 *If* $u \in R^n, u = (u_1, u_2, \ldots, u_n)$, *the norm of* u *can be introduced as*

$$\|u\| = \sqrt{u \cdot u} = \sqrt{u_1^2 + \cdots + u_n^2}.$$

The triangle inequality and Schwarz's inequality in R^n *become:*

ii-a) $\sqrt{(u_1 + v_1)^2 + \cdots + (u_n + v_n)^2} \leq \sqrt{\sum_{i=0}^{n} u_i^2} + \sqrt{\sum_{i=0}^{n} v_i^2}$;
   (Minkowski's inequality for sums)

iii-a) $|\sum_{i=0}^{n} u_i v_i| \leq (\sqrt{\sum_{i=0}^{n} u_i^2})(\sqrt{\sum_{i=0}^{n} v_i^2})$.
   (Cauchy-Schwarz inequality for sums)

Definition 24

  a) *The non-zero vectors* u *and* v *are* parallel *if there exists* $a \in R$ *such that* $u = av$.

  b) *The vectors* u *and* v *are* orthogonal *or* perpendicular *if their scalar product is zero, that is, if* $u \cdot v = 0$.

  c) *The* angle *between vectors* u *and* v *is* $\arccos(\frac{uv}{\|u\| \|v\|})$.

Vector Subspaces

Definition 25 *A non-empty subset S of a vector space V is a* subspace *of V if for any $u, v \in S$ and $a \in R$*

$$u + v \in S \qquad \text{and} \qquad au \in S.$$

Example 34 *V is a subset of itself. $\{0\}$ is also a subset of V. These subspaces are called* proper *subspaces.*

Example 35 *$L = \{(x, y)|y = mx + n\}$ where $m, n \in R$ and $m \neq 0$ is a subspace of R^2.*

Definition 26 *Let $u_1, u_2 \ldots u_k$ be vectors in a vector space V. The set S of all linear combinations of these vectors*

$$S = \{a_1 u_1 + a_2 u_2 + \cdots + a_k u_k | a_i \in R, i = 1, \cdots k\}$$

is called the subspace generated *or* spanned *by the vectors $u_1, u_2 \ldots, u_k$ and denoted as $sp(u_1, u_2 \cdots u_k)$.*

Proposition 14 *S is a subspace of V.*

Example 36 *Let $u_1 = (2, -1, 1)', u_2 = (3, 4, 0)'$. Then the subspace of R^3 generated by u_1 and u_2 is*

$$sp(u_1, u_2) = \{au_1 + bu_2 | a, b \in R\} = \{(2a + 3b, -a + 4b, a)' | a, b \in R\}.$$

Independence and Bases

Definition 27 *A set $\{u_1, u_2 \ldots u_k\}$ of vectors in a vector space V is* linearly dependent *if there exists the real numbers $a_1, a_2 \ldots a_k$, not all zero, such that $a_1 u_1 + a_2 u_2 + \ldots a_k u_k = 0$.*

In other words, the set of vectors in a vector space is linearly dependent if and only if one vector can be written as a linear combination of the others. ⚠

Example 37 *The vectors*

$$u_1 = (2, -1, 1)', \quad u_2 = (1, 3, 4)', \quad u_3 = (0, -7, -7)'$$

are linearly dependent since $u_3 = u_1 - 2u_2$.

Definition 28 *A set $\{u_1, u_2 \ldots u_k\}$ of vectors in a vector space V is linearly independent if $a_1 u_1 + a_2 u_2 + \cdots a_k u_k = 0$ implying $a_1 = a_2 = \cdots = a_k = 0$ (that is, they are not linearly dependent).*

In other words, the definition says that a set of vectors in a vector space is linearly independent if and only if none of the vectors can be written as a linear combination of the others.

Proposition 15 *Let $\{u_1, u_2 \ldots u_n\}$ be n vectors in R^n. The following conditions are equivalent:*
  i)   The vectors are independent;
  ii)   The matrix having these vectors as columns is non-singular;
  iii)   The vectors generate R^n.

Example 38 *The vectors*

$$u_1 = (1, 2, -2)', \quad u_2 = (2, 3, 1)', \quad u_3 = (-2, 0, 1)'$$

in R^3 are linearly independent since $\det \begin{pmatrix} 1 & 2 & -2 \\ 2 & 3 & 0 \\ -2 & 1 & 1 \end{pmatrix} = -17 \neq 0.$

Definition 29 *A set $\{u_1, u_2 \ldots u_k\}$ of vectors in V is a basis for V if it, first, generates V (that is, $V = sp(u_1, u_2 \ldots u_k)$), and, second, is linearly independent.*

Any set of n linearly independent vectors in R^n form a basis for R^n.

$\triangle$

Example 39 *The vectors from the preceding example*

$$u_1 = (1, 2, -2)', \quad u_2 = (2, 3, 1)', \quad u_3 = (-2, 0, 1)'$$

form a basis for R^3.

Example 40 *Consider the following vectors in R^n:*

$$e_i = (0, \ldots, 0, 1, 0, \ldots, 0)',$$

where 1 is in the ith position, $i = 1, \ldots, n$. The set $E_n = \{e_1, \ldots, e_n\}$ form a basis for R^n which is called the standard basis.

Definition 30 *Let V be a vector space and $B = \{u_1, u_2, \ldots, u_k\}$ a basis for V. Since B generates V, for any $u \in V$ there exists the real numbers $x_1, x_2, \ldots, x_n$ such that $u = x_1 u_1 + \ldots + x_n u_n$. The column vector $x = (x_1, x_2, \ldots, x_n)'$ is called the* vector of coordinates *of u with respect to B.*

Example 41 *Consider the vector space R^n with the standard basis E_n. For any $u = (u_1, \ldots, u_n)'$ we can represent u as $u = u_1 e_1 + \ldots u_n e_n$; therefore, $(u_1, \ldots, u_n)'$ is the vector of coordinates of u with respect to E_n.*

Example 42 *Consider the vector space R^2. Let us find the coordinate vector of $(-1, 2)'$ with respect to the basis $B = (1, 1)', (2, -3)'$ (i.e. find $(-1, 2)'_B$). We have to solve for a, b such that $(-1, 2)' = a(1, 1)' + b(2, -3)'$. Solving the system*

$$\begin{cases} a & +2b & = -1 \\ a & -3b & = 2 \end{cases}$$

we find $a = \frac{1}{5}, b = \frac{-3}{5}$. Thus, $(-1, 2)'_B = (\frac{1}{5}, \frac{-3}{5})'$.

Definition 31 *The* dimension *of a vector space V $dim(V)$ is the number of elements in any basis for V.*

Example 43 *The dimension of the vector space R^n with the standard basis E_n is $dim(R^n) = n$.*

Linear Transformations and Changes of Bases

Definition 32 *Let U, V be two vector spaces. A* linear transformation *of U into V is a mapping $T : U \to V$ such that for any $u, v \in U$ and any $a, b \in R$*

$$T(au + bv) = aT(u) + bT(v).$$

Example 44 *Let A be an $m \times n$ real matrix. Then the mapping $T : R^n \to R^m$ defined by $T(u) = Au$ is a linear transformation.*

Example 45 (Rotation of the Plane)
The function $T_R : R^2 \to R^2$ that rotates the plane counterclockwise through a positive angle α is a linear transformation.

To check this, first note that any two-dimensional vector $u \in R^2$ can be expressed in polar coordinates as $u = (r \cos \theta, r \sin \theta)'$ where $r = \|u\|$ and θ is the angle that u makes with the x-axis of the system of coordinates.

The mapping T_R is thus defined by

$$T_R(u) = (r \cos(\theta + \alpha), r \sin(\theta + \alpha))'.$$

Therefore,

$$T_R(u) = \begin{pmatrix} r(\cos \theta \cos \alpha - \sin \theta \sin \alpha) \\ r(\sin \theta \cos \alpha + \cos \theta \sin \alpha) \end{pmatrix},$$

or, alternatively,

$$T_R(u) = \begin{pmatrix} \cos \alpha & -\sin \alpha \\ \sin \alpha & \cos \alpha \end{pmatrix} \begin{pmatrix} r \cos \theta \\ r \sin \theta \end{pmatrix} = Au,$$

where

$$A = \begin{pmatrix} \cos \alpha & -\sin\alpha \\ \sin \alpha & \cos \alpha \end{pmatrix} \qquad \text{(the rotation matrix).}$$

From example 44 it follows that T_R is a linear transformation.

Proposition 16 Let U and V be two vector spaces, $B = (b_1, \ldots, b_n)$ a basis for U and $C = (c_1, \ldots, c_m)$ a basis for V.

- Any linear transformation T can be represented by an $m \times n$ matrix A_T whose ith column is the coordinate vector of $T(b_i)$ relative to C.

- If $x = (x_1, \ldots, x_n)'$ is the coordinate vector of $u \in U$ relative to B and $y = (y_1, \ldots, y_m)'$ is the coordinate vector of $T(u)$ relative to C then T defines the following transformation of coordinates:

$$y = A_T x \text{ for any } u \in U.$$

Definition 33 The matrix A_T is called the matrix representation of T relative to bases B, C.

Any linear transformation is uniquely determined by a transformation of coordinates.

Example 46 *Consider the linear transformation*

$$T : R^3 \to R^2, \quad T((x, y, z)') = (x - 2y, x + z)'$$

and bases

$$B = \{(1, 1, 1)', (1, 1, 0)', (1, 0, 0)'\} \text{ for } R^3$$

$$\text{and} \quad C = \{(1, 1)', (1, 0)'\} \text{ for } R^2.$$

How can we find the matrix representation of T relative to bases B, C?

<u>Solution:</u> *We have*

$$T((1, 1, 1)') = (-1, 2), \quad T((1, 1, 0)') = (-1, 1), \quad T((1, 0, 0)') = (1, 1).$$

The columns of A_T are formed by the coordinate vectors of $T((1, 1, 1)')$, $T((1, 1, 0)')$, $T((1, 0, 0)')$ relative to C. Applying the procedure developed in Example 42 we find

$$A_T = \begin{pmatrix} 2 & 1 & 1 \\ -3 & -2 & 0 \end{pmatrix}.$$

Definition 34 (Changes of Bases)

   Let V be a vector space of dimension n, B and C be two bases for V, and $I : V \to V$ be the identity transformation ($I(v) = v$ for all $v \in V$). The change-of-basis matrix D relative to B, C is the matrix representation of I relative to B, C.

Example 47 *For $u \in V$, let $x = (x_1, \ldots, x_n)'$ be the coordinate vector of u relative to B and $y = (y_1, \ldots, y_n)'$ is the coordinate vector of u relative to C. If D is the change-of-basis matrix relative to B, C then $y = Cx$. The change-of-basis matrix relative to C, B is D^{-1}.*

Example 48 *Given the following bases for R^2: $B = \{(1, 1)', (1, 0)'\}$ and $C = \{(0, 1)', (1, 1)'\}$, find the change-of-basis matrix D relative to B, C.*

   The columns of D are the coordinate vectors of $(1, 1)', (1, 0)'$ relative to C. Following Example 42 we find $D = \begin{pmatrix} 0 & -1 \\ 1 & 1 \end{pmatrix}.$

Proposition 17 *Let* $T : V \to V$ *be a linear transformation, and let* B, C *be two bases for* V. *If* A_1 *is the matrix representation of* T *in the basis* B, A_2 *is the matrix representation of* T *in the basis* C, *and* D *is the change-of-basis matrix relative to* C, B *then* $A_2 = D^{-1} A_1 D$.

Further Reading

- Bellman, R. E. (1997). *Introduction to Matrix Analysis* (2nd ed.). Philadelphia, PA: Society for Industrial and Applied Mathematics.

- Fraleigh, J. B. and Beauregard, R. A. (1995). *Linear Algebra* (3rd ed.). Addison-Wesley Publishing Company.

- Gantmacher, F. R. (1998). *The Theory of Matrices* (Vols. 1–2). American Mathematical Society.

- Lang, S. (2004). *Linear Algebra* (3rd ed.). New York: Springer.

Chapter 2

Calculus

2.1 The Concept of Limit

Definition 35 *The function $f(x)$ has a limit A (or tends to A as a limit) as x approaches a if for each given number $\varepsilon > 0$, no matter how small, there exists a positive number δ (that depends on ε) such that $|f(x) - A| < \varepsilon$ whenever $0 < |x - a| < \delta$.*

The standard notation is $\lim_{x \to a} f(x) = A$ or $f(x) \to A$ as $x \to a$.

Definition 36 *The function $f(x)$ has a left-side (or right-side) limit A as x approaches a from the left (or right),*

$$\lim_{x \to a_-} f(x) = A \qquad (\lim_{x \to a_+} f(x) = A),$$

if for each given number $\varepsilon > 0$ there exists a positive number δ such that $|f(x) - A| < \varepsilon$ whenever $a - \delta < x < a$ $(a < x < a + \delta)$.

Recipe 8 – How to Calculate Limits:
    We can apply the following basic rules for limits:
if $\lim_{x \to x_0} f(x) = A$ and $\lim_{x \to x_0} g(x) = B$ then

1. *if C is constant, then $\lim_{x \to x_0} C = C$;*

2. *$\lim_{x \to x_0}(f(x) + g(x)) = \lim_{x \to x_0} f(x) + \lim_{x \to x_0} g(x) = A + B$;*

3. *$\lim_{x \to x_0} f(x)g(x) = \lim_{x \to x_0} f(x) \cdot \lim_{x \to x_0} g(x) = A \cdot B$;*

4. *$\lim_{x \to x_0} \dfrac{f(x)}{g(x)} = \dfrac{\lim_{x \to x_0} f(x)}{\lim_{x \to x_0} g(x)} = \dfrac{A}{B}$;*

5. $\lim_{x \to x_0} (f(x))^n = (\lim_{x \to x_0} f(x))^n = A^n$;

6. $\lim_{x \to x_0} f(x)^{g(x)} = e^{\lim_{x \to x_0} g(x) \ln f(x)} = e^{B \ln A}$.

Some important results:

$$\lim_{x \to 0} \frac{\sin x}{x} = 1, \qquad \lim_{x \to \infty} \left(1 + \frac{1}{x}\right)^x = e = 2.718281828459\ldots,$$

$$\lim_{x \to 0} \frac{\ln(1 + x)}{x} = 1, \qquad \lim_{x \to 0} \frac{(1 + x)^p - 1}{x} = p,$$

$$\lim_{x \to 0} \frac{e^x - 1}{x} = 1, \qquad \lim_{x \to 0} \frac{a^x - 1}{x} = \ln a, \quad a > 0.$$

Example 49

a) $\lim_{x \to 1} \dfrac{x^3 - 1}{x - 1}$ $= \lim_{x \to 1} \dfrac{(x - 1)(x^2 + x + 1)}{x - 1}$

$\qquad\qquad\qquad = \lim_{x \to 1} (x^2 + x + 1) = 1 + 1 + 1 = 3.$

b) $\lim_{x \to 0} \dfrac{\sin^2 x}{x} = \lim_{x \to 0} (\dfrac{\sin x}{x} \sin x) = \lim_{x \to 0} \dfrac{\sin x}{x} \lim_{x \to 0} \sin x = 1 \cdot 0 = 0.$

c) $\lim_{x \to \infty} x^2 \ln \dfrac{\sqrt{x^2 + 1}}{x}$ $= \lim_{x \to \infty} x^2 \ln \sqrt{\dfrac{x^2 + 1}{x^2}}$

$\qquad\qquad\qquad = \lim_{x \to \infty} x^2 \dfrac{1}{2} \ln(1 + \dfrac{1}{x^2})$

$\qquad\qquad\qquad = \dfrac{1}{2} \lim_{x \to \infty} \dfrac{\ln(1 + 1/x^2)}{1/x^2} = \dfrac{1}{2}.$

d) $\lim_{x \to 0_+} x^{\frac{1}{1 + \ln x}} = e^{\lim_{x \to 0_+} \frac{1}{1 + \ln x} \ln x} = e^{\lim_{x \to 0_+} \frac{1}{1/\ln x + 1}} = e^1 = e.$

Another powerful tool for evaluating limits is *L'Hôpital's rule.*

Proposition 18 (L'Hôpital's Rule)

Suppose $f(x)$ and $g(x)$ are differentiable[1] in an interval (a, b) around x_0 except possibly at x_0, and suppose that $f(x)$ and $g(x)$ both approach

---

[1] For the definition of differentiability see the next section.

0 when x approaches x_0. If $g'(x) \neq 0$ for all $x \neq x_0$ in (a, b) and $\lim_{x \to x_0} \frac{f'(x)}{g'(x)} = A$ then

$$\lim_{x \to x_0} \frac{f(x)}{g(x)} = \lim_{x \to x_0} \frac{f'(x)}{g'(x)} = A.$$

The same rule applies if $f(x) \to \pm\infty$, $g(x) \to \pm\infty$. x_0 can be either finite or infinite.

Note that L'Hôpital's rule can be applied only if we have expressions of the form $\frac{0}{0}$ or $\frac{\pm\infty}{\pm\infty}$.   △

Example 50

a) $\lim_{x \to 0} \dfrac{x - \sin x}{x^3} = \lim_{x \to 0} \dfrac{1 - \cos x}{3x^2} = \lim_{x \to 0} \dfrac{\sin x}{6x} = \dfrac{1}{6} \lim_{x \to 0} \dfrac{\sin x}{x} = \dfrac{1}{6}.$

b) $\lim_{x \to 0} \dfrac{x - \sin 2x}{x^3} = \lim_{x \to 0} \dfrac{1 - 2\cos 2x}{3x^2} = \dfrac{\lim_{x \to 0}(1 - 2\cos 2x)}{\lim_{x \to 0} 3x^2}$

$$= \dfrac{-1}{\lim_{x \to 0} 3x^2} = -\infty.$$

2.2 Univariate Differentiation

Let us first discuss the basic principles of calculus in the case of one variable.

Definition 37 *A function $f(x)$ is continuous at $x = a$ if $\lim_{x \to a} f(x) = f(a)$.*

If $f(x)$ and $g(x)$ are continuous at a then:

- $f(x) \pm g(x)$ and $f(x)g(x)$ are continuous at a;

- if $g(a) \neq 0$ then $\frac{f(x)}{g(x)}$ is continuous at a;

- if $g(x)$ is continuous at a and $f(x)$ is continuous at $g(a)$ then $f(g(x))$ is continuous at a.

In general, any function built from continuous functions by additions, subtractions, multiplications, divisions, and compositions is continuous where defined.

If $\lim_{x \to a_+} f(x) = c_1 \neq c_2 = \lim_{x \to a_-} f(x)$, $|c_1|, |c_2| < \infty$, the function $f(x)$ is said to have a *jump discontinuity* at a. (Note that limit from above and limit from below, or right-hand side limit and left-hand side limit, at point a, sometimes are also denoted as $\lim_{h \downarrow a}$ and $\lim_{h \uparrow a}$, respectively.) If $\lim_{x \to a} f(x) = \pm \infty$, we call this type of discontinuity *infinite discontinuity*.

Suppose there is a functional relationship between x and y, $y = f(x)$. One of the natural questions one may ask is: How does y change if x changes? We can answer this question using the notion of the *difference quotient*. Denoting the change in x as $\Delta x = x - x_0$, the difference quotient is defined as

$$\frac{\Delta y}{\Delta x} = \frac{f(x_0 + \Delta x) - f(x_0)}{\Delta x}.$$

Taking the limit of the above expression, we arrive at the following definition.

Definition 38 *The derivative of $f(x)$ at x_0 is*

$$f'(x_0) = \lim_{x \to x_0} \frac{f(x_0 + \Delta x) - f(x_0)}{\Delta x}.$$

If the limit exists, f is called differentiable at x.

An alternative notation for the derivative often found in textbooks is

$$f'(x) = y' = \frac{df}{dx} = \frac{df(x)}{dx}.$$

The first and second derivatives with respect to time are usually denoted by dots (˙ and ¨, respectively), i.e if $z = z(t)$ then $\dot{z} = \frac{dz}{dt}$, $\ddot{z} = \frac{d^2 z}{dt^2}$.

The set of all continuously differentiable functions in the domain D (i.e. the argument of a function may take any value from D) is denoted by $C^{(1)}(D)$. For instance, $C^{(1)}(R)$ is the set of real-valued functions $f : R \to R$ such that the derivative f' of f exists and is continuous for any real argument.

Geometrically speaking, the derivative represents the slope of the tangent line to f at x. Using the derivative, the equation of the tangent line to $f(x)$ at x_0 can be written as

$$y = f(x_0) + f'(x_0)(x - x_0).$$

Note that the continuity of $f(x)$ is a necessary but NOT sufficient condition for its differentiability!   ⚠

Example 51 *For instance, $f(x) = |x - 2|$ is continuous at $x = 2$ but not differentiable at $x = 2$.*

The geometric interpretation of the derivative gives us the following formula (usually called Newton's approximation method), which allows us to find an approximate root of $f(x) = 0$. Let us define a sequence

$$x_{n+1} = x_n - \frac{f(x_n)}{f'(x_n)}.$$

If the initial value x_0 is chosen such that x_0 is reasonably close to an actual root, this sequence will converge to that root.

The derivatives of higher order $(2,3,...,n)$ can be defined in the same manner:

$$f''(x_0) = \frac{d}{dx} f'(x_0) = \lim_{x \to x_0} \frac{f'(x_0 + \Delta x) - f'(x_0)}{\Delta x}$$

$$\cdots \quad \cdots \quad \cdots$$

$$f^{(n)}(x_0) = \frac{d^n}{dx^n} f^{(n-1)}(x_0),$$

provided that these limits exist. If so, $f(x)$ is called n times continuously differentiable, $f \in C^{(n)}$. The symbol $\frac{d^n}{dx^n}$ denotes an operator of taking the nth derivative of a function with respect to x. The class of infinitely differentiable functions is denoted as $C^{(\infty)}$.

Example 52 *Discuss the continuity and differentiability of the five given functions. Please pay attention to the transition point when applicable.*

1. $f(x) = \begin{cases} +x^2, & x \geq 0 \\ -x^2, & x < 0 \end{cases}$

2. $f(x) = \begin{cases} +x^2 + 1, & x \geq 0 \\ -x^2 - 1, & x < 0 \end{cases}$

3. $f(x) = \begin{cases} x^3, & x \geq 1 \\ x, & x < 1 \end{cases}$

4. $f(x) = \begin{cases} x^3, & x < 1 \\ 3x - 2, & x \geq 1 \end{cases}$

5. $f(x) = x^{5/3}$

Solution:

1. $f'(x) = \begin{cases} 2x, & x \geq 0 \\ -2x, & x < 0 \end{cases}$ _As x converges to 0 both from above and below, $f'(x)$ converges to 0, so the function is C^1._

2. _The function is not continuous (and thus not differentiable). As x tends to 0 from above, $f(x)$ converges to 1, whereas $f(x)$ tends to -1 as x tends to 0 from below._

3. _The function is continuous, since $\lim_{x \to 1} f(x) = 1$ no matter how the limit is taken. However, it is not differentiable at $x = 1$, since $\lim_{h \downarrow 1}[f(1 + h) - f(1)]/h = 3$ and $\lim_{h \uparrow 1}[f(1 + h) - f(1)]/h = 1$. (Recall that from the geometric perspective, the function f' is continuous if the tangent line to the graph of f at $(x, f(x))$ changes continuously as x changes.)_

4. _The function is C^1 at $x = 1$ because the value of the derivative of $f(x)$ at $x = 1$ equals 3, no matter whether x tends to 1 from below or from above._

$$f'(x) = \begin{cases} 3x^2, & x < 1 \\ 3, & x \geq 1 \end{cases}$$

5. _The derivative $f'(x) = (5/3)x^{2/3}$, so $f(x)$ is C^1. But $x^{2/3}$ is not differentiable at $x = 0$, therefore f is not C^2 at $x = 0$. Anywhere else it is C^∞._

2.3 Rules of Differentiation

Suppose we have two differentiable functions of the same variable x, say $f(x)$ and $g(x)$. Then

- $(f(x) \pm g(x))' = f'(x) \pm g'(x)$;

- $(f(x)g(x))' = f'(x)g(x) + f(x)g'(x)$ (product rule);

- $\left(\dfrac{f(x)}{g(x)} \right)' = \dfrac{f'(x)g(x) - f(x)g'(x)}{g^2(x)}$ (quotient rule);

- if $f = f(y)$ and $y = g(x)$ then $\dfrac{df}{dx} = \dfrac{df}{dy}\dfrac{dy}{dx} = f'(y)g'(x)$ (chain rule or composite-function rule).

  This rule can be easily extended to the case of more than two functions involved.

  Example 53 (Application of the Chain Rule)
  Let $f(x) = \sqrt{x^2 + 1}$. *We can decompose as* $f = \sqrt{y}$, *where* $y = x^2 + 1$. *Therefore,*

  $$\frac{df}{dx} = \frac{df}{dy}\frac{dy}{dx} = \frac{1}{2\sqrt{y}} \cdot 2x = \frac{x}{\sqrt{y}} = \frac{x}{\sqrt{x^2 + 1}}.$$

- if $x = u(t)$ and $y = v(t)$ (i.e. x and y are *parametrically* defined) then

  $$\frac{dy}{dx} = \frac{dy/dt}{dx/dt} = \frac{\dot{v}(t)}{\dot{u}(t)} \qquad \text{(recall that } \dot{\ } \text{ means } \tfrac{d}{dt}\text{)},$$

  $$\frac{d^2y}{dx^2} = \frac{d}{dx}\left(\frac{\dot{v}(t)}{\dot{u}(t)} \right) = \frac{\frac{d}{dt}\left(\frac{\dot{v}(t)}{\dot{u}(t)} \right)}{dx/dt} = \frac{\ddot{v}(t)\dot{u}(t) - \dot{v}(t)\ddot{u}(t)}{(\dot{u}(t))^3}.$$

Example 54 *If* $x = a(t - \sin t)$, $y = a(1 - \cos t)$, *where a is a parameter, then*

$$\frac{dy}{dx} = \frac{\sin t}{1 - \cos t},$$

$$\frac{d^2y}{dx^2} = \frac{\cos t(1 - \cos t) - \sin^2 t}{a(1 - \cos t)^3} = \frac{\cos t - 1}{a(1 - \cos t)^2} = -\frac{1}{a(1 - \cos t)^2}.$$

Some special rules:

$$f(x) = constant \qquad\qquad \Rightarrow \quad f'(x) = 0$$
$$f(x) = x^a \text{ (}a\text{ is constant)} \qquad \Rightarrow \quad f'(x) = ax^{a-1}$$
$$f(x) = e^x \qquad\qquad \Rightarrow \quad f'(x) = e^x$$
$$f(x) = a^x \text{ (}a > 0\text{)} \qquad\qquad \Rightarrow \quad f'(x) = a^x \ln a$$
$$f(x) = \ln x \qquad\qquad \Rightarrow \quad f'(x) = \tfrac{1}{x}$$
$$f(x) = \log_a x \text{ (}a > 0, a \neq 1\text{)} \quad \Rightarrow \quad f'(x) = \tfrac{1}{x}\log_a e = \tfrac{1}{x \ln a}$$
$$f(x) = \sin x \qquad\qquad \Rightarrow \quad f'(x) = \cos x$$
$$f(x) = \cos x \qquad\qquad \Rightarrow \quad f'(x) = -\sin x$$
$$f(x) = \text{tg}x \qquad\qquad \Rightarrow \quad f'(x) = \tfrac{1}{\cos^2 x}$$
$$f(x) = \text{ctg}x \qquad\qquad \Rightarrow \quad f'(x) = -\tfrac{1}{\sin^2 x}$$
$$f(x) = \arcsin x \qquad\qquad \Rightarrow \quad f'(x) = \tfrac{1}{\sqrt{1-x^2}}$$
$$f(x) = \arccos x \qquad\qquad \Rightarrow \quad f'(x) = -\tfrac{1}{\sqrt{1-x^2}}$$
$$f(x) = \text{arctg}x \qquad\qquad \Rightarrow \quad f'(x) = \tfrac{1}{1+x^2}$$
$$f(x) = \text{arcctg}x \qquad\qquad \Rightarrow \quad f'(x) = -\tfrac{1}{1+x^2}$$

More hints:

- if $y = \ln f(x)$ then $y' = \dfrac{f'(x)}{f(x)}$.

- if $y = e^{f(x)}$ then $y' = f'(x)e^{f(x)}$.

- if $y = f(x)^{g(x)}$ then

$$
\begin{aligned}
y' \quad &= \quad (f(x)^{g(x)})' = (e^{g(x)\ln f(x)})' \\
&\overset{\text{(by chain rule)}}{=} \quad e^{g(x)\ln f(x)}(g(x)\ln f(x))' \\
&\overset{\text{(by product rule)}}{=} \quad e^{g(x)\ln f(x)}\left(g'(x)\ln f(x) + g(x)\frac{f'(x)}{f(x)}\right) \\
&= \quad f(x)^{g(x)}\left(g'(x)\ln f(x) + g(x)\frac{f'(x)}{f(x)}\right).
\end{aligned}
$$

- if a function f is a one-to-one mapping (or single-valued mapping),[2] it has the inverse f^{-1}. Thus, if $y = f(x)$ and $x = f^{-1}(y)$ then $\dfrac{dx}{dy} = \dfrac{1}{f'(x)} = \dfrac{1}{f'(f^{-1}(y))}$ or, in other words, $\dfrac{dx}{dy} = \dfrac{1}{dy/dx}$.

---

[2] In the one-dimensional case the set of all one-to-one mappings coincides with the set of strictly monotonic functions.

Example 55 *Given* $y = \ln x$, *its inverse is* $x = e^y$. *Therefore* $\dfrac{dx}{dy} =$
$\dfrac{1}{1/x} = x = e^y$.

- if a function f is a product (or quotient) of a number of other functions, the logarithmic function might be helpful while taking the derivative of f. If $y = f(x) = \dfrac{g_1(x)g_2(x)\dots g_k(x)}{h_1(x)h_2(x)\dots h_l(x)}$ then after taking the (natural) logarithms of both sides and rearranging the terms we get

$$\frac{dy}{dx} = y(x) \cdot \left(\frac{g_1'(x)}{g_1(x)} + \dots + \frac{g_k'(x)}{g_k(x)} - \frac{h_1'(x)}{h_1(x)} - \dots - \frac{h_l'(x)}{h_l(x)} \right).$$

Definition 39 *If* $y = f(x)$ *and* dx *is any number then the differential of* y *is defined as* $dy = f'(x)dx$.

The rules of differentials are similar to those of derivatives:

- If k is constant then $dk = 0$;

- $d(u \pm v) = du \pm dv$;

- $d(uv) = v \cdot du + u \cdot dv$;

- $d\left(\dfrac{u}{v}\right) = \dfrac{v \cdot du - u \cdot dv}{v^2}$.

Differentials of higher order can be found in the same way as derivatives.

2.4   Maxima and Minima of a Function of One Variable

Definition 40 *If* $f(x)$ *is continuous in a neighborhood* U *of a point* x_0, *it is said to have a local or relative maximum (minimum) at* x_0 *if for all* $x \in U$, $x \neq x_0$, $f(x) < f(x_0)$ $(f(x) > f(x_0))$.

Proposition 19 (Weierstrass's Theorem)

If f is continuous on a closed and bounded subset S of $\mathbf{R}^1$ (or, in the general case, of $\mathbf{R}^n$), then f reaches its maximum and minimum in S, i.e. there exist points $m, M \in S$ such that $f(m) \le f(x) \le f(M)$ for all $x \in S$.

Proposition 20 (Fermat's Theorem, or the Necessary Condition for Extremum)

If $f(x)$ is differentiable at x_0 and has a local extremum (minimum or maximum) at x_0 then $f'(x_0) = 0$.

Note that if the first derivative vanishes at some point, it does not imply that at this point f possesses an extremum. We can only state that f has a *stationary* point.

Some useful results:

Proposition 21 (Rolle's Theorem)

If f is continuous in $[a, b]$, differentiable in (a, b) and $f(a) = f(b)$, then there exists at least one point $c \in (a, b)$ such that $f'(c) = 0$.

Proposition 22 (Lagrange's Theorem, or the Mean Value Theorem)

If f is continuous in $[a, b]$ and differentiable in (a, b), then there exists at least one point $c \in (a, b)$ such that $f'(c) = \dfrac{f(b) - f(a)}{b - a}$.

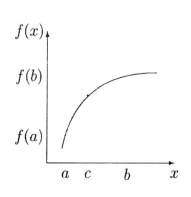

The geometric meaning of Lagrange's theorem:

the slope of a tangent line to $f(x)$ at the point c coincides with the slope of the straight line, connecting two points $(a, f(a))$ and $(b, f(b))$.

If we denote $a = x_0$, $b = x_0 + \Delta x$, then we may re-write Lagrange's theorem as $f(x_0 + \Delta x) - f(x_0) = \Delta x \cdot f'(x_0 + \theta \Delta x)$, where $\theta \in (0, 1)$.

This statement can be generalized in the following way:

Proposition 23 (Cauchy's Theorem, or the Generalized Mean Value Theorem)

If f and g are continuous in $[a, b]$ and differentiable in (a, b) then there exists at least one point $c \in (a, b)$ such that $(f(b) - f(a))g'(c) = (g(b) - g(a))f'(c)$.

Recall the meaning of the first and second derivatives of a function f. The sign of the first derivative tells us whether the *value* of the function increases ($f' > 0$) or decreases ($f' < 0$), whereas the sign of the second derivative tells us whether the *slope* of the function increases ($f'' > 0$) or decreases ($f'' < 0$). This gives us an insight into how to verify that at a stationary point we have a maximum or a minimum.

Assume that $f(x)$ is differentiable in a neighborhood of x_0 and it has a stationary point at x_0, i.e. $f'(x_0) = 0$.

Proposition 24 (The First-Derivative Test for Local Extremum)

If at a stationary point x_0 the first derivative of a function f	*a) changes its sign from positive to negative;* *b) changes its sign from negative to positive;* *c) does not change its sign;*

then the value of the function at x_0, $f(x_0)$, will be	*a) a local maximum;* *b) a local minimum;* *c) neither a local maximum nor a local minimum.*

Proposition 25 (Second-Derivative Test for Local Extremum)

A stationary point x_0 of $f(x)$ will be a local maximum if $f''(x_0) < 0$ and a local minimum if $f''(x_0) > 0$.

However, it may happen that $f''(x_0) = 0$, therefore the second-derivative test is not applicable. To compensate for this, we can extend the latter result and apply the following general test:

Proposition 26 (*n*th-Derivative Test)

If $f'(x_0) = f''(x_0) = \ldots = f^{(n-1)}(x_0) = 0$, $f^{(n)}(x_0) \neq 0$ and $f^{(n)}$ is continuous at x_0 then at point x_0 $f(x)$ has

a) an inflection point if n is odd;

b) a local maximum if n is even and $f^{(n)}(x_0) < 0$;

c) a local minimum if n is even and $f^{(n)}(x_0) > 0$.

To prove this statement we can use the *Taylor series*: If f is continuously differentiable enough at x_0, it can be expanded around a point x_0, i.e. this function can be transformed into a polynomial form in which the coefficient of the various terms are expressed in terms of the derivative values of f, all evaluated at the point of expansion x_0. More precisely,

$$f(x) = \sum_{i=0}^{\infty} \frac{f^{(i)}(x_0)}{i!}(x - x_0)^i,$$

where $k! = 1 \cdot 2 \cdot \ldots \cdot k$, $0! = 1$, $f^{(0)}(x_0) = f(x_0)$,

or

$$f(x) = \sum_{i=0}^{n} \frac{f^{(i)}(x_0)}{i!}(x - x_0)^i + R_n,$$

where

$$R_n = f^{(n+1)}(x_0 + \theta(x - x_0))\frac{(x - x_0)^{n+1}}{(n + 1)!}, \quad 0 < \theta < 1$$

is the remainder in the Lagrange form.

The Maclaurin series is the special case of the Taylor series when we set $x_0 = 0$.

Example 56 *Expand e^x around $x = 0$.*
Since $\frac{d^n}{dx^n}e^x = e^x$ for all n and $e^0 = 1$,

$$e^x = 1 + \frac{x}{1!} + \frac{x^2}{2!} + \frac{x^3}{3!} + \ldots = \sum_{i=0}^{\infty} \frac{x^i}{i!}.$$

The expansion of a function into Taylor series is useful as an approximation device. For instance, we can derive the following approximation:

$$f(x + dx) \approx f(x) + f'(x)dx, \qquad \text{when } dx \text{ is small.}$$

Note, however, that in general polynomial approximation is not the most efficient one.   ⚠

Recipe 9 – How to Find (Global) Maximum or Minimum Points:

If $f(x)$ has a maximum or minimum in a subset S (of $\mathbf{R}^1$ or, in general, $\mathbf{R}^n$), then we need to check the following points for maximum/minimum values of f:
- *interior points of S that are stationary;*
- *points at which f is not differentiable;*
- *extrema of f at the boundary of S.*

Note that the first-derivative test gives us only *relative* (or *local*) extrema. It may happen (due to Weierstrass's theorem) that f reaches ⚠ its global maximum at the border point.

Example 57 *Find the maximum of $f(x) = x^3 - 3x$ in $[-2, 3]$.*

The first-derivative test gives two stationary points: $f'(x) = 3x^2 - 3 = 0$ at $x = 1$ and $x = -1$. The second-derivative test guarantees that $x = -1$ is a local maximum. However, $f(-1) = 2 < f(3) = 18$. Therefore the global maximum of f in $[-2, 3]$ is reached at the border point $x = 2$.

Recipe 10 – How to Sketch the Graph of a Function:
Given f, you should perform the following actions:

1. *Check at which points the function is continuous, differentiable, etc.*

2. *Find its asymptotics, i.e. vertical asymptotes (finite x's at which $f(x) = \infty$) and non-vertical asymptotes. By definition, $y = ax + b$ is a non-vertical asymptote for the curve $y = f(x)$ if $\lim_{x\to\infty}(f(x) - (ax + b)) = 0$.*

   How to find an asymptote for the curve $y - f(x)$ as $x \to \infty$:

   - *Examine $\lim_{x\to\infty} f(x)/x$; if the limit does not exist, there is no asymptote as $x \to \infty$.*

   - *If $\lim_{x\to\infty} f(x)/x = a$, examine $\lim_{x\to\infty} f(x) - ax$; if the limit does not exist, there is no asymptote as $x \to \infty$.*

   - *If $\lim_{x\to\infty} f(x) - ax = b$, the straight line $y = ax + b$ is an asymptote for the curve $y = f(x)$ as $x \to \infty$.*

3. *Find all stationary points and check whether they are local minima, local maxima, or neither.*

4. *Find all points of inflection, i.e. points given by $f''(x) = 0$, at which the second derivative changes its sign.*

   Note that a zero first derivative value is not required for an inflection point. ⚠

5. *Find intervals in which f is increasing (decreasing).*

6. *Find intervals in which f is convex (concave).*

Definition 41 *f is called convex (concave) at x_0 if $f''(x_0) \geq 0$ ($f''(x_0)$ ≤ 0), and strictly convex (concave) if the inequalities are strict.*

Example 58 *Sketch the graph of the function $y = \frac{2x-1}{(x-1)^2}$.*
   a) $y(x)$ is defined for $x \in (-\infty, 1) \cup (1, +\infty)$.
   b) $y(x)$ is continuous in $(-\infty, 1) \cup (1, +\infty)$.
   c) $y(x) \to 0$ if $x \to \pm\infty$, therefore $y = 0$ is the horizontal asymptote.
   d) $y(x) \to \infty$ if $x \to \pm 1$, therefore $x = 1$ is the vertical asymptote.
   e) $y'(x) = -\dfrac{2x}{(x-1)^3}$. $y'(x)$ is not defined at $x = 1$, it is positive if $x \in (0, 1)$ (the function is increasing) and negative if $x \in (-\infty, 0) \cup (1, +\infty)$ (the function is decreasing).
   $x = 0$ is the unique extremum (minimum), $y(0) = -1$.
   f) $y''(x) = 2\dfrac{2x-1}{(x-1)^4}$. It is not defined at $x = 1$, positive at $x \in (-\frac{1}{2}, 1) \cup (1, +\infty)$ (the function is convex) and negative at $x \in (-\infty, -\frac{1}{2})$ (the function is concave).
   $x = -\dfrac{1}{2}$ is the unique point of inflection, $y(-\frac{1}{2}) = -\dfrac{8}{9}$.
   g) Intersections with the axes are given by $x = 0 \Rightarrow y(x) = -1$, $y = 0 \Rightarrow x = \dfrac{1}{2}$.
   Finally, the sketch of the graph looks like this:

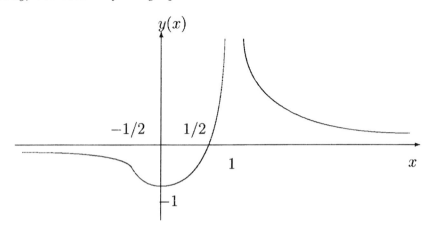

2.5   Integration

In this section, we focus mainly on a univariate case.

Definition 42 *Let $f(x)$ be a continuous function. The indefinite integral of f (denoted by $\int f(x)dx$) is defined as*

$$\int f(x)dx = F(x) + C,$$

where $F(x)$ is such that $F'(x) = f(x)$, and C is an arbitrary constant.

Rules of integration:

- $\int (af(x) + bg(x))dx = a \int f(x)dx + b \int g(x)dx$, a, b are constants (linearity of the integral);

- $\int f'(x)g(x)dx = f(x)g(x) - \int f(x)g'(x)dx$ (integration by parts);

- $\int f(u(t))\frac{du}{dt}dt = \int f(u)du$ (integration by substitution).

   Intuitively, integration by parts can be derived from the rule of differentiation of a product of two functions: Since $(fg)' = f'g - fg'$, or $f'g = (fg)' - fg'$, integration yields $\int f'gdx = fg - \int fg'dx$.

   Integration by substitution follows from the chain rule: Let $F(x)$ be such that $F'(x) = f(x)$. If $z = z(x)$ and $F = F(z(x))$ then $F'_x = F'_z \cdot \frac{dz}{dx}$, or $F'(X(z(x))) = f(z) \cdot \frac{dz}{dx}$. Therefore, by integrating both sides of the equation, $\int f(x)dx = \int f(z)\frac{dz}{dx}dx$.

Some special rules of integration:

$$\int \frac{f'(x)}{f(x)}dx = \ln|f(x)| + C, \qquad \int f'(x)e^{f(x)}dx = e^{f(x)} + C$$
$$\int \frac{1}{x}dx = \ln|x| + C, \qquad \int x^a dx = \frac{x^{a+1}}{a+1} + C, \quad a \neq -1$$
$$\int e^x dx = e^x + C, \qquad \int a^x dx = \frac{a^x}{\ln a} + C \quad a > 0$$

See any calculus reference book for tables of integrals of basic functions.

Example 59   a) $\displaystyle\int \frac{x^2 + 2x + 1}{x}dx = \int xdx + \int 2dx + \int \frac{1}{x}dx =$

$\dfrac{x^2}{2} + 2x + \ln|x| + C.$

b) $\displaystyle\int xe^{-x^2}dx = -\frac{1}{2}\int(-2x)e^{-x^2}dx \overset{\text{\textit{(substitution} } z = -x^2)}{=} -\frac{1}{2}\int e^z dz =$

$-\dfrac{e^{-x^2}}{2} + C.$

c) $\displaystyle\int xe^x dx$ $\overset{(by\ parts,\ f(x)\ =\ f'(x)\ =\ e^x,\ g(x)\ =\ x)}{=\!=}$ $\displaystyle xe^x - \int e^x dx =$
$xe^x - e^x + C.$

d) $\displaystyle\int \frac{dx}{x(x-2)} = \frac{1}{2}\int\left(\frac{1}{x-2} - \frac{1}{x}\right)dx = \frac{1}{2}(\ln|x-2| - \ln|x|) =$
$\ln\sqrt{|(x-2)/x|} + C.$

e) $\displaystyle\int (3x^2 - 2x^4)e^{-x^2}dx = \int 3x^2 e^{-x^2}dx - \int 2x^4 e^{-x^2}dx =$

$$\text{(second term by parts, } f'(x) = 2xe^{-x^2},\ f(x) = -e^{-x^2},\ g(x) = x^3)$$

$$= \int 3x^2 e^{-x^2}dx - \left[-x^3 e^{-x^2} + \int 3x^2 e^{-x^2}dx\right] = x^3 e^{-x^2} + C.$$

Definition 43 (the Newton-Leibniz formula)
   The definite integral of a continuous function f is

$$\int_a^b f(x)dx = F(x)|_a^b = F(b) - F(a), \qquad\qquad (2.1)$$

for $F(x)$ such that $F'(x) = f(x)$ for all $x \in [a,b]$

The indefinite integral is a function. The definite integral is a number!

⚠

We understand the definite integral in the Riemann sense:
Given a partition $a = x_0 < x_1 < \ldots < x_n = b$ and numbers $\zeta_i \in [x_i, x_{i+1}]$, $i = 0, 1, \ldots, n-1$,

$$\int_a^b f(x)dx = \lim_{\max_i(x_{i+1}-x_i)\to 0}\sum_{i=0}^{n-1} f(\zeta_i)(x_{i+1} - x_i).$$

Since every definite integral has a definite value, this value may be interpreted geometrically to be a particular area under a given curve defined by $y = f(x)$.

Note that if a curve lies below the x axis, this area will be negative.

⚠

Properties of definite integrals: For any arbitrary real numbers a, b, c, α, β

- $\int_a^b(\alpha f(x) + \beta g(x))dx = \alpha\int_a^b f(x)dx + \beta\int_a^b g(x)dx$

- $\int_a^b f(x)dx = -\int_b^a f(x)dx$

- $\int_a^a f(x)dx = 0$

- $\int_a^b f(x)dx = \int_a^c f(x)dx + \int_c^b f(x)dx$

- $|\int_a^b f(x)dx| \le \int_a^b |f(x)|dx$

- $\int_a^b f(x)g'(x)dx = \int_{g(a)}^{g(b)} f(u)du, \quad u = g(x)$ (change of variable)

- $\int_a^b f(x)g'(x)dx = f(x)g(x)|_a^b - \int_a^b f'(x)g(x)dx$

While taking definite integrals by substitution, do not forget to change the limits of integration accordingly. $\triangle$

Example 60 *It was shown earlier that by substitution $z(x) = -x^2$, an indefinite integral $\int xe^{-x^2}dx = -0.5\int e^z dz$. Thus its definite counterpart $\int_1^2 xe^{-x^2}dx = -0.5\int_{-1}^{-4} e^z dz$.*

Some more useful results: If λ is a real parameter,

- $\frac{d}{d\lambda}\int_{a(\lambda)}^{b(\lambda)} f(x)dx = f(b(\lambda))b'(\lambda) - f(a(\lambda))a'(\lambda)$

  In particular, $\frac{d}{dx}\int_a^x f(t)dt = f(x)$

  If $\int_a^b \frac{d}{d\lambda}f(x,\lambda)dx$ exists, then

- $\frac{d}{d\lambda}\int_a^b f(x,\lambda)dx = \int_a^b f'_\lambda(x,\lambda)dx$

Example 61 *We may apply this formula when we need to evaluate a definite integral, which cannot be integrated in elementary functions. For instance, let us find*

$$I(\lambda) = \int_0^1 xe^{-\lambda x}dx.$$

If we introduce another integral

$$J(\lambda) = \int_0^1 e^{-\lambda x}dx = -\frac{e^{-\lambda x}}{\lambda}|_0^1 = -\frac{e^{-\lambda} - 1}{\lambda},$$

then

$$I(\lambda) = -\int_0^1 \frac{d}{d\lambda}e^{-\lambda x}dx = -\frac{dJ(\lambda)}{d\lambda} = \frac{1 - e^{-\lambda}(1+\lambda)}{\lambda^2}.$$

We can combine the last two formulas to get:

- Leibniz's formula:

$$\frac{d}{d\lambda}\int_{a(\lambda)}^{b(\lambda)} f(x,\lambda)dx = f(b(\lambda),\lambda)b'(\lambda) - f(a(\lambda),\lambda)a'(\lambda)$$

$$+ \int_{a(\lambda)}^{b(\lambda)} f'_\lambda(x,\lambda)dx$$

Proposition 27 (Mean Value Theorem)

If $f(x)$ is continuous in $[a,b]$ then there exists at least one point $c \in (a,b)$ such that

$$\int_a^b f(x)dx = f(c)(b-a).$$

Intuitively, the integral mean value theorem can be derived from Lagrange's, or the calculus mean value theorem: Let $F(x)$ satisfy the conditions of Lagrange's theorem in the interval $[a,b]$ and $F'(x) = f(x)$. Then Lagrange's theorem implies $F(b) - F(a) = F'(c)(b-a)$ for some $c \in (a,b)$, and by the Newton-Leibniz formula, $\int_a^b f(x)dx = F'(c)(b - a) = f(c)(b-a)$.

Numerical methods of approximating the definite integrals:
a) The trapezoid formula: Denoting $y_i = f(a + i\frac{b-a}{n})$,

$$\int_a^b f(x)dx \approx \frac{b-a}{2n}(y_0 + 2y_1 + 2y_2 + \ldots + 2y_{n-1} + y_n),$$

n is an arbitrary integer number.
b) Simpson's formula: Denoting $y_i = f(a + i\frac{b-a}{2n})$,

$$\int_a^b f(x)dx \approx \frac{b-a}{6n}(y_0 + 4\sum_{i=1}^{n} y_{2i-1} + 2\sum_{i=1}^{n-1} y_{2i} + y_{2n}).$$

Improper integrals are integrals of one of the following forms:

- $\int_a^\infty f(x)dx = \lim_{b\to\infty}\int_a^b f(x)dx$ or $\int_{-\infty}^b f(x)dx = \lim_{a\to-\infty}\int_a^b f(x)dx.$

- $\int_a^b f(x)dx = \lim_{\epsilon\to 0}\int_{a+\epsilon}^b f(x)dx$ where $\lim_{x\to a} f(x) = \infty.$

- $\int_a^b f(x)dx = \lim_{\epsilon \to 0} \int_a^{b-\epsilon} f(x)dx$ where $\lim_{x \to b} f(x) = \infty$.

If these limits exist, the improper integral is said to be *convergent*, otherwise it is called *divergent*.

Example 62

$$\int_0^1 \frac{1}{x^p}dx = \lim_{\epsilon \to 0_+} \int_\epsilon^1 \frac{1}{x^p}dx = \begin{cases} \lim_{\epsilon \to 0_+}(\ln 1 - \ln \epsilon), & p = 1 \\ \lim_{\epsilon \to 0_+}(\frac{1}{1-p} - \frac{\epsilon^{1-p}}{1-p}), & p \neq 1 \end{cases}$$

Therefore the improper integral converges if $p < 1$ and diverges if $p \geq 1$.

Economics Application 3 (Optimal Subsidy and Compensated Demand Curve)

Consider the consumer's dual problem (recall that the dual to utility maximization is expenditure minimization: we minimize the objective function $M = p_x x + p_y y$, subject to the utility constraint $U = U(x, y)$, where M is income level, x, y and p_x, p_y are quantities and prices of two goods, respectively).

The solution to the expenditure-minimization problem under the utility constraint is the budget line with income M^*, $M^*(p_x, p_y) = p_x x^* + p_y y^*$.

If we hold utility and p_y constant, due to Hotelling's lemma

$$\frac{dM^*(p_x)}{dp_x} = x_c^*(p_x),$$

where $x_c^*(p_x)$ is the income-compensated (or simply compensated) demand curve for good x.

By rearranging terms, $dM^*(p_x) = x_c^*(p_x)dp_x$, where dM^* is the optimal subsidy for an infinitesimal change in price, we can "add up" a series of infinitesimal changes by taking integrals over those infinitesimal changes. Thus, if price increases from p_x^1 to p_x^2, the optimal subsidy S^* required to maintain a given level of utility would be the integral over all the infinitesimal income changes as price changes:

$$S^* = \int_{p_x^1}^{p_x^2} x_c^*(p_x)dp_x.$$

For example, if $p_x^1 = 1$, $p_x^2 = 4$, $x_c^*(p_x) = \frac{1}{\sqrt{p_x}}$ then

$$S^* = \int_1^4 \frac{1}{\sqrt{p_x}}dp_x = 2\sqrt{p_x}\big|_1^4 = 2 \cdot (2 - 1) = 2.$$

Multivariate definite integrals can be evaluated in a similar fashion. If Ω is a subset of R^n that contains all points $(x_1, \ldots, x_n)$ such that $a_i \leq x_i \leq b_i$ for all $i = 1, 2, \ldots, n$, and $f(x_1, \ldots, x_n)$ is a continuous function in Ω then the multiple (n-tiple) definite integral of f over Ω is defined as

$$\int \int \cdots \int_\Omega f(x_1, \ldots, x_n) dx_1 \cdots dx_n =$$

$$= \int_{a_n}^{b_n} \left(\int_{a_{n-1}}^{b_{n-1}} \cdots \left(\int_{a_1}^{b_1} f(x_1, \ldots, x_{n-1} x_n) dx_1 \right) \ldots dx_{n-1} \right) dx_n.$$

This formula suggests that the integration is to be performed first with respect to x_1, all other variables being kept constant, then with respect to x_2, treating the remaining variables $x_3, \ldots, x_n$ as constants, etc.

Example 63 *A random variable X is said to have standard normal distribution with zero mean and unit variance (this is a special case of the famous Gauss distribution, or normal distribution) if its density function $f(x) = \dfrac{1}{\sqrt{2\pi}} e^{-\frac{x^2}{2}}$. To check the important property of the density function $\int_{-\infty}^{\infty} f_X(x) dx = 1$, we compute*

$$\left(\int_{-\infty}^{\infty} \frac{1}{\sqrt{2\pi}} e^{-\frac{x^2}{2}} dx \right)^2 = \frac{1}{\sqrt{2\pi}} \int_{-\infty}^{\infty} \int_{-\infty}^{\infty} e^{-\frac{x^2}{2}} e^{-\frac{y^2}{2}} dx dy$$

$$= \frac{1}{\sqrt{2\pi}} \int_{-\infty}^{\infty} \int_{-\infty}^{\infty} e^{-\frac{x^2+y^2}{2}} dx dy$$

$$= \frac{1}{\sqrt{2\pi}} \int_0^{\infty} \int_0^{2\pi} e^{-\frac{\rho^2}{2}} |J| d\rho d\theta$$

$$= \frac{1}{\sqrt{2\pi}} \int_0^{\infty} \rho e^{-\frac{\rho^2}{2}} d\rho \int_0^{2\pi} d\theta$$

$$= -\int_0^{\infty} e^{-r} dr = 1.$$

Here we introduce polar coordinates $x = \rho \cos\theta$, $y = \rho \sin\theta$, with the Jacobian of the transformation equaling

$$|J| = \det \begin{pmatrix} x_\rho & y_\rho \\ x_\theta & y_\theta \end{pmatrix} = \det \begin{pmatrix} \cos\theta & \sin\theta \\ -\rho\sin\theta & \rho\cos\theta \end{pmatrix}$$

$$= \rho(\cos^2\theta + \sin^2\theta) = \rho.$$

2.6   Functions of More than One Variable

Let us consider a function $y = f(x_1, x_2, \ldots, x_n)$ where the variables x_i, $i = 1, 2, \ldots, n$ are all independent of one other, i.e. each varies without affecting others. The *partial derivative of y with respect to* x_i is defined as

$$
\begin{aligned}
\frac{\partial y}{\partial x_i} &= \lim_{\Delta x_i \to 0} \frac{\Delta y}{\Delta x_i} \\
&= \lim_{\Delta x_i \to 0} \frac{f(x_1, \ldots, x_i + \Delta x_i, \ldots, x_n) - f(x_1, \ldots, x_i, \ldots, x_n)}{\Delta x_i}
\end{aligned}
$$

for all $i = 1, 2, \ldots, n$. If these limits exist, our function is *differentiable* with respect to all arguments. Partial derivatives are often denoted with subscripts, e.g. $\frac{\partial f}{\partial x_i} = f_{x_i} = f_i\, (= f'_{x_i})$.

Example 64 *If* $y = 4x_1^3 + x_1 x_2 + \ln x_2$ *then* $\frac{\partial y}{\partial x_1} = 12x_1^2 + x_2$, $\frac{\partial y}{\partial x_2} = x_1 + 1/x_2$.

Again, $C^{(1)}(R^n)$ denotes the set of real-valued functions $f : R^n \to R^n$ such that all partial derivatives $f_1, \ldots, f_n$ of f exist and are continuous for all real arguments.

Definition 44 The total differential *of a function f is defined as*

$$
dy = \frac{\partial f}{\partial x_1} dx_1 + \frac{\partial f}{\partial x_2} dx_2 + \ldots + \frac{\partial f}{\partial x_n} dx_n.
$$

If $y = f(x_1, \ldots, x_n)$, $x_i = x_i(t_1, \ldots, t_m)$, $i = 1, 2, \ldots, n$, then for all $j = 1, 2, \ldots, m$

$$
\frac{\partial y}{\partial t_j} = \sum_{i=1}^{n} \frac{\partial f(x_1, \ldots, x_n)}{\partial x_i} \frac{\partial x_i}{\partial t_j}.
$$

This rule is called *the chain rule.*

Example 65 *If* $z = f(x, y)$, $x = u(t)$, $y = v(t)$ *then*

$$
\frac{dz}{dt} = \frac{\partial f}{\partial x} \frac{dx}{dt} + \frac{\partial f}{\partial y} \frac{dy}{dt}.
$$

(a special case of the chain rule)

Example 66 *If $z = f(x, y, t)$ where $x = u(t)$, $y = v(t)$ then*

$$\frac{dz}{dt} = \frac{\partial f}{\partial x}\frac{dx}{dt} + \frac{\partial f}{\partial y}\frac{dy}{dt} + \frac{\partial f}{\partial t}.$$

(dz/dt is sometimes called the total derivative).

The *second-order partial derivative* of $y = f(x_1, \ldots, x_n)$ is defined as

$$\frac{\partial^2 y}{\partial x_j \partial x_i} = f_{x_i x_j}(x_1, \ldots, x_n) = \frac{\partial}{\partial x_j} f_{x_i}(x_1, \ldots, x_n), \quad i, j = 1, 2, \ldots, n.$$

Proposition 28 (Schwarz's Theorem or Young's Theorem)
If at least one of the two partials is continuous, then

$$\frac{\partial^2 f}{\partial x_j \partial x_i} = \frac{\partial^2 f}{\partial x_i \partial x_j}, \quad i, j = 1, 2, \ldots, n.$$

Similar to the case of one variable, we can expand a function of several variables in a polynomial form of Taylor series, using partial derivatives of higher order.

Definition 45 *The second-order total differential of a function $y = f(x_1, x_2, \ldots, x_n)$ is defined as*

$$d^2 y = d(dy) = \sum_{i=1}^{n} \sum_{j=1}^{n} \frac{\partial^2 f}{\partial x_i \partial x_j} dx_i dx_j, \quad i, j = 1, 2, \ldots, n.$$

Example 67 *Let $z = f(x, y)$ and $f(x, y)$ satisfy the conditions of Schwarz's theorem (i.e. $f_{xy} = f_{yx}$). Then $d^2 z = f_{xx} dx^2 + 2f_{xy} dx dy + f_{yy} dy^2$*

The total differentials can also be used as an approximation device. For instance, if $y = f(x_1, x_2)$ then by definition

$$y - y_0 = \Delta f(x_1^0, x_2^0) \approx df(x_1^0, x_2^0).$$

Therefore,

$$
\begin{aligned}
f(x_1, x_2) &\approx f(x_1^0, x_2^0) + df(x_1^0, x_2^0) \\
&\approx f(x_1^0, x_2^0) + \frac{\partial f(x_1^0, x_2^0)}{\partial x_1}(x_1 - x_1^0) + \frac{\partial f(x_1^0, x_2^0)}{\partial x_2}(x_2 - x_2^0).
\end{aligned}
$$

Further, the accuracy of the approximation can be improved by using differentials of higher order. For instance, with second-order differentials involved,

$$y - y_0 \approx df(x_1^0, x_2^0) + \frac{d^2 f(x_1^0, x_2^0)}{2}.$$

In the multivariate case, second-order approximation becomes

$$f(\mathbf{x}) \approx f(\mathbf{x^0}) + \nabla f(\mathbf{x^0})\Delta\mathbf{x} + \frac{1}{2}\Delta\mathbf{x}'H(\mathbf{x^0})\Delta\mathbf{x},$$

where $\mathbf{x}$, $\mathbf{x^0}$ and $\Delta\mathbf{x} = \mathbf{x} - \mathbf{x^0}$ are n-dimensional vectors, H is the so-called Hessian matrix of second-order mixed partial derivatives, and the gradient vector $\nabla f = (\partial f/\partial x_1, \ldots, \partial f/\partial x_n)$.

Definition 46 *Function $f : R^n \longrightarrow R$ is homogeneous of degree r if for any vector $x = (x_1, \ldots, x_n)$ and any positive real number $t > 0$*

$$f(tx) = t^r f(x).$$

Proposition 29 *If f is homogeneous of degree r then f_i is homogeneous of degree $r - 1$.*

Proposition 30 (Euler's Theorem) *If f is homogeneous of degree r then for all x*

$$f_1(x)x_1 + \cdots + f_n(x)x_n = rf(x).$$

Euler's theorem can be inverted:

Proposition 31 *If there exists positive r such that for all x*

$$x_1 f_1(x) + x_2 f_2(x) + \cdots + x_n f_n(x) = rf(x)$$

then f is homogeneous of degree r.

Economics Application 4 *It can be shown that with homogeneous utility function $u(x)$, the marginal rate of substitution (MRS) is homogeneous of degree 0.*

Indeed, if $u(x)$ is homogeneous of degree r then

$$u_i(tx) = \frac{\partial}{\partial x_i}(u(tx)) = t^{r-1}u_i(x),$$

$$u_j(tx) = \frac{\partial}{\partial x_j}(u(tx)) = t^{r-1}u_j(x),$$

and

$$MRS_{ij}(tx) = \frac{u_i(tx)}{u_j(tx)} = \frac{t^{r-1}u_i(x)}{t^{r-1}u_j(x)} = t^0 \cdot \frac{u_i(x)}{u_j(x)}.$$

Thus MRS is homogeneous of degree 0.

Definition 47 *Function $f : R^n \longrightarrow R$ is said to be homothetic if there exists an increasing function $F : R \longrightarrow R$ and function $g : R^n \longrightarrow R$, which is homogeneous of degree 1, such that*

$$f(x) = F(g(x)).$$

2.7 Unconstrained Optimization in the Case of More than One Variable

Let $z = f(x_1, \ldots, x_n)$.

Definition 48 *z has a stationary point at $x^* = (x_1^*, \ldots, x_n^*)$ if $dz = 0$ at x^*, i.e. all f'_{x_i} should vanish at x^*.*

The conditions

$$f'_{x_i}(x^*) = 0, \quad \text{for all } i = 1, 2, \ldots, n$$

are called *the first-order necessary conditions (F.O.N.C.)*.

The *second-order necessary condition (S.O.N.C.)* requires sign semi-definiteness of the second-order total differential.

Definition 49 $z = f(x_1, \ldots, x_n)$ *has a local maximum (minimum) at* $x^* = (x_1^*, \ldots, x_n^*)$ *if* $f(x^*) - f(x) \geq 0 (\leq 0)$ *for all* x *in some neighborhood of* x^*.

The second-order sufficient condition (S.O.C.):
x^* is a maximum (minimum) if $d^2 z$ at x^* is negative definite (positive definite).

Recipe 11 – How to Check the Sign Definiteness of $d^2 z$:

Let us define the symmetric Hessian matrix H *of second-order partial derivatives as*

$$H = \begin{pmatrix} \frac{\partial^2 f}{\partial x_1 \partial x_1} & \cdots & \frac{\partial^2 f}{\partial x_1 \partial x_n} \\ \vdots & & \vdots \\ \frac{\partial^2 f}{\partial x_n \partial x_1} & \cdots & \frac{\partial^2 f}{\partial x_n \partial x_n} \end{pmatrix}.$$

Thus, the sign definiteness of $d^2 z$ *is equivalent to the sign definiteness of the quadratic form* H *evaluated at* x^*. *We can apply either the principal minors test or the eigenvalue test discussed in the previous chapter.*

Example 68 *Consider* $z = f(x, y)$. *If the first-order necessary conditions are met at* x^* *then*

$$x^* \text{ is a (local)} \quad \left\{ \begin{array}{c} minimum \\ maximum \end{array} \right\}$$

$$if \quad \left\{ \begin{array}{l} f''_{xx}(x^*) > 0, \; f''_{xx}(x^*) f''_{yy}(x^*) - (f''_{xy}(x^*))^2 > 0 \\ f''_{xx}(x^*) < 0, \; f''_{xx}(x^*) f''_{yy}(x^*) - (f''_{xy}(x^*))^2 > 0 \end{array} \right\}.$$

If $f''_{xx}(x^*) f''_{yy}(x^*) - (f''_{xy}(x^*))^2 < 0$, *we identify a stationary point as a saddle point.*

Note the sufficiency of this test. For instance, even if $f''_{xx}(x^*) f''_{yy}(x^*) - (f''_{xy}(x^*))^2 = 0$, we still may have an extremum at (x^*). ⚠

Example 69 (Classification of Stationary Points of a $C^{(2)}$-Function[3] of n Variables)

---

[3] Recall that $C^{(2)}$ means 'twice continuously differentiable'.

If $x^* = (x_1^*, \ldots, x_n^*)$ is a stationary point of $f(x_1, \ldots, x_n)$ and if $|H_k(x^*)|$ is the following determinant evaluated at x^*,

$$|H_k| = \det \begin{vmatrix} \dfrac{\partial^2 f}{\partial x_1 \partial x_1} & \cdots & \dfrac{\partial^2 f}{\partial x_1 \partial x_k} \\ \vdots & & \vdots \\ \dfrac{\partial^2 f}{\partial x_k \partial x_1} & \cdots & \dfrac{\partial^2 f}{\partial x_k \partial x_k} \end{vmatrix}, \quad k = 1, 2, \ldots, n$$

then

a) if $(-1)^k |H_k(x^*)| > 0$, $k = 1, 2, \ldots, n$, then x^* is a local maximum;

b) if $|H_k(x^*)| > 0$, $k = 1, 2, \ldots, n$, then x^* is a local minimum;

c) if $|H_n(x^*)| \neq 0$ and neither of the two conditions above is satisfied, then x^* is a saddle point.

Example 70 Let us find the extrema of the function $z(x, y) = x^3 - 8y^3 + 6xy + 1$.

$$\text{F.O.N.C.:} \quad z_x = 3(x^2 + 2y) = 0, \quad z_y = 6(-4y^2 + x) = 0.$$

Solving F.O.N.C. for x, y we obtain two stationary points: $x = y = 0$ and $x = 1$, $y = -1/2$.

S.O.C.: The Hessian matrices evaluated at stationary points are

$$H|_{(0,0)} = \begin{pmatrix} 6x & 6 \\ 6 & -48y \end{pmatrix}_{(0,0)} = \begin{pmatrix} 0 & 6 \\ 6 & 0 \end{pmatrix},$$

$$H|_{(1,-1/2)} = \begin{pmatrix} 6 & 6 \\ 6 & 24 \end{pmatrix},$$

therefore $(0,0)$ is a saddle point, $(1, -1/2)$ is a local minimum.

2.8   The Implicit Function Theorem

Proposition 32 (The Two-Dimensional Case)

If $F(x,y) \in C^{(k)}$ in a set D (i.e. F is k times continuously differentiable at any point in D) and (x_0, y_0) is an interior point of D, $F(x_0, y_0) = c$ (c is constant) and $F'_y(x_0, y_0) \neq 0$ then the equation $F(x,y) = c$ defines y as a $C^{(k)}$-function of x in some neighborhood of (x_0, y_0), i.e. $y = \psi(x)$ and

$$\frac{dy}{dx} = -\frac{F'_x(x,y)}{F'_y(x,y)}.$$

Example 71 *Let us consider the function* $\dfrac{1 - e^{-ax}}{1 + e^{ax}}$, *where a is a parameter, $a > 1$.*

What happens to the extremum value of x if a increases by a small number?

To answer this question, first write down the F.O.N.C. for a stationary point:

$$f'_x = \frac{a(e^{-ax} - e^{ax} + 2)}{((1 + e^{ax})^2} = 0.$$

If this expression vanishes at, say, $x = x^*$, we can apply the implicit function theorem to the F.O.N.C. and obtain

$$\frac{dx}{da} = -\frac{\partial(e^{-ax} - e^{ax} + 2)/\partial a}{\partial(e^{-ax} - e^{ax} + 2)/\partial x} = -\frac{x}{a} < 0$$

in some neighborhood of (x^*, a). The negativity of the derivative is due to the fact that x^* is positive, therefore, every x in a small neighborhood of x^* should be positive as well.

Let us formulate the general case of the implicit function theorem.

Consider the system of m equations with n *exogenous* variables, $x_1, \ldots, x_n$, and m *endogenous* variables, $y_1, \ldots, y_m$:

$$\begin{cases} f^1(x_1, \ldots, x_n, y_1, \ldots, y_m) = 0 \\ f^2(x_1, \ldots, x_n, y_1, \ldots, y_m) = 0 \\ \quad \cdots \ \cdots \ \cdots \\ f^m(x_1, \ldots, x_n, y_1, \ldots, y_m) = 0 \end{cases} \tag{2.2}$$

Our objective is to find $\dfrac{\partial y_j}{\partial x_i}$, $i = 1, \ldots, n$, $j = 1, \ldots, m$.

If we take the total differentials of all f^j, we get

$$\frac{\partial f^1}{\partial y_1} dy_1 + \ldots + \frac{\partial f^1}{\partial y_m} dy_m = -\left(\frac{\partial f^1}{\partial x_1} dx_1 + \ldots + \frac{\partial f^1}{\partial x_n} dx_n \right),$$

$$\ldots$$

$$\frac{\partial f^m}{\partial y_1} dy_1 + \ldots + \frac{\partial f^m}{\partial y_m} dy_m = -\left(\frac{\partial f^m}{\partial x_1} dx_1 + \ldots + \frac{\partial f^m}{\partial x_n} dx_n \right).$$

Now allow only x_i to vary (i.e. $dx_i \neq 0$, $dx_l = 0$, $l = 1, 2, \ldots, i-1, i+1, \ldots, n$) and divide each remaining term in the system above by dx_i. Thus we arrive at the following linear system with respect to $\dfrac{\partial y_1}{\partial x_i}, \ldots \dfrac{\partial y_m}{\partial x_i}$ (note that in fact $dy_1/dx_i, \ldots, dy_m/dx_i$ should be interpreted as partial derivatives with respect to x_i since we allow only x_i to vary, holding constant all other x_l):

$$\frac{\partial f^1}{\partial y_1} \frac{\partial y_1}{\partial x_i} + \ldots + \frac{\partial f^1}{\partial y_m} \frac{\partial y_m}{\partial x_i} = -\frac{\partial f^1}{\partial x_i},$$

$$\ldots$$

$$\frac{\partial f^m}{\partial y_1} \frac{\partial y_1}{\partial x_i} + \ldots + \frac{\partial f^m}{\partial y_m} \frac{\partial y_m}{\partial x_i} = -\frac{\partial f^m}{\partial x_i},$$

which can be solved for $\dfrac{\partial y_1}{\partial x_i}, \ldots \dfrac{\partial y_m}{\partial x_i}$ by, say, Cramer's rule or by the inverse matrix method.

Let us define the *Jacobian matrix* of $f^1, \ldots, f^m$ with respect to $y_1, \ldots, y_m$ as

$$\mathcal{J} = \begin{pmatrix} \frac{\partial f^1}{\partial y_1} & \cdots & \frac{\partial f^1}{\partial y_m} \\ \vdots & \ddots & \vdots \\ \frac{\partial f^m}{\partial y_1} & \cdots & \frac{\partial f^m}{\partial y_m} \end{pmatrix}.$$

Given the definition of Jacobian, we are in a position to formulate the general result:

Proposition 33 (General Implicit Function Theorem)

Suppose $f^1, \ldots, f^m$ are $C^{(k)}$-functions in a set $D \subset \mathbf{R}^{n+m}$.

Let $(x^*, y^*) = (x_1^*, \ldots, x_n^*, y_1^*, \ldots, y_m^*)$ be a solution to (2.2) in the interior of A.

Suppose also that $\det(\mathcal{J})$ *does not vanish at* (x^*, y^*). *Then (2.2) defines* $y_1, \ldots, y_m$ *as* $C^{(k)}$-*functions of* $x_1, \ldots, x_n$ *in some neighborhood of* (x^*, y^*) *and in that neighborhood, for* $j = 1, 2, \ldots, n,$

$$
\begin{pmatrix} \frac{\partial y_1}{\partial x_j} \\ \vdots \\ \frac{\partial y_m}{\partial x_j} \end{pmatrix} = -\mathcal{J}^{-1} \cdot \begin{pmatrix} \frac{\partial f^1}{\partial x_j} \\ \vdots \\ \frac{\partial f^m}{\partial x_j} \end{pmatrix}.
$$

Example 72 *Given the system of two equations* $x^2 + y^2 = z^2/2$ *and* $x + y + z = 2$, *find* $x' = x'(z)$, $y' = y'(z)$ *in the neighborhood of the point* $(1, -1, 2)$.

In this setup, $f^1(x, y, z) = x^2 + y^2 - z^2/2$, $f^2(x, y, z) = x + y + z - 2$, $f^1, f^2 \in C^\infty(\mathbf{R}^3)$, $f^1(1, -1, 2) = f^2(1, -1, 2) = 0$, *and at* $(1, -1, 2)$

$$
\det(\mathcal{J}) = \det \begin{pmatrix} \frac{\partial f^1}{\partial x} & \frac{\partial f^1}{\partial y} \\ \frac{\partial f^2}{\partial x} & \frac{\partial f^2}{\partial y} \end{pmatrix} = \det \begin{pmatrix} 2x & 2y \\ 1 & 1 \end{pmatrix}_{(1,-1,2)}
$$

$$
= \det \begin{pmatrix} 2 & -2 \\ 1 & 1 \end{pmatrix} = 4 \neq 0.
$$

Therefore the conditions of the implicit function theorem are satisfied and

$$
\begin{pmatrix} \frac{dx}{dz} \\ \frac{dy}{dz} \end{pmatrix} = -\frac{1}{2x - 2y} \begin{pmatrix} 1 & -2y \\ -1 & 2x \end{pmatrix} \begin{pmatrix} -z \\ 1 \end{pmatrix} \implies \begin{aligned} \frac{dx}{dz} &= \frac{z + 2y}{2x - 2y} \\ \frac{dy}{dz} &= -\frac{z + 2x}{2x - 2y} \end{aligned}
$$

The Jacobian matrix can also be applied to test whether functional (linear or nonlinear) dependence exists among a set of n functions in n variables. More precisely, if we have n functions of n variables $g_i = g_i(x_1, \ldots, x_n)$, $i = 1, 2, \ldots, n$, then the determinant of the Jacobian matrix of $g_1, \ldots, f_n$ with respect to $x_1, \ldots, x_m$ will be identically zero for all values of $x_1, \ldots, x_n$ if and only if the n functions $g_1, \ldots, g_n$ are functionally (linearly or nonlinearly) dependent.

Proposition 34 (Envelope Theorem)

Let $f \in C^2(R^{n+m})$, *and suppose that at* a^0 *there is a solution to the optimization problem*

$$
\max_x f(x, a),
$$

where x *is an n-dimensional vector of exogenous variables, and* a *is an m-dimensional vector of endogenous variables.*

Let us denote the solution by x^0, and assume that $f_{xx}(x^0, a^0)$ is negative definite. Then by the Implicit Function Theorem there exists a C^1 function $x^*(a)$, such that

$$f_x(x^*(a), a) = 0.$$

Furthermore, it can be shown that the matrix of second-order mixed partial derivatives $f_{xx}(x^*(a), a)$ is negative definite also for those a's which are sufficiently close to a^0. Thus a solution to the maximization problem exists in the neighborhood of a^0. If we define the optimal value function $V(\cdot)$ as

$$V(a) = max_x f(x, a)$$

then $V(a) = f(x^*(a), a)$.
The envelope theorem states the following result:

$$V_a(a^0) = \frac{\partial x^*(a^0)}{\partial a} f_x(x^*, a^0) + f_a(x^*, a^0) = f_a(x^*, a^0).$$

The latter follows directly from the first-order conditions $f_x(x^0, a^0) = 0$.

In particular, with $a = (a_1, \ldots, a_n)$,

$$\frac{dV(a)}{da_i} = \sum_{j=1}^{n} \frac{\partial f(x^*(a), a)}{\partial x_j^*} \cdot \frac{\partial x_j^*}{\partial a_i} + \frac{\partial f(x^*, a)}{\partial a_i}$$

$$= \frac{\partial f(x^*, a)}{\partial a_i} = \frac{\partial f(x, a)}{\partial a_i} \Big|_{x=x^*}.$$

Example 73 *Consider the following maximization problem:*

$$\max_{x, y} f(x, y; a, b) = ax^2 - x + by^2 - y$$

1. *Derive x^* and y^* that satisfy the first-order conditions.*

2. *Use the Hessian matrix to specify conditions for concavity and convexity. Under what restrictions on a and b will the stationary point (x^*, y^*) become a maximum?*

3. *Perform comparative statics (or sensitivity analysis): Compute marginal change in y^* as a and b marginally change (assume independence of a and b).*

4. *Find how the value of the function $f(x^*(a,b); y^*(a,b))$ varies as a changes. Do it two ways. First, plug in $x^*(a,b)$ and $y^*(a,b)$ into $f(\cdot, \cdot)$ and take the derivative of $f(x^*(a,b); y^*(a,b))$ with respect to a. Second, use the envelope theorem. (Both methods should produce the same result. Which method you think is easier to implement?)*

Solution:

1. $x^* = \frac{1}{2a}$ and $y^* = \frac{1}{2b}$.

2. *The Hessian matrix for $f(x,y;a,b)$ is $\mathcal{H} = \begin{pmatrix} 2a & 0 \\ 0 & 2b \end{pmatrix}$. The conditions for (strict) concavity are $a < 0$ and $4ab > 0$, which yields $a < 0$ and $b < 0$. Under these conditions the stationary point is maximum. The conditions for (strict) convexity are $a > 0$ and $4ab > 0$.*

3. $\frac{dy^*}{da} = 0$ and $\frac{dy^*}{db} = -\frac{1}{2b^2} < 0$

4. *Method 1: Plugging the values of x^* and y^* back into the original function implies $f(x^*, y^*; a, b) = -\frac{1}{4a} - \frac{1}{4b}$. Therefore, $\frac{\partial f(x^*, y^*; a, b)}{\partial a} = \frac{1}{4a^2} > 0$.*

   Method 2: The envelope theorem for unconstrained maximization can be alternatively set up as follows: Assume the function $f(\mathbf{x}; \mathbf{p})$ is maximized with respect to x. Consider the function f at the optimum, that is, $f(\mathbf{x}^*(\mathbf{p}), \mathbf{p})$. The total differential of this function with respect to p_i is then

$$\frac{df(\mathbf{x}^*(\mathbf{p}), \mathbf{p})}{dp_i} = \frac{\partial f(\mathbf{x}^*(\mathbf{p}), \mathbf{p})}{\partial p_i}.$$

   Thus, using the envelope theorem,

$$\frac{d}{da} f(x^*, y^*; a, b) = \frac{\partial}{\partial a} f(x, y; a, b) \Big|_{x=x^*, y=y^*}.$$

Computing the partial and evaluating at $x = x^*$, $y = y^*$ we find

$$\frac{\partial}{\partial a} f(x, y; a, b)\bigg|_{x=x^*, y=y^*} = x^2\big|_{x=x^*, y=y^*}$$

$$= (x^*)^2 = \left(\frac{1}{2a}\right)^2 = \frac{1}{4a^2}.$$

Economics Application 5 (Profit-Maximizing Firm)

Consider that a firm has the profit function $F(l, k) = pf(l, k) - wl - rk$ where f is the firm's (neoclassical) production function, p is the price of output, l, k are the amount of labor and capital employed by the firm (in units of output), w is the real wage and r is the real rental price of capital. The firm takes p, w and r as given. Assume that the Hessian matrix of f is negative definite.

a) Show that if the wage increases by a small amount, then the firm decides to employ less labor;

b) Show that if the wage increases by a small amount and the firm is constrained to maintain the same amount of capital k_0, then the firm will reduce l by less than it does in part a).

Solution:

a) The firm's objective is to choose l and k so that it maximizes its profits. The first-order conditions for maximization are:

$$\begin{cases} p\frac{\partial f}{\partial l}(l, k) - w = 0 \\ p\frac{\partial f}{\partial k}(l, k) - r = 0 \end{cases} \quad or, \quad \begin{cases} F^1(l, k; w) = 0 \\ F^2(l, k; r) = 0 \end{cases}$$

The Jacobian of this system of equations is:

$$|J| = \begin{vmatrix} \frac{\partial F^1}{\partial l} & \frac{\partial F^1}{\partial k} \\ \frac{\partial F^2}{\partial l} & \frac{\partial F^2}{\partial k} \end{vmatrix} = p^2 \begin{vmatrix} f_{11}(l, k) & f_{12}(l, k) \\ f_{12}(l, k) & f_{22}(l, k) \end{vmatrix} = p^2|H| > 0,$$

where f_{ij} is the second-order partial derivative of f with respect to the jth and the ith arguments and $|H|$ is the Hessian of f. The fact that H is negative definite means that $f_{11} < 0$ and $|H| > 0$ (which also implies $f_{22} < 0$).

Since $|J| \neq 0$ the implicit function theorem can be applied; thus, l and k are implicit functions of w. The first-order partial derivative of l with respect to w is:

$$\frac{\partial l}{\partial w} = -\frac{\begin{vmatrix} -1 & pf_{12}(l,k) \\ 0 & pf_{22}(l,k) \end{vmatrix}}{|J|} = \frac{f_{22}(l,k)}{p|H|} < 0.$$

Thus, the optimal l falls in response to a small increase in w.

b) If the firm's amount of capital is fixed at k_0, the profit function is $F(l) = pf(l, k_0) - wl - rk_0$ and the first-order condition is

$$p\frac{\partial f}{\partial l}(l, k_0) - w = 0.$$

Since $f_{11} \neq 0$, the above condition defines l as an implicit function of w. Taking the derivative of the F.O.C. with respect to l, we have:

$$pf_{11}(l, k_0)\frac{\partial l}{\partial w} = 1 \ or$$

$$\frac{\partial l}{\partial w} = \frac{1}{pf_{11}(l, k_0)}.$$

This change in l is smaller in absolute value than the change in l in part a) since

$$\frac{f_{22}(l, k_0)}{p|H|} < \frac{1}{pf_{11}(l, k_0)}$$

(Note that both ratios are negative.)

For another economic application see the section 'Constrained Optimization'.

2.9 Concavity, Convexity, Quasi-concavity and Quasi-convexity

Definition 50 *A function $z = f(x_1, \ldots, x_n)$ is concave if and only if for any pair of distinct points $u = (u_1, \ldots, u_n)$ and $v = (v_1, \ldots, v_n)$ in the domain of f and for any $\theta \in (0,1)$*

$$\theta f(u) + (1 - \theta)f(v) \leq f(\theta u + (1 - \theta)v)$$

and convex if and only if

$$\theta f(u) + (1 - \theta)f(v) \geq f(\theta u + (1 - \theta)v).$$

If f is differentiable, an alternative definition of concavity (or convexity) is

$$f(v) \leq f(u) + \sum_{i=1}^{n} f'_{x_i}(u)(v_i - u_i)$$

$$(or \ f(v) \geq f(u) + \sum_{i=1}^{n} f'_{x_i}(u)(v_i - u_i)).$$

If we change weak inequalities $\leq$ and $\geq$ to strict inequalities, we will get the definition of strict concavity and strict convexity, respectively.

Some useful results:

- The sum of concave (or convex) functions is a concave (or convex) function. If, additionally, at least one of these functions is strictly concave (or convex), the sum is also strictly concave (or convex).

- If $f(x)$ is a (strictly) concave function then $-f(x)$ is a (strictly) convex function and vice versa.

Recipe 12 – How to Check Concavity (or Convexity):
A twice continuously differentiable function $z = f(x_1, \ldots, x_n)$ is concave (or convex) if and only if $d^2 z$ is everywhere negative (or positive) semi-definite.

$z = f(x_1, \ldots, x_n)$ is strictly concave (or strictly convex) if (but not only if) $d^2 z$ is everywhere negative (or positive) definite.

Again, any test for the sign definiteness of a symmetric quadratic form is applicable.

Example 74 $z(xy) = x^2 + xy + y^2$ *is strictly convex.*

Economics Application 6 *Consider Cobb-Douglas production function $Y(K, L) = K^a L^b$ with $a, b \in (0, 1)$ and $a + b < 1$. It is strictly concave in the positive quadrant because $Y_{KK} = a(a - 1)K^{a-2}L^b < 0$, $Y_{LL} = b(b-1)K^a L^{b-2} < 0$, $Y_{KL} = abK^{a-1}L^{b-1}$, and $Y_{KK}Y_{LL} - (Y_{KL})^2 = K^{2a-2}L^{2b-2}ab[(a-1)(b-1) - ab] = K^{2a-2}L^{2b-2}ab[1 - (a+b)] > 0$. Thus the Hessian matrix is negative definite in the positive quadrant, implying strict concavity of Y.*

The conditions for concavity and convexity are necessary and sufficient, while those for strict concavity (strict convexity) are only sufficient.

Definition 51 *A set S in $\mathbf{R}^n$ is convex if for any $x, y \in S$ and any
$\theta \in [0, 1]$ the linear combination $\theta x + (1 - \theta)y \in S$.*

Definition 52 *A function $z = f(x_1, \ldots, x_n)$ is quasi-concave if and
only if for any pair of distinct points $u = (u_1, \ldots, u_n)$ and $v = (v_1, \ldots, v_n)$
in the convex domain of f and for any $\theta \in (0, 1)$*

$$f(v) \geq f(u) \Longrightarrow f(\theta u + (1 - \theta)v) \geq f(u)$$

and quasi-convex if and only if

$$f(v) \geq f(u) \Longrightarrow f(\theta u + (1 - \theta)v) \leq f(v).$$

The above definition can be modified as follows:
*A function $f : \mathcal{D} \to R$ is quasi-concave on $\mathcal{D}$ if and only if for all
$x_1, x_2 \in \mathcal{D}$ and for all $\lambda \in (0, 1)$,*

$$f[\lambda x_1 + (1 - \lambda)x_2] \geq \min\{f(x_1), f(x_2)\},$$

*and it is quasi-convex on $\mathcal{D}$ if and only if for all $x_1, x_2 \in \mathcal{D}$ and for all
$\lambda \in (0, 1)$,*

$$f[\lambda x_1 + (1 - \lambda)x_2] \leq \max\{f(x_1), f(x_2)\}.$$

*If f is differentiable, an alternative definition of quasi-concavity
(quasi-convexity) is*

$$f(v) \geq f(u) \implies \sum_{i=1}^{n} f'_{x_i}(u)(v_i - u_i) \geq 0 \quad (f(v) \geq f(u)$$

$$\implies \sum_{i=1}^{n} f'_{x_i}(v)(v_i - u_i) \geq 0).$$

*A change of weak inequalities $\leq$ and $\geq$ on the right to strict in-
equalities gives us the definitions of strict quasi-concavity and strict
quasi-convexity, respectively.*

Example 75 *The function $z(x, y) = xy$, $x, y \geq 0$ is quasi-concave.*

Example 76 *Let $f : R \rightarrow R$ be defined by*

$$\begin{cases} x^3, & x \in [0,1] \\ 1, & x \in (1,2] \\ x^3, & x > 2 \end{cases}$$

Since f is a non-decreasing function, it is both quasi-concave and quasi-convex on R.

⚠ Note that any convex (concave) function is quasi-convex (quasi-concave), but not *vice versa*.

Definition 53 *A function $f : A \longrightarrow R$ is semi-strictly quasi-concave on $A \subseteq R^n$ if it is quasi-concave on A and $(\forall x, y \in A)$, $(\forall \alpha \in [0,1])$*

$$f(x) > f(y) \Longrightarrow f(\alpha x + (1 - \alpha)y) > f(y)$$

Definition 54 *A differentiable function $f : A \longrightarrow R$ is pseudo-concave on $A \subseteq R^n$ if it is quasi-concave and $(\forall x, y \in A)$:*

$$f(x) > f(y) \Longrightarrow Df(y)(x - y)^T = \sum_{i=1}^{n} f_i(y)(x_i - y_i) > 0$$

2.10   Appendix: Matrix Derivatives

Matrix derivatives play an important role in economic analysis, especially in econometrics.

Certain conventions are used in differentiating matrices or differentiating with respect to matrices:

1. The derivative of a column (row) vector with respect to a scalar is also a column (row) vector;

2. The derivative of a scalar with respect to a column (row) vector is a row (column) vector;

3. The derivative of a scalar with respect to a $[m \times n]$ matrix is a $[n \times m]$ matrix;

4. The derivative of a vector with respect to a vector is a matrix.

If $\bar{x}$ is the column vector $\bar{x} = (x_1, x_2, \ldots, x_n)'$ and the scalar y is a differentiable function of the column vector $\bar{x}$,

$$y = f(\bar{x}) = f(x_1, x_2, \ldots, x_n),$$

then the vector of the first order partial derivatives (the *gradient* vector) is the row vector

$$\frac{\partial y}{\partial \bar{x}} = \left(\frac{\partial y}{\partial x_1}, \ldots, \frac{\partial y}{\partial x_n} \right).$$

If $\bar{x}$, $\bar{y}$ and $\bar{f}$ are column vectors,

$$\bar{x} = (x_1, x_2, \ldots, x_n)',$$

$$\bar{y} = (y_1, y_2, \ldots, y_m)',$$

$$\bar{f} = \bar{f}(\bar{x}) = (f_1(\bar{x}), f_2(\bar{x}), \ldots, f_m(\bar{x}))',$$

then the equation $\bar{y} = \bar{f}(\bar{x})$ defines a transformation $\bar{f}$ from $\mathbf{R}^n$ to $\mathbf{R}^m$, i.e.

$$
\begin{aligned}
y_1 &= f_1(x_1, \ldots, x_n), \\
\ldots \quad &\ldots \quad \ldots \\
y_m &= f_m(x_1, \ldots, x_n).
\end{aligned}
$$

The derivative of the vector function $\bar{f}$ of the vector variable $\bar{x}$ with respect to $\bar{x}$ is the *Jacobian matrix* of this transformation:

$$
\frac{\partial \bar{y}}{\partial \bar{x}} = \begin{pmatrix} \frac{\partial y_1(\bar{x})}{\partial x_1} & \cdots & \frac{\partial y_1(\bar{x})}{\partial x_n} \\ \vdots & & \vdots \\ \frac{\partial y_m(\bar{x})}{\partial x_1} & \cdots & \frac{\partial y_m(\bar{x})}{\partial x_n} \end{pmatrix}.
$$

If $\bar{a}$ and $\bar{x}$ are $[n \times 1]$-vectors and A is a $[m \times n]$ matrix, then

$$\frac{\partial}{\partial \bar{x}}(\bar{a}' \cdot \bar{x}) = \bar{a}', \qquad \frac{\partial}{\partial \bar{x}}(A\bar{x}) = A, \qquad \frac{\partial}{\partial \bar{a}}(\bar{a}' \cdot \bar{x}) = \frac{\partial}{\partial \bar{a}}(\bar{x}' \cdot \bar{a}) = \bar{x}'.$$

If $\bar{x}$ is $[n \times 1]$-vectors and A is a $[n \times n]$ matrix (a quadratic form), then

$$\frac{\partial}{\partial \bar{x}}(\bar{x}' A \bar{x}) = \bar{x}(A + A'), \qquad \frac{\partial^2}{\partial \bar{x} \partial \bar{x}'}(\bar{x}' A \bar{x}) = A + A'.$$

If y is a scalar function of a matrix $A = (a_{ij})_{[m \times n]}$, $y = y(A)$, then the derivative of the scalar y with respect to the matrix A is

$$\frac{\partial y}{\partial A} = \begin{pmatrix} \frac{\partial y}{\partial a_{11}} & \cdots & \frac{\partial y}{\partial a_{1n}} \\ \vdots & & \vdots \\ \frac{\partial y}{\partial a_{m1}} & \cdots & \frac{\partial y}{\partial a_{mn}} \end{pmatrix}.$$

If A is a $[n \times n]$ non-singular matrix, the derivative of its determinant with respect to A is given by

$$\frac{\partial}{\partial A}(\det(A)) = (C_{ij}),$$

where (C_{ij}) is the matrix of co-factors of A.

Example 77 (Matrix Derivatives in Econometrics: How to Find the Least Squares Estimator in a Multiple Regression Model)

Consider the multiple regression model:

$$y = X\beta + \epsilon$$

where $[n \times 1]$ vector y is the dependent variable, X is a $[n \times k]$ matrix of k explanatory variables with $rank(X) = k$, β is a $[k \times 1]$ vector of coefficients which are to be estimated, and ϵ is a $[n \times 1]$ vector of disturbances. We assume that the matrices of observations X and y are given.

Our goal is to find an estimator b for β using the least squares method. The least squares estimator of β is a vector b, which minimizes the residual sum of squares $E(b)$,

$$E(b) = e'e = (y - Xb)'(y - Xb) = y'y - y'Xb - b'X'y + b'X'Xb.$$

The first-order condition for extremum is:

$$\frac{dE(b)}{db} = 0.$$

To obtain the first-order conditions we use the following formulas:

- $\dfrac{d(a'b)}{db} = a,$

- $\dfrac{d(b'a)}{db} = a,$

- $$\frac{d(Mb)}{db} = M',$$

- $$\frac{d(b'Mb)}{db} = (M + M')b,$$

where a, b are $[k \times 1]$ vectors and M is a $[k \times k]$ matrix.
Using these formulas we obtain:

$$\frac{dE(b)}{db} = -2X'y + 2X'Xb$$

and the first-order condition implies:

$$X'Xb = X'y$$

or

$$b = (X'X)^{-1}X'y.$$

On the other hand, by the third derivation rule above, we have:

$$\frac{d^2 E(b)}{db^2} = (2X'X)' = 2X'X.$$

To check whether the solution b is indeed a minimum, we need to prove the positive definiteness of the matrix $X'X$.

First, notice that $X'X$ is a symmetric matrix. The symmetry is obvious:

$$(X'X)' = X'X.$$

To prove positive definiteness, we take an arbitrary $[k \times 1]$ vector z, $z \neq \mathbf{0}$ and check the following quadratic form:

$$z'(X'X)z = (Xz)'Xz.$$

The assumptions $\mathrm{rank}(X) = k$ and $z \neq 0$ imply $Xz \neq 0$. It follows that $(Xz)'Xz > 0$ for any $z \neq 0$ or, equivalently, $z'X'Xz > 0$ for any $z \neq 0$, which means that (by definition) $X'X$ is positive definite.

2.11   Appendix: Topological Structure and Its Implications

This section presents a collection of background mathematical facts widely used in economic analysis (e.g., in general equilibrium theory).

Basic Concepts

Definition 55 (topology) *If T is a subset of 2^A and satisfies the following properties:*

> 1. $\emptyset \in T, \qquad A \in T$;

> 2. *Any (finite or non-finite) union of sets in T belongs to T;*

> 3. *Any finite intersection of sets in T belongs to T;*

then T is called a topology defined on set A.

Definition 56 *A set with a topology defined on it is called a topological space.*

Definition 57 *A vector space with a norm defined on it is called a normed space.*

Definition 58 *Let V be a normed space. The set $B(x, r)$ defined as: $B(x, r) = \{y \in V : \|y - x\| < r\}$ is called an open ball with the center $x \in V$ and radius $r \in R, \quad r > 0$.*

Definition 59 *Any set $N \subseteq V$ such that there exists an open ball with the center x contained in N is called a neighborhood of a point $x \in V$.*

Definition 60 *A point $x \in A$ such that A is a neighborhood of x is called an interior point of a set $A \subseteq V$.*

Definition 61 *A set $A \subseteq V$ is open if all its points are interior.*

Example 78 *Let $V = R$, then a set $A = (0, 1)$ is an open set. To prove the latter, let us suppose a point x belongs to A, and define $r(x) = min\{\frac{x}{2}, \frac{1-x}{2}\}$. Then the following open ball is contained in A:*

$$
\begin{aligned}
B(x, r) &= \{x - r(x), x + r(x)\} \\
&= \left(x - min\left\{\frac{x}{2}, \frac{1-x}{2}\right\}, x + min\left\{\frac{x}{2}, \frac{1-x}{2}\right\}\right) \\
&= \begin{cases} \left(\frac{x}{2}, \frac{3x}{2}\right), & x \leq 0.5 \\ \left(\frac{3x-1}{2}, \frac{x+1}{2}\right), & x > 0.5 \end{cases}
\end{aligned}
$$

Definition 62 *A set whose complement is open is called a closed set.*

Example 79 *Let $V = R$. Then a set $A = [0, 1]$ is a closed set.*

Example 80 *Let $V = R$. Then $A = [0, 1)$ is neither a closed nor an open set.*

Definition 63 *A point $x \in V$ such that any open ball with center x contains a point of A other than x, is called a limit point of a set $A \subseteq V$.*

Definition 64 *The closure A^c of a set $A \subseteq V$ is the union of A and all limit points of A.*

Example 81 *Let $V = R$, $A = [0, 1) \cup \{3\}$. Then the closure of A is $A^c = [0, 1] \cup \{3\}$.*

Proposition 35 *An interior point of a set is always its limit point.*

Proposition 36 *A set A is closed $\iff A = A^c$ (i.e. A coincides with its closure).*

Definition 65 *A boundary point of a set $A \subseteq V$ is a point $x \in V$, such that any open ball with its center in x intersects with both A and A^c.*

Definition 66 *The set of all boundary points of $A \subseteq V$ is called the boundary of the set A.*

Example 82 *Let $V = R$, $A = [0, 1)$. A has two boundary points 0 and 1 and the boundary of A is equal to $\{0\} \cup \{1\}$.*

Proposition 37 *The boundaries of A and A^c coincide.*

Proposition 38   *i) Set A is closed $\iff$ set A contains its whole boundary.*

ii) Set A is open $\iff$ set A is disjoint from its boundary.

Bounded and Compact Sets

Definition 67 *The diameter of a set $A \subseteq V$ is a real number defined as*

$$diam(A) = \sup_{x,y \in A} \|y - x\|.$$

Note that for any A, $0 \leq diam(A) \leq +\infty$.

Definition 68 *A set with a finite diameter is called a bounded set.*

Proposition 39 *A set $A \subseteq V$ is bounded $\Longleftrightarrow \exists\ M \in R$ such that $\|x\| \leq M, \quad \forall x \in A$.*

Example 83 *Let $V = R$, $A = [0,1]$. The diameter of the set A is $diam(A) = 1$. Therefore, A is bounded since $\|x\| \leq 1\ \forall x \in A$.*

Definition 69 *A bounded and closed set is called a compact set.*

Definition 70 *A set $A \subseteq V$ is called a convex set if $\forall x, y \in A$ and for any real $\alpha \in [0,1]$ it follows that $\{\alpha x + (1 - \alpha)y\} \in A$.*

Definition 71 *The minimal convex set that contains $A \subseteq V$ is called the convex hull and is denoted $Co(A)$, that is the intersection of all convex sets that contain B.*

Separation Results

Definition 72 *Given $p \in R^N$, $p \neq 0$, and $c \in R$, the hyperplane generated by p and c is the set $H_{p,c} = \{z \in R^N : pz = c\}$.*

Definition 73 *The sets $\{z \in R^N : pz \geq c\}$ and $\{z \in R^N : pz \leq c\}$ are called, respectively, the half spaces above and the half spaces below the hyperplane $H_{p,c}$.*

Proposition 40 *Let $S \subseteq R^N$ be a non-empty, convex set, and $\bar{y} \in R^N \backslash S$. Then there exists a hyperplane that weakly separates $\bar{y}$ and S. If, in addition, S is compact then $\bar{y}$ and S are strongly separated.*

2.12   Appendix: Correspondences and Fixed-Point Theorems

This section briefly presents useful mathematical results that are applied in general equilibrium theory.

Definition 74 *A correspondence from set A to set Y is a function from X to the set of all subsets of Y (denoted as 2^Y).*

The difference between a correspondence from X to Y and a function from X to Y is that a function $f : X \longrightarrow Y$ assigns a unique element of Y to each element X, whereas a correspondence $F : X \longrightarrow 2^Y$ assigns a subset of Y to each element of X.   ⚠

Definition 75 *A correspondence $F : X \longrightarrow 2^Y$ is called non-empty valued if $\forall x \in X$, $F(x) \neq 0$.*

Definition 76 *A correspondence $F : X \longrightarrow 2^Y$ is called closed valued if $\forall x \in X$ $F(x)$ is a closed set in R^m.*

Definition 77 *A correspondence $F : X \longrightarrow 2^Y$ is called compact valued if $\forall x \in X$ $F(x)$ is a compact set in R^m.*

Definition 78 *A correspondence $F : X \longrightarrow 2^Y$ is called convex valued if $\forall x \in X$ $F(x)$ is a convex set in R^m.*

Definition 79 *A correspondence $F : X \longrightarrow Y$ is called upper hemicontinuous at x_0 if for any convergent sequence $\{x_r\}_{r=1}^{\infty}$ and for $\{y_r\}_{r=1}^{\infty} \subset Y$, conditions $x_r \longrightarrow x_0$, $y_r \longrightarrow y_0$ and $y_r \in F(x_r)$ for all r, imply that $y_0 \in F(x_0)$.*

Definition 80 *A correspondence $F : X \longrightarrow Y$ is called lower hemicontinuous at $x_0 \in X$ if for any sequence $\{x_r\} \in X$ such that $x_r \longrightarrow x_0$ and $\forall y_0 \in F(x_0)$ there exists a sequence $y_r \longrightarrow y_0$ where $y_r \in F(x_r)$, $\forall r$.*

Definition 81 *A correspondence $F : X \longrightarrow Y$ is upper hemicontinuous (lower hemicontinuous) on X if it is upper hemicontinuous (lower hemicontinuous) $\forall x \in X$.*

Definition 82 *A correspondence $F : X \longrightarrow Y$ is called continuous if it is upper and lower hemicontinuous on X.*

Definition 83 *A fixed point of a correspondence $F : X \longrightarrow X$ is a point $x \in X$ such that $x \in F(x)$.*

⚠ Recall that for a function $F : X \longrightarrow X$, a fixed point x^* satisfies the condition $f(x^*) = x^*$.

Proposition 41 (Brouwer's fixed point theorem)
    Let $f : x \longrightarrow X$ be a continuous function, and let X be a convex, non-empty, compact subset of R^N. Then f has a fixed point, i.e. $\exists x^* \in X$ such that $f(x^*) = x^*$.

⚠ Do not confuse the Brouwer fixed point theorem (which was proven in 1912 and deals with functions) with the more general Browder fixed point theorem (1967) that relates to fixed points of correspondences.

Proposition 42 (Browder Fixed-Point Theorem) *Let $Q : X \longrightarrow X$ be a correspondence where X is a compact convex subset of the Hausdorff topological vector space Y. Suppose Q is lower hemicontinuous, and $Q(x)$ is convex and non-empty $\forall x \in X$. Then $\exists x_0 \in X$ such that $x_0 \in Q(x_0)$.*

Proposition 43 (Kakutani's Fixed Point Theorem)
    Let X be a non-empty, compact and convex subset of R^N and $F : X \longrightarrow 2^X$ be a non-empty valued, compact valued and upper hemicontinuous correspondence. Then F has a fixed point.

Economics Application 7 (Existence Theorem)
    Suppose that $z(p)$ is a function defined for all strictly positive price vectors $p \in R_{++}^L$ and satisfying conditions. Then the system of equations $z(p) = 0$ has a solution. Hence, a Walrasian equilibrium exists in any pure exchange economy in which $\sum_i \omega_i \gg 0$ and every consumer has continuous, strictly convex, and strongly monotone preferences.

(The proof of the existence theorem is based on the Kakutani fixed point theorem. For further details see, e.g. MasCollel A., W.D. Whinston and J.R. Green, *Microeconomic Theory*.)

Further Reading

* Bartle, R. G. (1976). *The Elements of Real Analysis* (2nd ed.). New York: Wiley.

* MasCollel A., Whinston, W. D., and Green, J. R. (1995). *Microeconomic Theory*. New York: Oxford University Press.

* Edwards, C. N. and Penny, D.E. (1998). *Calculus and Analytical Geometry* (5th ed.). Englewood Cliffs: Prentice-Hall.

* Greene, W. H. (2008). *Econometric Analysis* (6th ed.). New Jersey: Prentice Hall.

* Rudin, W. (1976). *Principles of Mathematical Analysis* (3rd ed.). New York : McGraw-Hill.

* Searle, S. R. and Willett, L. S. (2001). *Matrix Algebra for Applied Economics*. New York: Wiley.

Chapter 3

Constrained Optimization

3.1 Optimization with Equality Constraints

Consider the problem[1]:

$$\text{extremize } f(x_1, \ldots, x_n) \qquad (3.1)$$
$$\text{subject to } g^j(x_1, \ldots, x_n) = b_j, \quad j = 1, 2, \ldots, m < n.$$

f is called the objective function, $g^1, g^2, \ldots, g^m$ are the constraint functions, $b_1, b_2, \ldots, b_m$ are the constraint constants. The difference $n - m$ is the number of *degrees of freedom* of the problem.

Note that n is strictly greater than m. ⚠

If it is possible to explicitly express (from the constraint functions) m independent variables as functions of the other $n - m$ independent variables, we can eliminate m variables in the objective function, thus the initial problem will be reduced to the unconstrained optimization problem with respect to $n - m$ variables. However, in many cases it is not technically feasible to explicitly express one variable as a function of the others.

Instead of the substitution and elimination method, we may resort to the easy-to-use and well-defined *method of Lagrange multipliers*.

Let f and $g^1, \ldots, g^m$ be $C^{(1)}$-functions and the Jacobian $\mathcal{J} = (\frac{\partial g^j}{\partial x_i})$, $i = 1, 2, \ldots, n$, $j = 1, 2, \ldots, m$, have the full rank, i.e. $\text{rank}(\mathcal{J}) = m$. We introduce the *Lagrangian function* as

---

[1] *"Extremize"* means to find either the minimum or the maximum of the objective function f.

$$L(x_1, \ldots, x_n, \lambda_1, \ldots, \lambda_m) =$$

$$= f(x_1, \ldots, x_n) + \sum_{j=1}^{m} \lambda_j (b_j - g^j(x_1, \ldots, x_n)),$$

where $\lambda_1, \lambda_2, \ldots, \lambda_m$ are constant (*Lagrange multipliers*).

Recipe 13 – What are the Necessary Conditions for the Solution to (3.1)?

Equate all partials of L with respect to $x_1, \ldots, x_n$, $\lambda_1, \ldots, \lambda_m$ to zero:

$$\frac{\partial L(x_1, \ldots, x_n, \lambda_1, \ldots, \lambda_m)}{\partial x_i} =$$

$$= \frac{\partial f(x_1, \ldots, x_n)}{\partial x_i} - \sum_{j=1}^{m} \lambda_j \frac{\partial g^j(x_1, \ldots, x_n)}{\partial x_i} = 0,$$

$$i = 1, 2, \ldots, n,$$

$$\frac{\partial L(x_1, \ldots, x_n, \lambda_1, \ldots, \lambda_m)}{\partial \lambda_j} = b_j - g^j(x_1, \ldots, x_n) = 0,$$

$$j = 1, 2, \ldots, m.$$

Solve these equations for $x_1, \ldots, x_n$ and $\lambda_1, \ldots, \lambda_m$.

In the end we will get a set of stationary points of the Lagrangian funciton. If $x^* = (x_1^*, \ldots, x_n^*)$ is a solution to (3.1), it should be a stationary point of L.

Intuitively, the first-order necessary conditions emerge as follows (take a two-dimensional case for simplicity): At the point of optimality $x^* = (x_1^*, x_2^*)$, the utility function and the constraint are tangent, that is, the slopes of tangent lines are equal to certain constant λ

$$\frac{f_1(x^*)}{f_2(x^*)} = \frac{g_1(x^*)}{g_2(x^*)} = \lambda.$$

The latter implies $f_1 - \lambda g_1 = 0$ and $f_2 - \lambda g_2 = 0$ at x^*, which are the F.O.C of optimization of the Lagrangian function $L = f(x_1, x_2) + \lambda(b - g(x_1, x_2))$.

It is important that rank$(\mathcal{J}) = m$, and the functions are continuously differentiable. (Otherwise, the Lagrange multiplier method may not identify a true optimal solution or may fail to work at all.)

Example 84 *This example shows that the Lagrange multiplier method does not always provide the solution to an optimization problem. Consider the problem:*

$$\min \ x^2 + y^2$$
$$\text{subject to } (x-1)^3 - y^2 = 0.$$

Notice that the solution to this problem is $(1,0)$. The restriction $(x-1)^3 = y^2$ implies $x \geq 1$. If $x = 1$, then $y = 0$ and $x^2 + y^2 = 1$. If $x > 1$, then $y > 0$ and $x^2 + y^2 > 1$. Thus, the minimum of $x^2 + y^2$ is attained for $(1,0)$.

The Lagrangian function is

$$L \ = \ x^2 + y^2 - \lambda[(x-1)^3 - y^2]$$

The first-order conditions are

$$\begin{cases} \frac{\partial L}{\partial x} &= 0, \\ \frac{\partial L}{\partial y} &= 0, \quad or \\ \frac{\partial L}{\partial \lambda} &= 0, \end{cases} \qquad \begin{cases} 2x - 3\lambda(x-1)^2 &= 0, \\ 2y + 2\lambda y &= 0, \\ (x-1)^3 - y^2 &= 0. \end{cases}$$

Clearly, the first equation is not satisfied for $x = 1$. Thus, the Lagrange multiplier method fails to detect the solution $(1,0)$.

Example 85 *Another example that illustrates the constraint qualification failure when the Lagrange multiplier method does not work properly (in this particular example matrix $\mathcal{J}$ does not have the full rank at the point of optimality).*

Let us minimize $f(x,y,z) = x^2 + (y-1)^2 + z^2$, subject to $x + y = \sqrt{2}$ and $x^2 + y^2 = 1$. Obviously, the problem has a unique solution $(\sqrt{2}/2, \sqrt{2}/2, 0)$. On the other hand, the F.O.C for the Lagrangian function $L = x^2 + (y-1)^2 + z^2 + \lambda(\sqrt{2} - x - y) + \mu(1 - x^2 - y^2)$ yield

$$\lambda \ = \ 2x(\mu - 1), \tag{3.2}$$
$$\lambda - 2 \ = \ 2y(\mu - 1). \tag{3.3}$$

Computing $((3.2) + (3.3))$ we find $2\lambda - 2 = 2\sqrt{2}(\mu - 1)$, or, taking squares, $(\lambda-1)^2 = 2(\mu-1)^2$. But the computation of squared first-order conditions $(3.2)^2 + (3.3)^2$ brings a contradictory result $(\lambda - 1)^2 + 1 = 2(\mu - 1)^2$.

If we need to check whether a stationary point is a maximum or minimum, the following *local sufficient conditions* can be applied:

Proposition 44 *Let us introduce a bordered Hessian* $|\bar{H}_r|$ *as*

$$|\bar{H}_r| = \det \begin{pmatrix} 0 & \cdots & 0 & \frac{\partial g^1}{\partial x_1} & \cdots & \frac{\partial g^1}{\partial x_r} \\ \vdots & \ddots & \vdots & \vdots & \ddots & \vdots \\ 0 & \cdots & 0 & \frac{\partial g^m}{\partial x_1} & \cdots & \frac{\partial g^m}{\partial x_r} \\ \frac{\partial g^1}{\partial x_1} & \cdots & \frac{\partial g^m}{\partial x_1} & \frac{\partial^2 L}{\partial x_1 \partial x_1} & \cdots & \frac{\partial^2 L}{\partial x_1 \partial x_r} \\ \vdots & \ddots & \vdots & \vdots & \ddots & \vdots \\ \frac{\partial g^1}{\partial x_r} & \cdots & \frac{\partial g^m}{\partial x_r} & \frac{\partial^2 L}{\partial x_r \partial x_1} & \cdots & \frac{\partial^2 L}{\partial x_r \partial x_r} \end{pmatrix},$$

$r = 1, 2, \ldots, n.$

Let f and $g^1, \ldots, g^m$ be $C^{(2)}$-functions and let x^* satisfy the necessary conditions for problem (3.1). Let $|\bar{H}_r(x^*)|$ be the bordered Hessian determinant evaluated at x^*. Then

- *if $(-1)^m|\bar{H}_r(x^*)| > 0$, $r = m+1, \ldots, n$, then x^* is a local minimum point for problem (3.1).*

- *if $(-1)^r|\bar{H}_r(x^*)| > 0$, $r = m+1, \ldots, n$, then x^* is a local maximum point for problem (3.1).*

Example 86 (Local Second-Order Conditions for the Case with one Constraint)
Suppose that $x^* = (x_1^*, \ldots, x_n^*)$ satisfies the necessary conditions for the problem
$$\max(\min) f(x_1, \ldots, x_n) \text{ subject to } g(x_1, \ldots, g_n) = b,$$
i.e. all partial derivatives of the Lagrangian function are zero at x^*. Define

$$|\bar{H}_r| = \det \begin{pmatrix} 0 & \frac{\partial g}{\partial x_1} & \cdots & \frac{\partial g}{\partial x_r} \\ \frac{\partial g}{\partial x_1} & \frac{\partial^2 L}{\partial x_1 \partial x_1} & \cdots & \frac{\partial^2 L}{\partial x_1 \partial x_r} \\ \vdots & \vdots & \ddots & \vdots \\ \frac{\partial g}{\partial x_r} & \frac{\partial^2 L}{\partial x_r \partial x_1} & \cdots & \frac{\partial^2 L}{\partial x_r \partial x_r} \end{pmatrix}, \qquad r = 1, 2, \ldots, n.$$

Let $|\bar{H}_r(x^*)|$ be $|\bar{H}_r|$ evaluated at x^*. Then

- x^* *is a local minimum point of* f *subject to the given constraint if* $|\bar{H}_r(x^*)| < 0$ *for all* $r = 2, \ldots, n$.

- x^* *is a local maximum point of* f *subject to the given constraint if* $(-1)^r|\bar{H}_r(x^*)| > 0$ *for all* $r = 2, \ldots, n$.

Example 87 (Comparative Statics of Constrained Optimization Problem)

Consider a constrained maximization problem

$$\max_x f(x, a)$$

subject to $\quad g(x, a) = 0,$

where x *is a n-dimensional vector of endogenous variables and* $a \in R^m$ *is a vector of parameters (exogenous variables).*

The Lagrangian for this problem is $L(x, \lambda, a) = F(x, a) + \lambda g(x, a)$, *with the first-order conditions*

$$L_x(x, \lambda, a) = f_x(x, a) + \lambda g_x(x, a),$$
$$L_\lambda(x, \lambda, a) = g(x, a).$$

Let us apply the implicit function theorem to the first-order conditions: Denote

$$F(x, \lambda, a) = \left(\begin{array}{c} f_x(x, a) + \lambda g_x(x, a) \\ g(x, a) \end{array} \right)$$

and suppose that there is a point such that $F(x^0, \lambda^0, a^0) = 0$, *and* $F_{x\lambda}(x^0, \lambda^0, a^0)$ *is invertible (which holds true, for instance, if the sufficient second-order conditions are satisfied). Then there are functions* $x^*(a)$ *and* $\lambda^*(a)$ *and real* $\epsilon > 0$ *such that*

(i) $x^*(a^0) = x^0, \lambda^*(a^0) = \lambda^0;$

(ii) $(\forall a | a - a^0| < \epsilon) \ F(x^*(a), \lambda^*(a), a) = 0;$

(iii) $\left(\begin{array}{c} x_a^*(a) \\ \lambda_a^*(a) \end{array} \right) = -[F_{x\lambda}(x^0, \lambda^0, a^0)]^{-1} F_a(x^0, \lambda^0, a^0).$

For the constrained maximization problem, $F_{x\lambda}$ is evaluated as

$$F_{x\lambda}(x^0, \lambda^0, a^0) = \begin{pmatrix} L_{xx}(x^0, \lambda^0, a^0) & L_{x\lambda}(x^0, \lambda^0, a^0) \\ L_{\lambda x}(x^0, \lambda^0, a^0) & L_{\lambda\lambda}(x^0, \lambda^0, a^0) \end{pmatrix}.$$

Finally, derivation of x^* and λ^* is completed by finding

$$F_a(x^0, \lambda^0, a^0) = \begin{pmatrix} f_{xa}(x^0, a^0) + \lambda^0 g_{xa}(x^0, a^0) \\ g_a(x^0, a^0) \end{pmatrix} = \begin{pmatrix} L_{xa} \\ L_{\lambda a} \end{pmatrix}.$$

Economics Application 8 (Utility Maximization)

In order to illustrate how the Lagrange-multiplier method can be applied, let us consider the following optimization problem:

The preferences of a consumer over two goods x and y are given by the utility function

$$U(x, y) = (x + 1)(y + 1) = xy + x + y + 1.$$

The prices of goods x and y are 1 and 2, respectively, and the consumer's income is 30. What bundle of goods will the consumer choose?

The consumer's budget constraint is $x + 2y \leq 30$ which, together with the conditions $x \geq 0$ and $y \geq 0$, determines his budget set (the set of all affordable bundles of goods). He will choose the bundle from his budget set that maximizes his utility. Since $U(x, y)$ is an increasing function in both x and y (over the domain $x \geq 0$ and $y \geq 0$), the budget constraint should be satisfied with equality. The consumer's optimization problem can be stated as follows:

$$\begin{aligned} &\max \ U(x, y) \\ &\text{subject to } x + 2y = 30. \end{aligned}$$

The Lagrangian function is

$$\begin{aligned} L &= U(x, y) - \lambda(x + 2y - 30) \\ &= xy + x + y + 1 - \lambda(x + 2y - 30). \end{aligned}$$

The first-order necessary conditions are

$$\begin{cases} \frac{\partial L}{\partial x} = 0 \\ \frac{\partial L}{\partial y} = 0 \\ \frac{\partial L}{\partial \lambda} = 0 \end{cases} \quad or \quad \begin{cases} y + 1 - \lambda = 0 \\ x + 1 - 2\lambda = 0 \\ x + 2y - 30 = 0 \end{cases}$$

From the first equation, we get $\lambda = y + 1$; substituting $y + 1$ for λ in the second equation, we obtain $x - 2y - 1 = 0$. This condition and the third condition above lead to $x = \frac{31}{2}$, $y = \frac{29}{4}$.

To check whether this solution really maximizes the objective function, let us apply the second-order sufficient conditions.

The second-order conditions involve the bordered Hessian:

$$\det \bar{H} = \begin{vmatrix} 0 & 1 & 2 \\ 1 & 0 & 1 \\ 2 & 1 & 0 \end{vmatrix} = 4 > 0.$$

Thus, $(x = \frac{31}{2}, y = \frac{29}{4})$ is indeed a maximum and represents the bundle demanded by the consumer.

Economics Application 9 (Applications of the Implicit Function Theorem)

Consider the problem of maximizing the function $f(x, y) = ax + y$ subject to the constraint $x^2 + ay^2 = 1$ where $x > 0$, $y > 0$ and a is a positive parameter. Given that this problem has a solution, find how the optimal values of x and y change if a increases by a very small amount.

Solution:

Differentiation of the Lagrangian function $L = ax + y - \lambda(x^2 + ay^2 - 1)$ with respect to x, y, λ gives the first-order conditions:

$$\begin{cases} a - 2\lambda x &= 0 \\ 1 - 2a\lambda y &= 0 \\ x^2 + ay^2 - 1 &= 0 \end{cases} \text{ or, } \begin{cases} F^1(x, \lambda; a) &= 0 \\ F^2(y, \lambda; a) &= 0 \\ F^3(x, y; a) &= 0 \end{cases} \tag{3.4}$$

The Jacobian of this system of three equations is:

$$J = \begin{pmatrix} \frac{\partial F^1}{\partial x} & \frac{\partial F^1}{\partial y} & \frac{\partial F^1}{\partial \lambda} \\ \frac{\partial F^2}{\partial x} & \frac{\partial F^2}{\partial y} & \frac{\partial F^2}{\partial \lambda} \\ \frac{\partial F^3}{\partial x} & \frac{\partial F^3}{\partial y} & \frac{\partial F^3}{\partial \lambda} \end{pmatrix} = \begin{pmatrix} -2\lambda & 0 & -2x \\ 0 & -2a\lambda & -2ay \\ 2x & 2ay & 0 \end{pmatrix},$$

$$|J| = -8a\lambda(x^2 + ay^2) = -8a\lambda.$$

From the first F.O.C. we can see that $\lambda > 0$, thus $|J| < 0$ at the optimal point. Therefore, in a neighborhood of the optimal solution, given by the system (3.4), the conditions of the implicit function theorem are met, and in this neighborhood x and y can be expressed as functions of a.

The first-order partial derivatives of x and y with respect to a are evaluated as:

$$\frac{\partial x}{\partial a} = -\frac{\begin{vmatrix} \frac{\partial F^1}{\partial a} & \frac{\partial F^1}{\partial y} & \frac{\partial F^1}{\partial \lambda} \\ \frac{\partial F^2}{\partial a} & \frac{\partial F^2}{\partial y} & \frac{\partial F^2}{\partial \lambda} \\ \frac{\partial F^3}{\partial a} & \frac{\partial F^3}{\partial y} & \frac{\partial F^3}{\partial \lambda} \end{vmatrix}}{|J|} = -\frac{\begin{vmatrix} 1\lambda & 0 & -2x \\ -2\lambda y & -2a\lambda & -2ay \\ y^2 & 2ay & 0 \end{vmatrix}}{|J|} =$$

$$= \frac{4ay^2(\lambda x + a)}{8a\lambda} = \frac{3xy^2}{2} > 0.$$

$$\frac{\partial y}{\partial a} = -\frac{\begin{vmatrix} \frac{\partial F^1}{\partial x} & \frac{\partial F^1}{\partial a} & \frac{\partial F^1}{\partial \lambda} \\ \frac{\partial F^2}{\partial x} & \frac{\partial F^2}{\partial a} & \frac{\partial F^2}{\partial \lambda} \\ \frac{\partial F^3}{\partial x} & \frac{\partial F^3}{\partial a} & \frac{\partial F^3}{\partial \lambda} \end{vmatrix}}{|J|} = -\frac{\begin{vmatrix} -2\lambda & 1 & -2x \\ 0 & -2\lambda y & -2ay \\ 2x & y^2 & 0 \end{vmatrix}}{|J|} =$$

$$= -\frac{-4xy(\lambda x + a) - 4\lambda y(x^2 + ay^2)}{-8a\lambda} = -\left(\frac{3xy}{4\lambda} + \frac{y}{2a}\right) < 0.$$

In conclusion, a small increase in a will lead to a rise in the optimal x and a fall in the optimal y.

Economics Application 10 *The economic interpretation of Lagrange multipliers is as follows:* λ_i *denotes the* shadow price *of the corresponding constraint constant (in other words, the available quantity of a certain resource)* b_j. *Technically speaking, if we maximize* $f(x)$ *s.t.* $g(x) = b$ *(here x is a n-dimensional vector and b parameter), then at the optimal* $x^* = x^*(b)$ *and* $\lambda^* = \lambda^*(b)$,

$$\frac{df(x^*(b), \lambda^*(b))}{db} = \frac{dL}{db}\Big|_{x=x^*, \lambda=\lambda^*} = \lambda^*.$$

Proposition 45 (Envelope Theorem for Constrained Problem)
    Let functions $f, g : R^{n+m} \longrightarrow R$, *and consider a constrained optimization problem*

$$\max_x f(x, a),$$

subject to $g(x, a) = 0$,

where x is an n-dimensional vector of exogenous variables, and a is an m-dimensional vector of endogenous variables.

Let $V(a)$ be the optimal value function associated with this problem, i.e., $V(a) = max_x f(x, a)$ subject to the constraint. The Lagrangian function for $V(a)$ is

$$L(x, \lambda, a) = f(x, a) + \lambda g(x, a).$$

If f and g are C^2 functions, and the sufficient second-order conditions are satisfied at the optimal solution (x^*, λ^*) to the constrained maximization, then

$$V_a(a) = L_a(x^*, \lambda^*, a) = f_a(x^*, a) + \lambda^* g_a(x^*, a).$$

If the constraint is presented in the form $g(x, a) = b$ where $a = (a_1, \ldots, a_m)$ and b is a scalar, then the envelope theorem can be rewritten as

$$\frac{dV}{da_i} = \sum_{j=1}^{n} \frac{\partial L^*}{\partial x_j^*} \cdot \frac{\partial x_j^*}{\partial a_i} + \frac{\partial L^*}{\partial \lambda^*} \cdot \frac{\partial \lambda^*}{\partial a_i} + \frac{\partial L^*}{\partial a_i} = \frac{\partial L}{\partial a_i}\Big|_{x=x^*(a,b), \lambda=\lambda^*(a,b)},$$

$$\frac{dV}{db} = \frac{\partial L}{\partial b}\Big|_{x=x^*(a,b), \lambda=\lambda^*(a,b)}.$$

Here we use the fact that at the point of optimality (x^*, λ^*), $\lambda^*(b - g(x^*, a)) = 0$ – thus $f(x^*, a) = L^* = f(x^*, a) + \lambda^*(b - g(x^*, a))$, – and $L_{x_j^*}^* = L_{\lambda^*}^* = 0$.

Economics Application 11 (Roy's Identity)

Consider the utility maximization problem with linear budget constraint

$$\max U(x_1, \ldots, x_n)$$
$$\text{subject to } p_1 x_1 + \ldots + p_n x_n = M.$$

Denote $x^* = (x_1^*, \ldots, x_n^*)$ and $p = (p_1, \ldots, p_n)$. The optimal solution $x^* = x^*(p, M)$ is called the Marshallian demand function. The maximum value function $U^* = U^*(p, M) = U(x^*(p, M))$ is called the indirect utility function. With the Lagrangian function $L = U(x_1, \ldots, x_n) + \lambda(M - p_1 x_1 - \ldots - p_n x_n)$, the envelope theorem implies

$$\frac{\partial U^*}{\partial M} = \frac{\partial U(x^*(p, M))}{\partial M} = \frac{\partial L(x, p, M)}{\partial M}\Big|_{x=x^*(p,M), \lambda=\lambda^*(p,M)} = \lambda^*, (3.5)$$

$$\frac{\partial U^*}{\partial p_i} = \frac{\partial L}{\partial p_i}\Big|_{x=x^*(p,M),\lambda=\lambda^*(p,M)} = -\lambda^* x_i^*. \tag{3.6}$$

Equation (3.5) evaluates the marginal utility of money λ^*. Equation (3.6) is called Roy's Identity. *It says that the marginal disutility of a price increase is proportional to the optimal demanded quantity, with λ^* (marginal utility of money) being the coefficient of proportionality. Sometimes, (3.6) also reads as*

$$x^*(p, M) = -\frac{\partial U^*/\partial p_i}{\partial U^*/\partial M}.$$

Finally, let us present a case in which the first-order necessary conditions also become sufficient.

Proposition 46 *Suppose that f and $g^1, \ldots, g^m$ are defined on an open convex set $S \subset \mathbf{R}^n$. Let x^* be a stationary point of the Lagrangian function. Then,*

- *if the Lagrangian is concave, x^* maximizes (3.1);*

- *if the Lagrangian is convex, x^* minimizes (3.1).*

3.2   The Case of Inequality Constraints

Classical methods of optimization (the method of Lagrange multipliers) deal with optimization problems with equality constraints in the form of $g(x_1, \ldots, x_n) = c$. Non-classical optimization, also known as *mathematical programming*, tackles problems with inequality constraints like $g(x_1, \ldots, x_n) \leq c$.

Mathematical programming includes *linear programming* and *non-linear programming*. In linear programming, the objective function and all inequality constraints are linear. When either the objective function or an inequality constraint is non-linear, we face a problem of non-linear programming.

In the following, we restrict our attention to non-linear programming. The problem of linear programming – also called a *linear program* – is discussed in the Appendix to this chapter.

3.2.1   Non-Linear Programming

The non-linear programming problem is that of choosing non-negative values of certain variables so as to maximize or minimize a given (non-linear) function subject to a given set of (non-linear) inequality constraints.

The non-linear programming maximum problem is

$$\max f(x_1, \ldots, x_n) \qquad (3.7)$$
$$\text{subject to } g^j(x_1, \ldots, x_n) \leq b_j, \quad j = 1, 2, \ldots, m.$$
$$x_1 \geq 0, \ldots, x_n \geq 0.$$

Similarly, the minimization problem is

$$\min f(x_1, \ldots, x_n)$$
$$\text{subject to } g^j(x_1, \ldots, x_n) \geq b_j, \quad j = 1, 2, \ldots, m.$$
$$x_1 \geq 0, \ldots, x_n \geq 0.$$

First, note that there are no restrictions on the relative size of m and n, unlike the case of equality constraints. Second, note that the direction of the inequalities ($\leq$ or $\geq$) at the constraints is only a convention, because the inequality $g^j \leq b_j$ can be easily converted to the $\geq$ inequality by multiplying it by -1, yielding $-g^j \geq -b_j$. Third, note that an equality constraint, say $g^k = b_k$, can be replaced by the two inequality constraints, $g^k \leq b_k$ and $-g^k \leq -b_k$.

Definition 84 *A constraint $g^j \leq b_j$ is called binding (or active) at $x^0 = (x_1^0, \ldots, x_n^0)$ if $g^j(x^0) = b_j$.*

Example 88 *The following is an example of a non-linear program:*

$$\max \quad \pi = x_1(10 - x_1) + x_2(20 - x_2)$$
$$\text{subject to} \quad 5x_1 + 3x_2 \leq 40$$
$$x_1 \leq 5$$
$$x_2 \leq 10$$
$$x_1 \geq 0, x_2 \geq 0.$$

Example 89 (How to Solve a Non-linear Program Graphically)

An intuitive way to understand a low-dimensional non-linear program is to represent the feasible set and the objective function graphically.

For instance, we may try to represent graphically the optimization problem from Example 88, which can be re-written as

$$\max \quad \pi = 125 - (x_1 - 5)^2 - (x_2 - 10)^2$$
$$\text{subject to} \quad 5x_1 + 3x_2 \leq 40$$
$$x_1 \leq 5$$
$$x_2 \leq 10$$
$$x_1 \geq 0, x_2 \geq 0;$$

or

$$\min \quad C = (x_1 - 5)^2 + (x_2 - 10)^2$$
$$\text{subject to} \quad 5x_1 + 3x_2 \leq 40$$
$$x_1 \leq 5$$
$$x_2 \leq 10$$
$$x_1 \geq 0, x_2 \geq 0.$$

Region OABCD below shows the feasible set of the non-linear program.

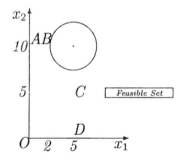

The value of the objective function can be interpreted as the square of the distance from the point $(5, 10)$ to the point (x_1, x_2). Thus, the non-linear program requires us to find the point in the feasible region which minimizes the distance to the point $(5, 10)$.

As can be seen in the figure above, the optimal point (x_1^*, x_2^*) lies on the line $5x_1 + 3x_2 = 40$ and it is also the tangency point with a circle centered at $(5, 10)$. These two conditions are sufficient to determine the optimal solution.

Consider $(x_1 - 5)^2 + (x_2 - 10)^2 - C = 0$. From the implicit function theorem, we have

$$\frac{dx_2}{dx_1} = -\frac{x_1 - 5}{x_2 - 10}.$$

Thus, the tangent to the circle centered at $(5, 10)$ which has the slope $-\frac{5}{3}$ is given by the equation

$$-\frac{5}{3} = -\frac{x_1 - 5}{x_2 - 10}$$

or

$$3x_1 - 5x_2 + 35 = 0.$$

Solving the system of equations

$$\begin{cases} 5x_1 + 3x_2 = 40 \\ 3x_1 - 5x_2 = -35 \end{cases}$$

we find the optimal solution $(x_1^*, x_2^*) = \left(\frac{95}{34}, \frac{295}{34}\right)$.

3.2.2 Kuhn-Tucker Conditions

For the purpose of ruling out certain irregularities on the boundary of the feasible set, a restriction on the constrained functions is imposed. This restriction is called *the constraint qualification*.

Definition 85 *Let $\bar{x} = (x_1, \ldots, x_n)$ be a point on the boundary of the feasible region of a non-linear program.*

  i) *A vector $dx = (dx_1, \ldots, dx_n)$ is a* test vector *of $\bar{x}$ if it satisfies the conditions*

  - $\bar{x}_i = 0 \implies dx_i \geq 0$;
  - $g^j(\bar{x}) = b_j \implies dg^j(\bar{x}) \leq (\geq)0$ *for a maximization (minimization) program.*

  ii) *A* qualifying arc *for a test vector of $\bar{x}$ is an arc with the point of origin $\bar{x}$, contained entirely in the feasible region and tangent to the test vector.*

  iii) *The* constraint qualification *is a condition which requires that, for any boundary point $\bar{x}$ of the feasible region and for any test vector of $\bar{x}$, there exists a qualifying arc for that test vector.*

The constraint qualification condition is often imposed in the following way:

- the gradients at x^* of those g^j-constraints which are binding (active) at x^* are linearly independent.

In other words, the Jacobian of binding constraints at x^* should have the full rank.

Example 90 *Consider a problem of maximizing $f(x, y) = xy$, subject to constraints $(x + y - 2)^2 \leq 0$, $x \geq 0$, $y \geq 0$. Obviously, the solution is such that $x + y - 2 = 0$, or $x^* = y^* = 1$. On the other hand, with the Lagrangian function $L = xy - \lambda(x + y - 2)^2$, the K.-T. conditions read as follows:*

$$L_x = y - 2\lambda(x + y - 2) \leq 0, \qquad L_y = x - 2\lambda(x + y - 2) \leq 0,$$

$$xL_x = x(y - 2\lambda(x + y - 2)) = 0, \ \ yL_x = y(x - 2\lambda(x + y - 2)) = 0,$$

$$L_\lambda = -(x + y - 2)^2 \geq 0, \qquad \lambda L_\lambda = -\lambda(x + y - 2)^2 = 0.$$

If $\lambda = 0$ then the constraint is inactive, thus from the first two F.O.C. it follows that $x = y = 0$.
If $\lambda \neq 0$ then $(x+y-2)^2 = 0$, i.e. x and y cannot vanish simultaneously. However, the first two constraints require $x = y = 0$, which leads to a contradiction.
The reason for K.-T. conditions to fail is that at the point of optimality, $rank(\nabla g(x^*, y^*) = (2(x^* + y^* - 20, 2(x^* + y^* - 2)) = (0, 0) = 0$.

Assume that the constraint qualification condition is satisfied and the functions in (3.7) are differentiable. If x^* is the optimal solution to (3.7), it should satisfy the following conditions (*the Kuhn-Tucker necessary maximum conditions* or *K.-T. conditions*):

$$\frac{\partial L}{\partial x_i} \leq 0, \ x_i \geq 0 \quad \text{and} \quad x_i \frac{\partial L}{\partial x_i} = 0, \quad i = 1, \ldots, n,$$

$$\frac{\partial L}{\partial \lambda_j} \geq 0, \ \lambda_j \geq 0 \quad \text{and} \quad \lambda_j \frac{\partial L}{\partial \lambda_j} = 0, \quad j = 1, \ldots, m$$

where

$$L(x_1, \ldots, x_n, \lambda_1, \ldots, \lambda_m) =$$

$$= f(x_1, \ldots, x_n) + \sum_{j=1}^{m} \lambda_j (b_j - g^j(x_1, \ldots, x_n))$$

is the Lagrangian function of a non-linear program.

The minimization version of Kuhn-Tucker necessary conditions is

$$\frac{\partial L}{\partial x_i} \geq 0, \ x_i \geq 0 \quad \text{and} \quad x_i \frac{\partial L}{\partial x_i} = 0, \quad i = 1, \ldots, n,$$

$$\frac{\partial L}{\partial \lambda_j} \leq 0, \ \lambda_j \geq 0 \quad \text{and} \quad \lambda_j \frac{\partial L}{\partial \lambda_j} = 0, \quad j = 1, \ldots, m$$

Recipe 14 – How do the K.-T. Conditions Emerge? An Intuitive Approach:

I. *Unconstrained optimization:* $\max f(x_1, \ldots, x_n)$
   $F.O.C.: \frac{\partial f}{\partial x_i} = 0,$
   $\quad i = 1, \ldots, n.$

II. *Inequality constraints optimization:* $\max f(x_1, \ldots, x_n)$
   (only non-negativity constraints) $s.t. \ x_i \geq 0, \ i = 1, \ldots, n$
   $F.O.C.: \frac{\partial f}{\partial x_i} \leq 0, \ x_i \geq 0,$
   $\quad x_i \cdot \frac{\partial f}{\partial x_i} = 0,$
   $\quad i = 1, \ldots, n$
   if $x_i > 0 \Rightarrow \frac{\partial f}{\partial x_i} = 0$
   (Same condition as in I)
   if $x_i = 0 \Rightarrow$ *Corner solution.*

III. *Inequality constraints optimization:* $\max f(x_1, \ldots, x_n)$
   (general case) $s.t. \ x_i \geq 0, \ i = 1, \ldots, n$
   $\quad g^j(x_1, \ldots, x_n) \leq b_j,$
   $\quad j = 1, \ldots, m$

The Lagrangian function associated with this problem is
$$\mathcal{L} = f + \sum_{j=1}^{m} \lambda_j (b_j - g^j)$$

IIIa.   *If we introduce "dummy" variables $s_j = b_j - g^j$ then III can*
        be modified to a constrained optimization problem
        with non-negativity constraints and equality constraints:

$$\max f(x_1, \ldots, x_n)$$
$$\text{s.t. } x_i \geq 0, \ i = 1, \ldots, n$$
$$s_j \geq 0, \ j = 1, \ldots, m$$
$$s_j + g^j(x_1, \ldots, x_n) = b_j,$$
$$j = 1, \ldots, m$$

The Lagrangian associated with this modified problem is

$$\mathbf{L} = f + \sum_{j=1}^{m} \lambda_j(b_j - g^j - s_j)$$

But we already know the F.O.C.
for the equality constrained optimization problem:

$$\frac{\partial \mathbf{L}}{\partial \lambda_j} = 0 \qquad \Rightarrow \qquad \frac{\partial \mathbf{L}}{\partial \lambda_j} = b_j - g^j - s_j = 0$$

Since $s_j \geq 0$,
we have $b_j - g^j \geq 0$

Therefore, $\boxed{\dfrac{\partial \mathcal{L}}{\partial \lambda_j} = b_j - g^j \geq 0}$

Non-negativity of x_i and s_j implies

$$\frac{\partial \mathbf{L}}{\partial x_i} \leq 0, \ x_i \geq 0, \ x_i \cdot \frac{\partial \mathbf{L}}{\partial x_i} = 0 \Rightarrow \boxed{\frac{\partial \mathcal{L}}{\partial x_i} \leq 0, \ x_i \geq 0, \ x_i \cdot \frac{\partial \mathcal{L}}{\partial x_i} = 0}$$

$$\frac{\partial \mathbf{L}}{\partial s_j} \leq 0, \ s_j \geq 0, \ s_j \cdot \frac{\partial \mathbf{L}}{\partial s_j} = 0 \Rightarrow \frac{\partial \mathbf{L}}{\partial s_j} = -\lambda_j \leq 0.$$

Thus, $\boxed{\lambda_j \geq 0}$

$$s_j \cdot \frac{\partial \mathbf{L}}{\partial s_j} = \lambda_j \cdot s_j = 0.$$

Therefore, $\lambda_j(b_j - g^j) = 0$

or $\boxed{\lambda_j \cdot \dfrac{\partial \mathcal{L}}{\partial \lambda_j} = 0}$

Conditions derived in the boxes are called

Kuhn–Tucker conditions.

Note that, in general, the K.-T. conditions are neither necessary nor sufficient for a local optimum. However, if certain assumptions are satisfied, the K.-T. conditions become necessary and even sufficient. ⚠

Example 91 *The Lagrangian function of the non-linear program in Example 88 is:*

$$L = x_1(10 - x_1) + x_2(20 - x_2)$$
$$+\lambda_1(30 - 5x_1 - 3x_2) + \lambda_2(5 - x_1) + \lambda_3(10 - x_2).$$

The K.-T. conditions are:

$$\frac{\partial L}{\partial x_1} = 10 - 2x_1 - 5\lambda_1 - \lambda_2 \leq 0$$

$$\frac{\partial L}{\partial x_2} = 20 - 2x_2 - 3\lambda_1 - \lambda_3 \leq 0$$

$$\frac{\partial L}{\partial \lambda_1} = -(5x_1 + 3x_2 - 40) \geq 0$$

$$\frac{\partial L}{\partial \lambda_2} = -(x_1 - 5) \geq 0$$

$$\frac{\partial L}{\partial \lambda_3} = -(x_2 - 10) \geq 0$$

$$x_1 \geq 0, x_2 \geq 0$$

$$\lambda_1 \geq 0, \lambda_2 \geq 0, \lambda_3 \geq 0$$

$$x_1 \frac{\partial L}{\partial x_1} = 0, x_2 \frac{\partial L}{\partial x_2} = 0$$

$$\lambda_j \frac{\partial L}{\partial \lambda_j} = 0, \ j = 1, 2, 3.$$

Proposition 47 (Necessity Theorem)
The K.-T. conditions are necessary conditions for a local optimum if the constraint qualification is satisfied.

Note that ⚠

- The failure of the constraint qualification signals certain irregularities at the boundary kinks of the feasible set. Only if the optimal solution occurs in such a kink may the K.-T. conditions not be satisfied.

- If all constraints are linear, the constraint qualification is always satisfied.

Example 92 *The constraint qualification for the non-linear program in Example 88 is satisfied since all constraints are linear. Therefore, the optimal solution $(\frac{95}{34}, \frac{295}{34})$ must satisfy the K.-T. conditions in Example 91.*

Example 93 *The following example illustrates a case where the Kuhn-Tucker conditions are not satisfied in the solution to an optimization problem. Consider the problem:*

$$\begin{aligned} \max \quad & y \\ \text{subject to} \quad & x + (y-1)^3 \leq 0, \\ & x \geq 0, y \geq 0. \end{aligned}$$

The solution to this problem is $(0,1)$. (If $y > 1$, then the restriction $x + (y-1)^3 \leq 0$ implies $x < 0$.)
The Lagrangian function is

$$L = y + \lambda[-x - (y-1)^3].$$

One of the Kuhn-Tucker conditions requires

$$\frac{\partial L}{\partial y} \leq 0,$$

or

$$1 - 3\lambda(y-1)^2 \leq 0.$$

As can be observed, this condition is not verified at the point $(0,1)$.

Proposition 48 (Kuhn-Tucker Sufficiency Theorem – 1) *If*

a) *f is differentiable and concave in the non-negative orthant;*

b) *each constraint function is differentiable and convex in the non-negative orthant;*

c) *the point x^* satisfies the Kuhn-Tucker necessary maximum conditions*

then x^* gives a global maximum of f.

To adapt this theorem for minimization problems, we need to interchange the two words "concave" and "convex" in a) and b) and use the Kuhn-Tucker necessary minimum condition in c).

Proposition 49 (Sufficiency Theorem – 2)
The K.-T. conditions are sufficient conditions for x^* to be a local optimum of a maximization (minimization) program if the following assumptions are satisfied:

- *the objective function f is differentiable and quasi-concave (or quasi-convex);*

- *each constraint g^i is differentiable and quasi-convex (or quasi-concave);*

- *any one of the following is satisfied:*

  - *there exists j such that $\frac{\partial f(x^*)}{\partial x_j} < 0(> 0)$;*

  - *there exists j such that $\frac{\partial f(x^*)}{\partial x_j} > 0(< 0)$ and x_j can take a positive value without violating the constraints;*

  - *f is concave (or convex).*

The problem of finding the non-negative vector

$$(x^*, \lambda^*): \quad x^* = (x_1^*, \ldots, x_n^*), \ \lambda^* = (\lambda_1^*, \ldots, \lambda_m^*)$$

which satisfies Kuhn-Tucker necessary conditions and for which

$$L(x, \lambda^*) \le L(x^*, \lambda^*) \le L(x^*, \lambda)$$

for all $x = (x_1, \ldots, x_n) \ge 0, \ \lambda = (\lambda_1, \ldots, \lambda_m) \ge 0$

is known as the *saddle point problem.*

Proposition 50 *If (x^*, λ^*) solves the saddle point problem then (x^*, λ^*) solves the problem (3.7).*

Economics Application 12 (Economic Interpretation of a Non-linear Program and Kuhn-Tucker Conditions)
   A non-linear program can be interpreted much like a linear program. A maximization program in the general form, for example, is the production problem facing a firm which has to produce n goods such that it maximizes its revenue subject to m resource (factor) constraints.
   The variables have the following economic interpretations:

- *x_i is the amount produced of the ith product;*

- *b_j is the amount of the jth resource available;*

- *f is the profit (revenue) function;*

- g^j is a function which shows how the jth resource is used in producing the n goods.

The optimal solution to the maximization program indicates the optimal quantities of each good the firm should produce.

In order to interpret the Kuhn-Tucker conditions, we first have to note the meanings of the following variables:

- $f_i = \frac{\partial f}{\partial x_i}$ is the marginal profit (revenue) of product i;

- λ_j is the shadow price of resource j;

- $g_i^j = \frac{\partial g^j}{\partial x_i}$ is the amount of resource j used in producing a marginal unit of product i;

- $\lambda_j g_i^j$ is the imputed cost of resource j incurred in the production of a marginal unit of product i.

The K.-T. condition $\frac{\partial L}{\partial x_i} \leq 0$ can be written as $f_i \leq \sum_{j=1}^{m} \lambda_j g_i^j$ and it says that the marginal profit of the ith product cannot exceed the aggregate marginal imputed cost of the ith product.

The condition $x_i \frac{\partial L}{\partial x_i} = 0$ implies that, in order to produce good i ($x_i > 0$), the marginal profit of good i must be equal to the aggregate marginal imputed cost ($\frac{\partial L}{\partial x_i} = 0$). The same condition shows that good i is not produced ($x_i = 0$) if there is an excess imputation ($x_i \frac{\partial L}{\partial x_i} < 0$).

The K.-T. condition $\frac{\partial L}{\partial \lambda_j} \geq 0$ is simply a restatement of constraint j, which states that the total amount of resource j used in producing all the n goods should not exceed total amount available b_j.

The condition $\frac{\partial L}{\partial \lambda_j} = 0$ indicates that if a resource is not fully used in the optimal solution ($\frac{\partial L}{\partial \lambda_j} > 0$), then its shadow price will be 0 ($\lambda_j = 0$). On the other hand, a fully used resource ($\frac{\partial L}{\partial \lambda_j} = 0$) has a strictly positive price ($\lambda_j > 0$).

Example 94 *Let us find an economic interpretation for the maximization program given in Example 88:*

$$\begin{aligned}
\max \quad & R = x_1(10 - x_1) + x_2(20 - x_2) \\
\text{subject to} \quad & 5x_1 + 3x_2 \leq 40 \\
& x_1 \leq 5 \\
& x_2 \leq 10 \\
& x_1 \geq 0, x_2 \geq 0.
\end{aligned}$$

A firm has to produce two goods using three kinds of resources available in the amounts 40,5,10 respectively. The first resource is used in the production of both goods: five units are necessary to produce one unit of good 1, and three units to produce one unit of good 2. The second resource is used only in producing good 1 and the third resource is used only in producing good 2.

The sale prices of the two goods are given by the linear inverse demand equations $p_1 = 10 - x_1$ and $p_2 = 20 - x_2$. The problem the firm faces is how much to produce of each good in order to maximize revenue $R = x_1 p_1 + x_2 p_2$. The solution $(2, 10)$ gives the optimal amounts the firm should produce.

Economics Application 13 (Sales-Maximizing Firm)

Suppose that a firm's objective is to maximize its sales revenue subject to the constraint that the profit is not less than a certain value. Let Q denote the amount of the good supplied on the market. If the revenue function is $R(Q) = 20Q - Q^2$, the cost function is $C(Q) = Q^2 + 6Q + 2$ and the minimum profit is 10, find the sales-maximizing quantity.

The firm's problem is

$$\max \ R(Q)$$
$$\text{subject to } R(Q) - C(Q) \geq 10 \text{ and } Q \geq 0$$

or

$$\max \ R(Q)$$
$$\text{subject to } 2Q^2 - 14Q + 2 \leq -10 \text{ and } Q \geq 0.$$

The Lagrangian function is

$$Z = 20Q - Q^2 - \lambda(2Q^2 - 14Q + 12)$$

and the Kuhn-Tucker conditions are

$$\frac{\partial Z}{\partial Q} \leq 0, \quad Q \geq 0, \quad Q\frac{\partial Z}{\partial Q} = 0, \tag{3.8}$$

$$\frac{\partial Z}{\partial \lambda} \geq 0, \quad \lambda \geq 0, \quad \lambda\frac{\partial Z}{\partial \lambda} = 0. \tag{3.9}$$

Explicitly, the first inequalities in each row above are

$$-Q + 10 - \lambda(2Q - 7) \leq 0, \tag{3.10}$$

$$-Q^2 + 7Q - 6 \geq 0. \tag{3.11}$$

The second inequality in (3.8) and inequality (3.11) imply that $Q > 0$ (if Q were 0, (3.11) would not be satisfied). From the third condition in (3.8) it follows that $\frac{\partial Z}{\partial Q} = 0$.

If λ were 0, the condition $\frac{\partial Z}{\partial Q} = 0$ would lead to $Q = 10$, which is not consistent with (3.11); thus, we must have $\lambda > 0$ and $\frac{\partial Z}{\partial \lambda} = 0$.

The quadratic equation $-Q^2 + 7Q - 6 = 0$ has the roots $Q_1 = 1$ and $Q_2 = 6$. Q_1 implies $\lambda = -\frac{9}{5}$, which contradicts $\lambda > 0$. The solution $Q_2 = 6$ leads to a positive value for λ. Thus, the sales-maximizing quantity is $Q = 6$.

3.3   Appendix: Linear Programming

The Setup of the Problem

There are two general types of linear programs:

- the maximization program in n variables subject to $m + n$ constraints:

$$
\begin{array}{ll}
\text{Maximize} & \pi = \sum_{j=1}^{n} c_j x_j \\
\text{subject to} & \sum_{j=1}^{n} a_{ij} x_j \leq r_i \quad (i = 1, 2, \ldots m) \\
\text{and} & x_j \geq 0 \quad\quad\quad\; (j = 1, 2, \ldots n).
\end{array}
$$

- the minimization program in n variables subject to $m + n$ constraints:

$$
\begin{array}{ll}
\text{Minimize} & C = \sum_{j=1}^{n} c_j x_j \\
\text{subject to} & \sum_{j=1}^{n} a_{ij} x_j \geq r_i \quad (i = 1, 2, \ldots m) \\
\text{and} & x_j \geq 0 \quad\quad\quad\; (j = 1, 2, \ldots n).
\end{array}
$$

Note that without loss of generality all constraints in a linear program may be written as either $\leq$ inequalities or $\geq$ inequalities because any $\leq$ inequality can be transformed into a $\geq$ inequality by multiplying both sides by -1.

A more concise way to express a linear program is by using matrix notation:

- the maximization program in n variables subject to $m + n$ constraints:

$$\begin{aligned} \text{Maximize} \quad & \pi = c'x \\ \text{subject to} \quad & Ax \le r \\ \text{and} \quad & x \ge 0. \end{aligned}$$

- the minimization program in n variables subject to $m + n$ constraints:

$$\begin{aligned} \text{Minimize} \quad & C = c'x \\ \text{subject to} \quad & Ax \ge r \\ \text{and} \quad & x \ge 0. \end{aligned}$$

where $c = (c_1, \ldots, c_n)'$, $x = (x_1, \ldots, x_n)'$, $r = (r_1, \ldots, r_m)'$,

$$A = \begin{pmatrix} a_{11} & a_{12} & \cdots & a_{1n} \\ a_{21} & a_{22} & \cdots & a_{2n} \\ & & \ddots & \\ a_{m1} & a_{m2} & \cdots & a_{mn} \end{pmatrix}$$

Example 95 *The following is an example of a linear program:*

$$\begin{aligned} \text{max} \quad & \pi = 10x_1 + 30x_2 \\ \text{subject to} \quad & 5x_1 + 3x_2 \le 40 \\ & x_1 \le 5 \\ & x_2 \le 10 \\ & x_1 \ge 0, x_2 \ge 0. \end{aligned}$$

Proposition 51 (Globality Theorem)

If the feasible set F of an optimization problem is a closed convex set, and if the objective function is a continuous concave (convex) function over the feasible set, then

i) *any local maximum (minimum) will also be a global maximum (minimum) and*

ii) *the points in F at which the objective function is optimized will constitute a convex set.*

Note that any linear program satisfies the assumptions of the globality theorem. Indeed, the objective function is linear; thus, it is both a concave and a convex function. On the other hand, each inequality restriction defines a closed halfspace, which is also a convex set. Since the intersection of a finite number of closed convex sets is also a closed convex set, it follows that the feasible set F is closed and convex.

The theorem says that for a linear program there is no difference between a local optimum and a global optimum, and if a pair of optimal solutions to a linear program exists, then any convex combination of the two must also be an optimal solution.

Definition 86 *An* extreme point *of a linear program is a point in its feasible set which cannot be derived from a convex combination of any other two points in the feasible set.*

Example 96 *For the problem in example 95 the extreme points are* $(0,0)$, $(0,10)$, $(2,10)$, $(5,5)$, $(5,0)$. *They are just the corners of the feasible region represented in the $x_1 0 x_2$ system of coordinates.*

Proposition 52 *The optimal solution to a linear program is an extreme point.*

The Simplex Method

The next issue is to develop an algorithm which will enable us to solve a linear program. Such an algorithm is called *the simplex method.*

The simplex method is a systematic procedure for finding the optimal solution to a linear program. Loosely speaking, using this method we move from one extreme point of the feasible region to another until the optimal point is attained.

The following two examples illustrate how the simplex method can be applied:

Example 97 *Consider again the maximization problem in Example 95.*

Step 1. *Transform each inequality constraint into an equality constraint by inserting a dummy variable (slack variable) in each constraint.*

After applying this step, the problem becomes:

$$\max \quad \pi = 10x_1 + 30x_2$$
$$\text{subject to} \quad 5x_1 + 3x_2 + d_1 = 40$$
$$x_1 + d_2 = 5$$
$$x_2 + d_3 = 10$$
$$x_1 \geq 0, x_2 \geq 0, d_1 \geq 0, d_2 \geq 0, d_3 \geq 0.$$

The extreme points of the new feasible set are called basic feasible solutions *(BFS). For each extreme point of the original feasible set there is a BFS. For example, the BFS corresponding to the extreme point* $(x_1, x_2) = (0, 0)$ *is* $(x_1, x_2, d_1, d_2, d_3) = (0, 0, 40, 5, 10)$.

Step 2. *Find a BFS to be able to start the algorithm.*

A BFS is easy to find when $(0, 0)$ *is in the feasible set, as it is in our example. Here, the BFS are* $(0, 0, 40, 5, 10)$. *In general, finding a BFS requires the introduction of additional artificial variables. Example 98 illustrates this situation.*

Step 3. *Set up a simplex tableau like the following one:*

Row	π	x_1	x_2	d_1	d_2	d_3	Constant
0	1	-10	-30	0	0	0	0
1	0	5	3	1	0	0	40
2	0	1	0	0	1	0	5
3	0	0	1	0	0	1	10

Row 0 contains the coefficients of the objective function. Rows $1, 2, 3$ include the coefficients of the constraints as they appear in the matrix of coefficients.

In row 0 we can see the variables which have non-zero values in the current BFS. They correspond to the zeros in row 0, in this case, the non-zero variables are d_1, d_2, d_3. These non-zero variables always correspond to a basis in R^m, where m is the number of constraints (see the general form of a linear program). The current basis is formed by the column coefficient vectors which appear in the tableau in rows $1, 2, 3$ under d_1, d_2, d_3 $(m = 3)$.

The rightmost column shows the non-zero values of the current BFS (in rows $1, 2, 3$) and the value of the objective function for the current BFS (in row 0). That is, $d_1 = 40$, $d_2 = 5$, $d_3 = 10$, $\pi = 0$.

Step 4. *Choose the pivot.*

The simplex method finds the optimal solution by moving from one BFS to another BFS until the optimum is reached. The new BFS is chosen such that the value of the objective function is improved (that is, it is larger for a maximization problem and lower for a minimization problem).

In order to switch to another BFS, we need to replace a variable currently in the basis by a variable which is not in the basis. This amounts to choosing a pivot *element which will indicate both the new variable and the exit variable.*

The pivot element is selected in the following way. The new variable (and, correspondingly, the pivot column) is that variable having the lowest negative value in row 0 (largest positive value for a minimization problem). Once we have determined the pivot column, the pivot row is found in this way:

- *pick the strictly positive elements in the pivot column except for the element in row 0;*

- *select the corresponding elements in the constant column and divide them by the strictly positive elements in the pivot column;*

- *the pivot row corresponds to the minimum of these ratios.*

In our example, the pivot column is the column of x_2 (since $-30 < -10$) and the pivot row is row 3 (since $min\{\frac{40}{3}, \frac{10}{1}\} = \frac{10}{1}$); thus, the pivot element is the 1 which lies at the intersection of column of x_2 and row 3.

Step 5. *Make the appropriate transformations to reduce the pivot column to a unit vector (that is, a vector with 1 in the place of the pivot element and 0 in the rest).*

Usually, first we have to divide the pivot row by the pivot element to obtain 1 in the place of the pivot element. However, this is not necessary in our case since we already have 1 in that position.

Then the new version of the pivot row is used to get zeros in the other places of the pivot column. In our example, we multiply the pivot row by 30 and add it to row 0; then we multiply the pivot row by -3 and add it to row 1.

Hence we obtain a new simplex tableau:

Row	π	x_1	x_2	d_1	d_2	d_3	Constant
0	1	-10	0	0	0	30	300
1	0	5	0	1	0	-3	10
2	0	1	0	0	1	0	5
3	0	0	1	0	0	1	10

Step 6. *Check that row 0 still contains negative values (positive values for a minimization problem). If it does, proceed to step 4. If it does not, the optimal solution must be found.*

In our example, we still have a negative value in row 0 (-10). According to step 4, the column of x_1 is the pivot column (x_1 enters the base). Since $min\{\frac{10}{5}, \frac{5}{1}\} = \frac{10}{5}$, it follows that row 1 is the pivot row. Thus, 5 in row 1 is the new pivot element. Applying step 5, we obtain a third simplex tableau:

Row	π	x_1	x_2	d_1	d_2	d_3	Constant
0	1	0	0	2	0	24	320
1	0	1	0	$\frac{1}{5}$	0	$\frac{-3}{5}$	2
2	0	0	0	$\frac{-1}{5}$	1	$\frac{3}{5}$	3
3	0	0	1	0	0	1	10

There is no negative element left in row 0 and the current basis consists of x_1, x_2, d_2. Thus, the optimal solution can be read from the rightmost column: $x_1 = 2$, $x_2 = 10$, $d_2 = 3$, $\pi = 320$. The correspondence between the optimal values and the variables in the current basis is made using the 1's in the column vectors of x_1, x_2, d_2.

Example 98 *Consider the minimization problem:*

$$\text{min} \quad C = x_1 + x_2$$
$$\text{subject to} \quad x_1 - x_2 \geq 0$$
$$-x_1 - x_2 \geq -6$$
$$x_2 \geq 1$$
$$x_1 \geq 0, x_2 \geq 0.$$

After applying step 1 of the simplex algorithm, we obtain:

$$\text{min} \quad C = x_1 + x_2$$
$$\text{subject to} \quad x_1 - x_2 - d_1 = 0$$
$$-x_1 - x_2 - d_2 = -6$$
$$x_2 - d_3 = 1$$
$$x_1 \geq 0, x_2 \geq 0, d_1 \geq 0, d_2 \geq 0, d_3 \geq 0.$$

In this example, $(0,0)$ is not in the feasible set. One way of finding a starting BFS is to transform the problem by adding 3 more artificial variables f_1, f_2, f_3 to the linear program as follows:

$$
\begin{aligned}
\min \quad & C = x_1 + x_2 + M(f_1 + f_2 + f_3) \\
\text{subject to} \quad & x_1 - x_2 - d_1 + f_1 = 0 \\
& x_1 + x_2 + d_2 + f_2 = 6 \\
& x_2 - d_3 + f_3 = 1 \\
& x_1 \geq 0, x_2 \geq 0, d_1 \geq 0, d_2 \geq 0, d_3 \geq 0 \\
& f_1 \geq 0, f_2 \geq 0, f_3 \geq 0.
\end{aligned}
$$

Before adding the artificial variables, the second constraint is multiplied by -1 to obtain a positive number on the right-hand side.

As long as M is high enough, the optimal solution will contain $f_1 = 0$, $f_2 = 0$, $f_3 = 0$ and thus the optimum is not affected by including the artificial variables in the linear program. In our case, we choose $M = 100$. (In maximization problems M is chosen very low.)

Now, the starting BFS is given by $(x_1, x_2, d_1, d_2, d_3, f_1, f_2, f_3) = (0, 0, 0, 0, 0, 0, 6, 1)$ and the first simplex tableau is:

Row	C	x_1	x_2	d_1	d_2	d_3	f_1	f_2	f_3	Constant
0	1	-1	-1	0	0	0	-100	-100	-100	0
1	0	1	-1	-1	0	0	1	0	0	0
2	0	1	1	0	1	0	0	1	0	6
3	0	0	1	0	0	-1	0	0	1	1

In order to obtain unit vectors in the columns of f_1, f_2 and f_3, we add $100(row1+row2+row3)$ to row 0:

Row	C	x_1	x_2	d_1	d_2	d_3	f_1	f_2	f_3	Constant
0	1	199	99	-100	100	-100	0	0	0	700
1	0	1	-1	-1	0	0	1	0	0	0
2	0	1	1	0	1	0	0	1	0	6
3	0	0	1	0	0	-1	0	0	1	1

Proceeding with the other steps of the simplex algorithm will finally lead to the optimal solution $x_1 = 1$, $x_2 = 1$, $C = 2$.

Sometimes the starting BFS (or the starting base) can be found in a simpler way than by using the general method that was just illustrated.

For example, our linear program can be written as:

$$\min \quad C = x_1 + x_2$$
$$\text{subject to} \quad -x_1 + x_2 + d_1 = 0$$
$$x_1 + x_2 + d_2 = 6$$
$$x_2 - d_3 = 1$$
$$x_1 \geq 0, x_2 \geq 0, d_1 \geq 0, d_2 \geq 0, d_3 \geq 0.$$

Taking $x_1 = 0$, $x_2 = 0$ we get $d_1 = 0$, $d_2 = 6$, $d_3 = -1$. Only d_3 contradicts the constraint $d_3 \geq 0$. Thus, we need to add only one artificial variable f to the linear program (so that the third constraint becomes $x_2 - d_3 + f = 1$, and the objective function is $\min C = x_1 + x_2 + Mf$, $M = 100$) and the variables in the starting base will be d_1, d_2, f.

The complete sequence of simplex tableaux leading to the optimal solution is given below.

Row	C	x_1	x_2	d_1	d_2	d_3	f	Constant
0	1	-1	-1	0	0	0	-100	0
1	0	-1	1	1	0	0	0	0
2	0	1	1	0	1	0	0	6
3	0	0	1	0	0	-1	1	1
0	1	-1	99	0	0	-100	0	100
1	0	-1	1*	1	0	0	0	0
2	0	1	1	0	1	0	0	6
3	0	0	1	0	0	-1	1	1
0	1	98	0	-99	0	-100	0	100
1	0	-1	1	1	0	0	0	0
2	0	2	0	-1	1	0	0	6
3	0	1*	0	-1	0	-1	1	1
0	1	0	0	-1	0	-2	-98	2
1	0	0	1	0	0	-1	1	1
2	0	0	0	1	1	2	-2	4
3	0	1	0	-1	0	-1	1	1

(The asterisks mark the pivot elements.)

As can be seen from the last tableau, the optimum is $x_1 = 1$, $x_2 = 1$, $d_2 = 4$, $C = 2$.

Economics Application 14 (Economic Interpretation of a Maximization Program)

The maximization program in the general form is the production problem facing a firm which has to produce n goods such that it maximizes its revenue subject to m resource (factor) constraints.

The variables have the following economic interpretations:

- *x_j is the amount produced of the jth product;*

- *r_i is the amount of the ith resource available;*

- *a_{ij} is the amount of the ith resource used in producing a unit of the jth product;*

- *c_j is the price of a unit of product j.*

The optimal solution to the maximization program indicates the optimal quantities of each good the firm should produce.

Example 99 *Let us find an economic interpretation of the maximization program given in Example 95:*

$$\begin{aligned}
\max \quad & \pi = 10x_1 + 30x_2 \\
\text{subject to} \quad & 5x_1 + 3x_2 \leq 40 \\
& x_1 \leq 5 \\
& x_2 \leq 10 \\
& x_1 \geq 0, x_2 \geq 0.
\end{aligned}$$

A firm has to produce two goods using three kinds of resources available in the amounts 40,5,10 respectively. The first resource is used in the production of both goods: five units are necessary to produce one unit of good 1, and three units to produce 1 unit of good 2. The second resource is used only in producing good 1 and the third resource is used only in producing good 2.

The question is, how much of each good should the firm produce in order to maximize its revenue if the sale prices of goods are 10 and 30, respectively? The solution $(2, 10)$ gives the optimal amounts the firm should produce.

Economics Application 15 (Economic Interpretation of a Minimization Program)

The minimization program (see the general form) is also connected to the firm's production problem. This time the firm wants to produce a fixed combination of m goods using n resources (factors). The problem lies in choosing the amount of each resource such that the total cost of all resources used in the production of the fixed combination of goods is minimized.

The variables have the following economic interpretations:

- x_j *is the amount of the jth resource used in the production process;*

- r_i *is the amount of the good i the firm decides to produce;*

- a_{ij} *is the marginal productivity of resource j in producing good i;*

- c_j *is the price of a unit of resource j.*

The optimal solution to the minimization program indicates the optimal quantities of each factor the firm should employ.

Example 100 *Consider again the linear program in Example 98:*

$$\begin{aligned} \min \quad & C = x_1 + x_2 \\ \text{subject to} \quad & x_1 - x_2 \geq 0 \\ & x_1 + x_2 \leq 6 \\ & x_2 \geq 1 \\ & x_1 \geq 0, x_2 \geq 0. \end{aligned}$$

What economic interpretation can be assigned to this linear program?

A firm has to produce two goods, at most six units of the first good and one unit of the second good. Two factors are used in the production process and the price of both factors is one. The marginal productivity of both factors in producing good 1 is one and the marginal productivity of the factors in producing good 2 are zero and one, respectively (good 2 uses only the second factor). The production process also requires that the amount of the second factor not exceed the amount of the first factor.

The question is how much of each factor the firm should employ in order to minimize the total factor cost. The solution $(1, 1)$ gives the optimal factor amounts.

Duality

As a matter of fact, the maximization and the minimization programs are closely related. This relationship is called *duality*.

Definition 87 *The* dual (program) *of the maximization program*

> *Maximize* $\pi = c'x$
> *subject to* $Ax \le r$
> *and* $x \ge 0.$

is the minimization program

> *Minimize* $\pi^* = r'y$
> *subject to* $A'y \ge c$
> *and* $y \ge 0.$

Definition 88 *The* dual (program) *of the minimization program*

> *Minimize* $C = c'x$
> *subject to* $Ax \ge r$
> *and* $x \ge 0.$

is the maximization program

> *Maximize* $C^* = r'y$
> *subject to* $A'y \le c$
> *and* $y \ge 0.$

Definition 89 *The original program from which the dual is derived is called* the primal (program).

Example 101 *The dual of the maximization program in Example 95 is*

> min $C^* = 40y_1 + 5y_2 + 10y_3$
> subject to $5y_1 + y_2 \ge 10$
> $3y_1 + y_3 \ge 30$
> $y_1 \ge 0, y_2 \ge 0, y_3 \ge 0.$

Example 102 *The dual of the minimization program in Example 98 is*

> max $\pi^* = -6y_2 + y_3$
> subject to $y_1 - y_2 \le 1$
> $-y_1 - y_2 + y_3 \le 1$
> $y_1 \ge 0, y_2 \ge 0, y_3 \ge 0.$

The following duality theorems clarify the relationship between the dual and the primal.

Proposition 53 *The optimal values of the primal and the dual objective functions are always identical, if the optimal solutions exist.*

Proposition 54

1. *If a certain choice variable in a primal (dual) program is optimally non-zero then the corresponding dummy variable in the dual (primal) must be optimally zero;*

2. *If a certain dummy variable in a primal (dual) program is optimally non-zero then the corresponding choice variable in the dual (primal) must be optimally zero.*

The last simplex tableau of a linear program offers not only the optimal solution to that program but also the optimal solution to the dual program. ⚠

Recipe 15 – How to Solve the Dual Program:
The optimal solution to the dual program can be read from row 0 of the last simplex tableau of the primal in the following way:

1. *the absolute values of the elements in the columns corresponding to the primal dummy variables are the values of the dual choice variables;*

2. *the absolute values of the elements in the columns corresponding to the primal choice variables are the values of the dual dummy variables.*

Example 103 *From the last simplex tableau in Example 95 we can infer the optimal solution to the problem in Example 101. It is $y_1 = 2$, $y_2 = 0$, $y_3 = 24$, $\pi^* = 320$.*

Example 104 *The last simplex tableau in Example 98 implies that the optimal solution to the problem in Example 102 is $y_1 = 1$, $y_2 = 0$, $y_3 = 2$, $C = 2$.*

Further reading

- Intriligator, M. (2002). *Mathematical Optimization and Economic Theory.* Society for Industrial and Applied Mathematics.

- Luenberger, D. G. (1984) *Introduction to Linear and Nonlinear Programming* (2nd ed.). Massachusetts: Addison-Wesley Inc.

Chapter 4

Dynamics

4.1 Differential Equations

Definition 90 *An equation $F(x, y, y', \ldots, y^{(n)}) = 0$ which assumes a functional relationship between independent variable x and dependent variable y and includes x, y and derivatives of y with respect to x is called an ordinary differential equation of order n.*

The order of a differential equation is determined by the order of the highest derivative in the equation.

A function $y = y(x)$ is said to be a solution to this differential equation, in some open interval I, if $F(x, y(x), y'(x), \ldots, y^{(n)}(x)) = 0$ for all $x \in I$.

4.1.1 Differential Equations of the First Order

Let us consider the first-order differential equations, $y' = f(x, y)$.

The solution $y(x)$ is said to solve Cauchy's problem (or the initial value problem) with the initial values (x_0, y_0), if $y(x_0) = y_0$.

Example 105 *Consider a simple model of population growth over time at a constant growth rate n. If we denote the population size at time t as $N(t)$ then the growth of the population at t, $\dot{N}(t)$, should be proportional to the size of the population $N(t)$, with the proportionality coefficient n, that is, $\dot{N}(t) = nN(t)$, or $d(\ln N(t))/dt = n$. The latter equation immediately integrates to $\ln N(t) = nt + \tilde{C}$, or $N(t) = Ce^{nt}$, where C is an integration constant. If we know that at time $t = 0$ the population* *size was N_0 then these initial values give rise to a solution (that is unique due to the Uniqueness Theorem) of the form $N(t) = N_0 e^{nt}$.*

Proposition 55 (Existence Theorem)
    If f is continuous in an open domain D, then for any given pair $(x_0, y_0) \in D$ there exists a solution $y(x)$ of $y' = f(x, y)$ such that $y(x_0) = y_0$.

Proposition 56 (Uniqueness Theorem)
    If f and $\frac{\partial f}{\partial y}$ are continuous in an open domain D, then given any $(x_0, y_0) \in D$ there exists a unique solution $y(x)$ of $y' = f(x, y)$ such that $y(x_0) = y_0$.

Definition 91 *A differential equation of the form $y' = f(y)$ is said to be autonomous (i.e. y' is determined by y alone).*

Example 106 *Consider an autonomous differential equation $y' = 2\sqrt{y}$. It always solves Cauchy's problem (the existence theorem holds), but the conditions of the Uniqueness Theorem are violated, because $d(2\sqrt{y})/dy = 1/\sqrt{y}$ is not continuous at $y = 0$. Indeed, one can check that there are two distinct solutions passing through the points on the horizontal axis in the $(X - Y)$-plane: $y_1(x) = (x + C)^2$, C is an arbitrary constant, and $y_2(x) \equiv 0$.*

Recipe 16 – How to Sketch Solution Curves without Solving the Differential Equation Itself:
    Note that the differential equation $y' = f(x, y)$ gives us the slope of the solution curves at all points of the region D. Thus, in particular, all the solution curves cross the curve $f(x, y) = k$ (k is a constant) with slope k. This curve is called the isocline of slope k. If we draw the set of isoclines obtained by taking different real values of k, we can schematically sketch the family of solution curves in the $(x - y)$-plane.

Recipe 17 – How to Solve a Differential Equation of the First Order:
    There is no general method of solving differential equations. However, in some cases the solution can be easily obtained:

- **The Variables Separable Case**

    If $y' = f(x)g(y)$ then $\frac{dy}{g(y)} = f(x)dx$. By integrating both parts of the latter equation, we will get a solution.

Example 107 *Solve Cauchy's problem* $(x^2 + 1)y' + 2xy^2 = 0$, *given* $y(0) = 1$.

If we rearrange the terms, this equation becomes equivalent to

$$\frac{dy}{y^2} = -\frac{2xdx}{x^2 + 1}.$$

The integration of both parts yields

$$\frac{1}{y} = \ln(x^2 + 1) + C,$$

or

$$y(x) = \frac{1}{\ln(x^2 + 1) + C}.$$

Since we should satisfy the condition $y(0) = 1$, we can evaluate the constant C, $C = 1$.

Economics Application 16 *Consider a model of economic growth:*

Output $Y = Y(t) = A(K(t))^1 - a(L(t))^a$,

Investment $\dot{K} = sY$,

Population growth $L = L_0 e^{nt}$,

Initial value $K(0) = K_0$.

To find the time path of K, we solve the differential equation

$$\dot{K} = sY = sAK^{1-a}L^a = aAK^{1-a}L_0^a e^{ant}.$$

Upon integration

$$\int K^{a-1}dK = sAL_0^a \int e^{ant}dt,$$

the solution is found as

$$\frac{1}{a}K^a = \frac{sAL_0^a}{an}e^{ant} + \tilde{C},$$

or

$$K^a = \frac{sAL_0^a}{n}e^{ant} + C.$$

The value of C is chosen such that $K(0) = K_0$, that is $C = K_0^a - sAL_0^a/n$. Finally,

$$K(t) = \left[K_0^a + \frac{sAL_0^a}{n}(e^{ant} - 1) \right]^{1/a}.$$

- **Differential Equations with Homogeneous Coefficients**

    Definition 92 *A function $f(x,y)$ is called homogeneous of degree n, if for any λ $f(\lambda x, \lambda y) = \lambda^n f(x,y)$.*

    If we have the differential equation $M(x,y)dx + N(x,y)dy = 0$ and $M(x,y)$, $N(x,y)$ are homogeneous of the same degree, then the change of variable $y = tx$ reduces this equation to the variables separable case.

Example 108 *Solve $(y^2 - 2xy)dx + x^2dy = 0$.*
    The coefficients are homogeneous of degree 2. Note that $y \equiv 0$ is a special solution. If $y \neq 0$, let us change the variable: $y = tx$, $dy = tdx + xdt$. Thus

$$x^2(t^2 - 2t)dx + x^2(tdx + xdt) = 0.$$

Dividing by x^2 (since $y \neq 0$, $x \neq 0$ as well) and rearranging the terms, we get a variables separable equation

$$\frac{dx}{x} = -\frac{dt}{t^2 - t}.$$

Taking the integrals of both parts,

$$\int \frac{dx}{x} = \ln|x|,$$

$$\int -\frac{dt}{t^2 - t} = \int \left(\frac{1}{t} - \frac{1}{t-1}\right) dt = \ln|t| - \ln|t-1| + C,$$

therefore

$$x = \frac{Ct}{t-1}, \quad \text{or, in terms of } y, \ x = \frac{Cy}{y-x}.$$

We can single out y as a function of x, $y = \dfrac{x^2}{x - C}$.

- **Exact Differential Equations**

    If we have the differential equation $M(x,y)dx + N(x,y)dy = 0$ and $\frac{\partial M(x,y)}{\partial y} \equiv \frac{\partial N(x,y)}{\partial x}$ then all the solutions of this equation are given by $F(x,y) = C$, C is a constant, where $F(x,y)$ is such that $\frac{\partial F}{\partial x} = M(x,y)$, $\frac{\partial F}{\partial y} = N(x,y)$.

Example 109 *Solve* $\frac{y}{x}dx + (y^3 + \ln x)dy = 0.$

$\frac{\partial}{\partial y}\left(\frac{y}{x}\right) = \frac{\partial}{\partial x}(y^3 + \ln x),$ *so we have checked that we are dealing with an exact differential equation. Therefore, we need to find* $F(x,y)$ *such that*

$$F'_x = \frac{y}{x} \implies F(x,y) = \int \frac{y}{x}dx = y\ln x + \phi(y),$$

and

$$F'_y = y^3 + \ln x.$$

To find $\phi(y)$, *note that*

$$F'_y = y^3 + \ln x = \ln x + \phi'(y) \implies \phi(y) = \int y^3 dy = \frac{1}{4}y^4 + C.$$

Therefore, all solutions of the equation are given by the implicit function $4y\ln x + y^4 = C.$

- **Linear Differential Equations of the First Order**

  If we need to solve $y' + p(x)y = q(x),$ *then all the solutions are given by the formula*

$$y(x) = e^{-\int p(x)dx}\left(C + \int q(x)e^{\int p(x)dx}dx\right).$$

Example 110 *Solve* $xy' - 2y = 2x^4.$

In other words, we need to solve $y' - \frac{2}{x}y = 2x^3.$ *Therefore,*

$$\begin{aligned}
y(x) &= e^{\int \frac{2}{x}dx}\left(C + \int 2x^3 e^{-\int \frac{2}{x}dx}sd\right) \\
&= e^{\ln x^2}\left(C + \int 2x^3 e^{\ln \frac{1}{x^2}}sd\right) = x^2(C + x^2).
\end{aligned}$$

4.1.2 Qualitative Theory of First-Order Differential Equations

Consider an autonomous differential equation $\dot{x}(t) = f(x(t))$.

Definition 93 *A solution x^* to the equation $\dot{x}(t) = f(x(t))$ is called an* equilibrium, *or* stationary state, *or* steady state, *if $x^* \equiv Const \; \forall t$ and $f(x^*) = 0$.*

Other solutions starting in a neighborhood of x^* may either approach the equilibrium x^* as time goes on (i.e., x^* is locally stable), or, on the contrary, diverge from it (that is, x^* is unstable).

Proposition 57 *Let x^* be the steady state of the equation $\dot{x}(t) = f(x(t))$. If $f'(x^*) < 0$ then x^* is locally asymptotically stable. If $f'(x^*) > 0$ then x^* is unstable.*

Economics Application 17 (Solow-Swan Growth Model)
   Consider a neoclassical production function $Y = F(K, L)$ with the following properties:

- *Homogeneity of degree 1;*

- $F_K > 0, \; F_L > 0, \; F_{KK} < 0, \; F_{LL} < 0;$

- $F(0, L) = F(K, 0) = 0;$

- *(Inada conditions) With $k = K/L$, $f'(k) = (F(K,L)/L)' = F'(k, 1) \to \infty$ as $k \to 0$, and $f'(k) \to 0$ as $k \to \infty$.*

Assume that output is split between saving $S = \alpha Y$, $\alpha \in (0,1)$, and consumption $C = Y - S$, given that all saving is invested, i.e., $S = I$. Capital stock thus accumulates due to excess of the investment over the capital stock depreciation with the depreciation rate δ:

$$\dot{K} = I - \delta K = S - \delta K = \alpha Y - \delta K.$$

In per capita terms, output $y = \dfrac{Y}{L} = F\left(\dfrac{K}{L}, 1\right) = f(k)$, where per capita capital $k = \dfrac{K}{L}$. Note that $f'(k) > 0$ and $f''(k) < 0$. Also, assume *constant population growth at a rate n, that is $\dfrac{\dot{L}}{L} = n$, or $\dot{L} = nL$, or*

$L(t) = L_0 e^{nt}$.

Since $K = kL$, the time derivative $\dot{K} = k\dot{L} + L\dot{k} = kn\dot{L} + L\dot{k}$. Therefore, the equation of capital growth becomes

$$knL + L\dot{k} = \alpha Y - \delta kL.$$

Dividing by L and rearranging terms, we arrive at per capita capital accumulation equation

$$\dot{k} = \alpha f(k) - (\delta + n)k.$$

This equation has the unique equilibrium k^* such that $\alpha f(k^*) - (\delta + n)k^* = 0$. The equilibrium is stable because $\alpha f'(k^*) - (\delta + n) < 0$ (at the point of intersection of $g_1(k) = \alpha f(k)$ and $g_2(k) = (\delta + n)k$ the slope of $g_2(k)$ is steeper than that of $g_1(k)$ due to Inada conditions.)

If the social planner opts for maximization of consumption in the steady state, (s)he can use the saving rate α as a policy variable. On the one hand, α determines the equilibrium level of capital k^*. On the other hand, per capita consumption in the steady state $c^* = (Y^* - S^*)/L = (1 - \alpha)Y^*/L = (1 - \alpha)f(k^*) = f(k^*) - \alpha f(k^*) = f(k^*) - (\delta + n)k^*$. Therefore, the necessary condition for c^* to be maximized as a function of α is

$$\frac{dc^*(\alpha)}{d\alpha} = 0 = \frac{df(k(\alpha))}{dk} \cdot \frac{dk(\alpha)}{d\alpha}\Big|_{k=k^*(\alpha)} - (\delta + n),$$

or

$$f'(k^*) = \delta + n.$$

The latter identity is called the Golden Rule of Capital Accumulation.

4.1.3 Linear Differential Equations of a Higher Order with Constant Coefficients

These are the equations of the form

$$y^{(n)} + a_1 y^{(n-1)} + \ldots + a_{n-1}y' + a_n y = f(x).$$

If $f(x) \equiv 0$ the equation is called *homogeneous*, otherwise it is called *non-homogeneous*.

Recipe 18 – How to find the general solution $y_g(x)$ of the homogeneous equation: *The general solution is a sum of basic solutions* $y_1, \ldots, y_n$,

$$y_g(x) = C_1 y_1(x) + \ldots + C_n y_n(x),$$

$C_1, \ldots, C_n$ *are arbitrary constants.*

These arbitrary constants, however, can be defined in a unique way, once we set up the initial value Cauchy problem. Find $y(x)$ *such that*

$$y(x) = y_{0_0}, y'(x) = y_{0_1}, \ldots, y^{(n-1)}(x) = y_{0_{n-1}} \text{ at } x = x_0,$$

where $x_0, y_{0_0}, y_{0_1}, \ldots, y_{0_{n-1}}$ *are given initial values (real numbers).*
    To find all basic solutions, we proceed as follows:

1. *Solve the characteristic equation*

$$\lambda^n + a_1 \lambda^{n-1} + \ldots + a_{n-1}\lambda + a_n = 0$$

   for λ. *The roots of this equation are* $\lambda_1, \ldots, \lambda_n$. *Some of these roots may be complex numbers. We may also have repeated roots.*

2. *If* λ_i *is a simple real root, the basic solution corresponding to this root is* $y_i(x) = e^{\lambda_i x}$.

3. *If* λ_i *is a repeated real root of degree* k, *it generates* k *basic solutions*

$$y_{i_1}(x) = e^{\lambda_i x}, \; y_{i_2}(x) = xe^{\lambda_i x}, \; \ldots, \; y_{i_k}(x) = x^{k-1}e^{\lambda_i x}.$$

4. *If* λ_j *is a simple complex root,* $\lambda_j = \alpha_j + i\beta_j$, $i = \sqrt{-1}$, *then the complex conjugate of* λ_j, *say* $\lambda_{j+1} = \alpha_j - i\beta_j$, *is also a root of the characteristic equations. Therefore the pair* λ_j, λ_{j+1} *give rise to two basic solutions:*[1]

$$y_{j_1} = e^{\alpha_j x} \cos \beta_j x, \quad y_{j_2} = e^{\alpha_j x} \sin \beta_j x.$$

---

[1] These solutions can be derived by applying a well-known formula $e^z = e^{a+ib} = e^a \cdot e^{ib} = e^a(\cos(b) + i\sin(b))$, where $z = a + ib$ is a complex number.

5. *If λ_j is a repeated complex root of degree l, $\lambda_j = \alpha_j + i\beta_j$, then the complex conjugate of λ_j, say $\lambda_{j+1} = \alpha_j - i\beta_j$, is also a repeated root of degree l. These $2l$ roots generate $2l$ basic solutions*

$$y_{j_1} = e^{\alpha_j x} \cos \beta_j x,$$

$$y_{j_2} = x e^{\alpha_j x} \cos \beta_j x, \quad \ldots, \quad y_{j_l} = x^{l-1} e^{\alpha_j x} \cos \beta_j x,$$

$$y_{j_{l+1}} = e^{\alpha_j x} \sin \beta_j x,$$

$$y_{j_{l+2}} = x e^{\alpha_j x} \sin \beta_j x, \quad \ldots, \quad y_{j_{2l}} = x^{l-1} e^{\alpha_j x} \sin \beta_j x.$$

Recipe 19 – How to Solve the Non-Homogeneous Equation:

The general solution to the non-homogeneous solution is $y_{nh}(x) = y_g(x) + y_p(x)$, where $y_g(x)$ is the general solution to the homogeneous equation and $y_p(x)$ is a particular solution to the non-homogeneous equation, i.e. any function which solves the non-homogeneous equation.

Recipe 20 – How to Find a Particular Solution to the Non-Homogeneous Equation:

1. *If $f(x) = P_k(x)e^{bx}$, $P_k(x)$ is a polynomial of degree k, then a particular solution is*

$$y_p(x) = x^s Q_k(x)e^{bx},$$

where $Q_k(x)$ is a polynomial of the same degree k. If b is not a root of the characteristic equation, $s = 0$; if b is a root of the characteristic polynomial of degree m then $s = m$.

2. *If $f(x) = P_k(x)e^{px} \cos qx + Q_k(x)e^{px} \sin qx$, $P_k(x), Q_k(x)$ are polynomials of degree k, then a particular solution can be found in the form*

$$y_p(x) = x^s R_k(x)e^{px} \cos qx + x^s T_k(x)e^{px} \sin qx,$$

where $R_k(x), T_k(x)$ are polynomials of degree k. If $p + iq$ is not a root of the characteristic equation, $s = 0$; if $p + iq$ is a root of the characteristic polynomial of degree m then $s = m$.

3. *The general method for finding a particular solution to a non-homogeneous equation is called the* variation of parameters *or the* method of undetermined coefficients.

Suppose we have the general solution to the homogeneous equation

$$y_g = C_1 y_1(x) + \ldots + C_n y_n(x),$$

where $y_i(x)$ are basic solutions. Treating constants $C_1, \ldots, C_n$ as functions of x, for instance, $u_1(x), \ldots, u_n(x)$, we can express a particular solution to the non-homogeneous equation as

$$y_p(x) = u_1(x) y_1(x) + \ldots + u_n(x) y_n(x),$$

where $u_1(x), \ldots, u_n(x)$ are solutions of the system

$$u_1'(x) y_1(x) + \ldots + u_n'(x) y_n(x) = 0,$$
$$u_1'(x) y_1'(x) + \ldots + u_n'(x) y_n'(x) = 0,$$

$$\ldots \quad \ldots \quad \ldots$$

$$u_1'(x) y_1^{(n-2)}(x) + \ldots + u_n'(x) y_n^{(n-2)}(x) = 0,$$
$$u_1'(x) y_1^{(n-1)}(x) + \ldots + u_n'(x) y_n^{(n-1)}(x) = f(x)$$

4. *If $f(x) = f_1(x) + f_2(x) + \ldots + f_r(x)$ and $y_{p1}(x), \ldots, y_{pr}(x)$ are particular solutions corresponding to $f_1(x), \ldots, f_r(x)$ respectively, then*

$$y_p(x) = y_{p1}(x) + \ldots + y_{pr}(x).$$

Example 111 *Solve $y'' - 5y' + 6y = x^2 + e^x - 5$.*
The characteristic roots are $\lambda_1 = 2$, $\lambda_2 = 3$, therefore the general solution to the homogeneous equation is

$$y_h(x) = C_1 e^{2x} + C_2 e^{3x}.$$

We search for a particular solution in the form

$$y_p(x) = ax^2 + bx + c + d e^x.$$

To find coefficients a, b, c, d, let us substitute the particular solution into the initial equation:

$$2a + d e^x - 5(2ax + b + d e^x) + 6(ax^2 + bx + c + d e^x) = x^2 - 5 + e^x.$$

Equating term-by-term the coefficients on the left-hand side to the right-hand side, we obtain

$$6a = 1, \ -5 \cdot 2a + 6 \cdot b = 0, \ 2a - 5b + 6c = -5, \ d - 5d + 6d = 1.$$

Thus $d = 1/2, \ a = 1/6, \quad b = 10/36, \quad c = -71/108.$

Finally, the general solution to the nonhomogeneous equation is

$$
\begin{aligned}
y(x) &= y_{nh}(x) = y_h(x) + y_p(x) \\
&= C_1 e^{2t} + C_2 e^{3t} + \frac{x^2}{6} + \frac{5x}{18} - \frac{71}{108} + \frac{e^x}{2}.
\end{aligned}
$$

Example 112 *Solve* $y'' - 2y' + y = \dfrac{e^x}{x}.$

The general solution to the homogeneous equation is $y_g(x) = C_1 x e^x + C_2 e^x$ *(the characteristic equation has the repeated root* $\lambda = 1$ *of degree 2). We look for a particular solution to the non-homogeneous equation in the form*

$$y_p(x) = u_1(x) x e^x + u_2(x) e^x.$$

We have

$$u_1'(x) x e^x + u_2'(x) e^x = 0,$$
$$u_1'(x)(e^x + x e^x) + u_2'(x) e^x = \frac{e^x}{x}.$$

Therefore $u_1'(x) = \frac{1}{x}, u_2'(x) = -1.$ *Integrating, we find that* $u_1(x) = \ln|x|, \ u_2(x) = -x.$ *Thus* $y_p(x) = \ln|x| x e^x + x e^x,$ *and*

$$y_n(x) = y_g(x) + y_p(x) = \ln|x| x e^x + C_1 x e^x + C_2 e^x.$$

4.1.4 Systems of First-Order Linear Differential Equations

The general form of a system of first-order differential equations is:

$$\dot{x}(t) = A(t)x(t) + b(t), \quad x(0) = x_0$$

where t is the independent variable ("time"), $x(t) = (x_1(t), \ldots, x_n(t))'$ is a vector of dependent variables, $A(t) = (a_{ij}(t))_{[n \times n]}$ is a real $[n \times n]$ matrix with variable coefficients, $b(t) = (b_1(t), \ldots, b_n(t))'$ a variant n-vector.

In what follows, we concentrate on the case when A and b do not depend on t, which is the case in constant coefficients differential equations systems:

$$\dot{x}(t) = Ax(t) + b, \quad x(0) = x_0. \tag{4.1}$$

We also assume that A is non-singular.

The solution to system (4.1) can be determined in two steps:

- First, we consider the homogeneous system (corresponding to $b = 0$):

$$\dot{x}(t) = Ax(t), \quad x(0) = x_0. \tag{4.2}$$

  The solution $x_c(t)$ of this system is called the complementary solution.

- Second, we find a particular solution x_p of the system (4.1), which is called the particular integral. The constant vector x_p is simply the solution to $Ax_p = -b$, i.e. $x_p = -A^{-1}b$. [2]

Given $x_c(t)$ and x_p, the general solution to the system (4.1) is simply:

$$x(t) = x_c(t) + x_p.$$

We can find the solution to the homogeneous system (4.2) in two different ways.

First, we can eliminate $n - 1$ unknowns and reduce the system to one linear differential equation of degree n.

Example 113 *Let*

$$\begin{cases} \dot{x} = 2x + y, \\ \dot{y} = 3x + 4y. \end{cases}$$

Taking the derivative of the first equation and eliminating y and $\dot{y}$ ($\dot{y} = 3x + 4y = 3x + 4\dot{x} - 4 \cdot 2x$), we arrive at the second-order homogeneous linear differential equation

$$\ddot{x} - 6\dot{x} + 5x = 0,$$

which has the general solution $x(t) = C_1 e^t + C_2 e^{5t}$. Since $y(t) = \dot{x} - 2x$, we find that $y(t) = -C_1 e^t + 3C_2 e^{5t}$.

---

[2] In a sense, x_p can be viewed as an affine shift, which restores the origin to a unique equilibrium of (4.1).

The second way is to write the solution to (4.2) as

$$x(t) = e^{At} x_0$$

where (by definition)

$$e^{At} = I + At + \frac{A^2 t^2}{2!} + \cdots$$

Unfortunately, this formula is of little practical use. In order to find a feasible formula for e^{At}, we distinguish three main cases.

Case 1: A has real and distinct eigenvalues

The fact that the eigenvalues are real and distinct implies that any corresponding eigenvectors are linearly independent. Consequently, A is diagonalizable, i.e.

$$A = P\Lambda P^{-1}$$

where $P = [v_1, v_2, \dots, v_n]$ is a matrix composed by the eigenvectors of A and Λ is a diagonal matrix whose diagonal elements are the eigenvalues of A. Therefore, $e^A = Pe^\Lambda P^{-1}$.

Thus, the solution to the system (4.2) can be written as:

$$
\begin{aligned}
x(t) &= Pe^{\Lambda t}P^{-1}x_0 \\
&= Pe^{\Lambda t}c \\
&= c_1 v_1 e^{\lambda_1 t} + \dots + c_n v_n e^{\lambda_n t}
\end{aligned}
$$

where $c = (c_1, c_2, \dots, c_n)$ is a vector of constants determined from the initial conditions $(c = P^{-1}x_0)$.

Case 2: A has real and repeated eigenvalues

First, consider the simpler case when A has only one eigenvalue λ which is repeated m times (m is called the algebraic multiplicity of λ). Generally, in this situation the maximum number of independent eigenvectors corresponding to λ is less than m, meaning that we cannot construct the matrix P of linearly independent eigenvectors and, therefore, A is not diagonalizable.[3]

---

[3] If A is a real and symmetric matrix, then A is always diagonalizable and the formula given for case 1 can be applied.

The solution in this case has the form:

$$x(t) = \sum_{i=1}^{m} c_i h_i(t)$$

where $h_i(t)$ are quasi-polynomials and c_i are constants determined by the initial conditions. For example, if $m = 3$, we have:

$$\begin{aligned}
h_1(t) &= e^{\lambda t} v_1 \\
h_2(t) &= e^{\lambda t}(tv_1 + v_2) \\
h_3(t) &= e^{\lambda t}(t^2 v_1 + 2tv_2 + 3v_3)
\end{aligned}$$

where v_1, v_2, v_3 are determined by the conditions:

$$(A - \lambda I)v_i = v_{i-1}, v_0 = 0.$$

If A happens to have several eigenvalues which are repeated, the solution to (4.2) is obtained by finding the solution corresponding to each eigenvalue, and then adding up the individual solutions.

Case 3: A has complex eigenvalues

Since A is a real matrix, the complex eigenvalues always appear in pairs, i.e. if A has the eigenvalue $\alpha + \beta i$, then it also must have the eigenvalue $\alpha - \beta i$.

Now consider the simpler case in which A has only one pair of complex eigenvalues, $\lambda_1 = \alpha + \beta i$ and $\lambda_2 = \alpha - \beta i$. Let v_1 and v_2 denote the eigenvectors corresponding to λ_1 and λ_2; then $v_2 = \bar{v}_1$, where $\bar{v}_1$ is the conjugate of v_1. The solution can be expressed as:

$$\begin{aligned}
x(t) &= e^{At} x_0 \\
&= P e^{\Lambda t} P^{-1} x_0 \\
&= P e^{\Lambda t} c \\
&= c_1 v_1 e^{(\alpha+\beta i)t} + c_2 v_2 e^{(\alpha-\beta i)t} \\
&= c_1 v_1 e^{\alpha t}(\cos \beta t + i \sin \beta t) + c_2 v_2 e^{\alpha t}(\cos \beta t - i \sin \beta t) \\
&= (c_1 v_1 + c_2 v_2)e^{\alpha t} \cos \beta t + i(c_1 v_1 - c_2 v_2)e^{\alpha t} \sin \beta t \\
&= h_1 e^{\alpha t} \cos \beta t + h_2 e^{\alpha t} \sin \beta t,
\end{aligned}$$

where $h_1 = c_1 v_1 + c_2 v_2$, $h_2 = i(c_1 v_1 - c_2 v_2)$ are real vectors.

Note that if A has more pairs of complex eigenvalues, then the solution to (4.2) is obtained by finding the solution corresponding to each eigenvalue, and then adding up the individual solutions. The same remark holds true when we encounter any combination of the three cases discussed above.

Example 114 *Consider*

$$A = \begin{pmatrix} 5 & 2 \\ -4 & -1 \end{pmatrix}.$$

Solving the characteristic equation $\det(A - \lambda I) = \lambda^2 - 4\lambda + 3 = 0$ *we find eigenvalues of* A*:* $\lambda_1 = 1$*,* $\lambda_2 = 3$*.*

From the condition $(A - \lambda_1 I)v_1 = 0$ *we find an eigenvector corresponding to* λ_1*:* $v_1 = \begin{pmatrix} -1 \\ 2 \end{pmatrix}.$

Similarly, the condition $(A - \lambda_2 I)v_2 = 0$ *gives an eigenvector corresponding to* λ_2*:* $v_2 = \begin{pmatrix} -1 \\ 1 \end{pmatrix}.$

Therefore, the general solution to the system $\dot{x}(t) = Ax(t)$ *is*

$$x(t) = c_1 v_1 e^{\lambda_1 t} + c_2 v_2 e^{\lambda_2 t}$$

or

$$x(t) = \begin{pmatrix} x_1(t) \\ x_2(t) \end{pmatrix} = \begin{pmatrix} -c_1 e^t - c_2 e^{3t} \\ 2c_1 e^t + c_2 e^{3t} \end{pmatrix}.$$

Example 115 *Consider*

$$A = \begin{pmatrix} 3 & 1 \\ -1 & 1 \end{pmatrix}.$$

$\det(A - \lambda I) = \lambda^2 - 4\lambda + 4 = 0$ *gives the eigenvalue* $\lambda = 2$ *repeated twice.*

From the condition $(A - \lambda_1 I)v_1 = 0$ *we find* $v_1 = \begin{pmatrix} -1 \\ 1 \end{pmatrix}$

and from $(A - \lambda_1 I)v_2 = v_1$ *we obtain* $v_2 = \begin{pmatrix} 1 \\ -2 \end{pmatrix}.$

The solution to the system $\dot{x}(t) = Ax(t)$ *is*

$$\begin{aligned} x(t) &= c_1 h_1(t) + c_2 h_2(t) \\ &= c_1 v_1 e^{\lambda t} + c_2 (t v_1 + v_2) e^{\lambda t} \end{aligned}$$

or

$$x(t) = \begin{pmatrix} x_1(t) \\ x_2(t) \end{pmatrix} = \begin{pmatrix} -c_1 + c_2 - c_2 t \\ c_1 - 2c_2 + c_2 t \end{pmatrix} e^{2t}.$$

Example 116 *Consider*

$$A = \begin{pmatrix} 3 & 2 \\ -1 & 1 \end{pmatrix}.$$

The equation $\det(A - \lambda I) = \lambda^2 - 4\lambda + 5 = 0$ *leads to the complex eigenvalues* $\lambda_1 = 2 + i$, $\lambda_2 = 2 - i$.
For λ_1 *we set the equation* $(A - \lambda_1 I)v_1 = 0$ *and obtain:*

$$v_1 = \begin{pmatrix} 1 + i \\ -1 \end{pmatrix}, \quad v_2 = \bar{v}_1 = \begin{pmatrix} 1 - i \\ -1 \end{pmatrix}.$$

The solution of the system $\dot{x}(t) = Ax(t)$ *is*

$$\begin{aligned} x(t) &= c_1 v_1 e^{\lambda_1 t} + c_2 v_2 e^{\lambda_2 t} \\ &= c_1 v_1 e^{(2+i)t} + c_2 v_2 e^{(2-i)t} \\ &= c_1 v_1 e^{2t}(\cos t + i \sin t) + c_2 v_2 e^{2t}(\cos t - i \sin t). \end{aligned}$$

Performing the computations, we finally obtain:

$$x(t) = (c_1 + c_2)e^{2t} \begin{pmatrix} \cos t - \sin t \\ -\cos t \end{pmatrix} + i(c_1 - c_2)e^{2t} \begin{pmatrix} \cos t + \sin t \\ -\sin t \end{pmatrix}.$$

Now let vector b be different from 0. To specify the solution to the system (4.1), we need a particular solution x_p. This is found from the condition $\dot{x} = 0$; thus, $x_p = -A^{-1}b$. Therefore, the general solution to the non-homogeneous system is

$$x(t) = x_c(t) + x_p = Pe^{\Lambda t}P^{-1}c - A^{-1}b$$

where the vector of constants c is determined by the initial condition $x(0) = x_0$ as $c = x_0 + A^{-1}b$.

Example 117 *Consider*

$$A = \begin{pmatrix} 5 & 2 \\ -4 & -1 \end{pmatrix}, b = \begin{pmatrix} 1 \\ 2 \end{pmatrix} \text{ and } x_0 = \begin{pmatrix} -1 \\ 2 \end{pmatrix}.$$

In Example 114 we obtained

$$x_c(t) = \begin{pmatrix} -c_1 e^t - c_2 e^{3t} \\ 2c_1 e^t + c_2 e^{3t} \end{pmatrix}.$$

In addition, we have

$$c = \begin{pmatrix} c_1 \\ c_2 \end{pmatrix} = x_0 + A^{-1}b = \begin{pmatrix} -1 \\ 2 \end{pmatrix} + \begin{pmatrix} \frac{-5}{3} \\ \frac{14}{3} \end{pmatrix} = \begin{pmatrix} \frac{-8}{3} \\ \frac{20}{3} \end{pmatrix}$$

and

$$-x_p = A^{-1}b = \begin{pmatrix} \frac{-1}{3} & \frac{-2}{3} \\ \frac{4}{3} & \frac{5}{3} \end{pmatrix} \begin{pmatrix} 1 \\ 2 \end{pmatrix} = \begin{pmatrix} \frac{-5}{3} \\ \frac{14}{3} \end{pmatrix}.$$

The solution to the system $\dot{x}(t) = Ax(t) + b$ is

$$x(t) = x_c(t) + x_p = \begin{pmatrix} \frac{8}{3}e^t - \frac{20}{3}e^{3t} + \frac{5}{3} \\ \frac{-16}{3}e^t + \frac{20}{3}e^{3t} + \frac{-14}{3} \end{pmatrix}.$$

Example 118 *Solve*

$$\begin{cases} \dot{x} &= 4x - y - z, \\ \dot{y} &= x + 2y - z, \\ \dot{z} &= x - y + 2z, \end{cases}$$

with the initial values $x(0) = 3$, $y(0) = 2$, $z(0) = 1$.

Solution:

Matrix $A = \begin{pmatrix} 4 & -1 & -1 \\ 1 & 2 & -1 \\ 1 & -1 & 2 \end{pmatrix}$ has eigenvalues $\lambda_1 = \lambda_2 = 3$, $\lambda_3 = 2$,

and the corresponding eigenvectors

$$V_1 = \begin{pmatrix} 1 \\ 1 \\ 0 \end{pmatrix}, \quad V_2 = \begin{pmatrix} 1 \\ 0 \\ 1 \end{pmatrix}, \quad V_3 = \begin{pmatrix} 1 \\ 1 \\ 1 \end{pmatrix}.$$

Therefore, the solution is

$$\begin{pmatrix} x(t) \\ y(t) \\ z(t) \end{pmatrix} = (C_1 V_1 + C_2 V_2)e^{3t} + C_3 V_3 e^{2t}$$

$$= \begin{pmatrix} (C_1 + C_2)e^{3t} + C_3 e^{2t} \\ C_1 e^{3t} + C_3 e^{2t} \\ C_2 e^{3t} + C_3 e^{2t} \end{pmatrix}.$$

Finally, undetermined coefficients C_1, C_2 and C_3 are found as a solution to a system of linear equations

$$\begin{pmatrix} x(0) \\ y(0) \\ z(0) \end{pmatrix} = \begin{pmatrix} 3 \\ 2 \\ 1 \end{pmatrix} = \begin{pmatrix} C_1 + C_2 + C_3 \\ C_1 + C_3 \\ C_2 + C_3 \end{pmatrix}.$$

After some algebra, $C_1 = 2$, $C_2 = 1$, $C_3 = 0$.

Economics Application 18 (General Equilibrium)

Consider a market for n goods $x = (x_1, x_2, \ldots, x_n)'$. Let $p = (p_1, p_2, \ldots, p_n)'$ denote the vector of prices and $D_i(p), S_i(p)$ demand and supply for good $i, i = 1, \ldots, n$. The general equilibrium is obtained for a price vector p^* which satisfies the conditions: $D_i(p^*) = S_i(p^*)$ for all $i = 1, \ldots, n$.

The dynamics of price adjustment for the general equilibrium can be described by a system of first-order differential equations of the form:

$$\dot{p} = W[D(p) - S(p)] = WAp, \qquad p(0) = p_0$$

where $W = diag(w_i)$ is a diagonal matrix having the speeds of adjustment $w_i, i = 1, \ldots, n$, $D(p) = [D_1(p), \ldots, D_n(p)]'$ on the diagonal and $S(p) = [S_1(p), \ldots, S_n(p)]'$ are linear functions of p and A is a constant $[n \times n]$ matrix.

We can solve this system by applying the methods discussed above. For instance, if $WA = P\Lambda P^{-1}$, then the solution is

$$p(t) = Pe^{\Lambda t}P^{-1}p_0.$$

Economics Application 19 (Dynamic IS-LM Model)

A simplified dynamic IS-LM model can be specified by the following system of differential equations:

$$\begin{aligned} \dot{Y} &= w_1(I - S) \\ \dot{r} &= w_2[L(Y, r) - M], \end{aligned}$$

with initial conditions

$$Y(0) = Y_0, r(0) = r_0,$$

where:

- Y is national income;

- r is the interest rate;

- $I = I_0 - \alpha r$ is the investment function (α is a positive constant);

- $S = s(Y - T) + (T - G)$ is national savings; (T, G and s represent taxes, government purchases and the savings rate, respectively);

- $L(Y, r) = a_1 Y - a_2 r$ is money demand (a_1 and a_2 are positive constants);

- M is money supply;

- w_1, w_2 are speeds of adjustment.

For simplicity, consider $w_1 = 1$ and $w_2 = 1$. The original system can be re-written as:

$$
\begin{pmatrix} \dot{Y} \\ \dot{r} \end{pmatrix} = \begin{pmatrix} -s & -\alpha \\ a_1 & -a_2 \end{pmatrix} \begin{pmatrix} Y \\ r \end{pmatrix} + \begin{pmatrix} I_0 - (1-s)T + G \\ -M \end{pmatrix}.
$$

Assume that $A = \begin{pmatrix} -s & -\alpha \\ a_1 & -a_2 \end{pmatrix}$ is diagonalizable, i.e. $A = P\Lambda P^{-1}$, and has distinct eigenvalues λ_1, λ_2 with corresponding eigenvectors v_1, v_2. (λ_1, λ_2 are the solutions of the equation $r^2 + (s + a_2)r + sa_2 + \alpha a_1 = 0$.)
The solution becomes:

$$
\begin{pmatrix} Y(t) \\ r(t) \end{pmatrix} = P e^{\Lambda t} P^{-1} (x_0 + A^{-1}b) - A^{-1}b
$$

where

$$
x_0 = \begin{pmatrix} Y_0 \\ r_0 \end{pmatrix}, b = \begin{pmatrix} I_0 - (1-s)T + G \\ -M \end{pmatrix}.
$$

We have

$$
A^{-1} = \frac{1}{\alpha a_1 + sa_2} \begin{pmatrix} -a_2 & \alpha \\ -a_1 & -s \end{pmatrix}
$$

and

$$
A^{-1}b = \frac{1}{\alpha a_1 + sa_2} \begin{pmatrix} -a_2[I_0 - (1-s)T + G] - \alpha M \\ -a_1[I_0 - (1-s)T + G] + sM \end{pmatrix}.
$$

Assume that $P^{-1} = \begin{pmatrix} \tilde{v}_1' \\ \tilde{v}_2' \end{pmatrix}$. *Then, we have:*

$$Pe^{\Lambda t}P^{-1} = [v_1, v_2] \begin{pmatrix} e^{\lambda_1 t} & 0 \\ 0 & e^{\lambda_2 t} \end{pmatrix} \begin{pmatrix} \tilde{v}_1' \\ \tilde{v}_2' \end{pmatrix} = v_1\tilde{v}_1'e^{\lambda_1 t} + v_2\tilde{v}_2'e^{\lambda_2 t}.$$

Finally, the solution is:

$$\begin{pmatrix} Y(t) \\ r(t) \end{pmatrix} = [v_1\tilde{v}_1'e^{\lambda_1 t} + v_2\tilde{v}_2'e^{\lambda_2 t}] \begin{pmatrix} Y_0 + \frac{-a_2 S - \alpha M}{\alpha a_1 + sa_2} \\ r_0 + \frac{-a_1 S + sM}{\alpha a_1 + sa_2} \end{pmatrix} -$$

$$- \begin{pmatrix} \frac{-a_2 S - \alpha M}{\alpha a_1 + sa_2} \\ \frac{-a_1 S + sM}{\alpha a_1 + sa_2} \end{pmatrix},$$

where $S = I_0 - (1 - s)T + G$.

Economics Application 20 (Solow-Swan Model Revisited)

Extending the model developed in Economics Application 17, the Solow model is the basic model of economic growth which relies on the following key differential equation explaining the dynamics of capital per unit of effective labor:

$$\dot{k}(t) = sf(k(t)) - (n + g + \delta)k(t), \qquad k(0) = k_0$$

where

- *k denotes capital per effective labor;*

- *s is the savings rate;*

- *n is the rate of population growth;*

- *g is the rate of technological progress;*

- *δ is the depreciation rate.*

Let us consider two types of production function.

Case 1: *The production function is linear:* $f(k(t)) = ak(t)$ *where a is a positive constant.*

The equation for $\dot{k}$ becomes:

$$\dot{k}(t) = (sa - n - g - \delta)k(t).$$

This linear differential equation has the solution

$$k(t) = k(0)e^{(sa-n-g-\delta)t} = k_0 e^{(sa-n-g-\delta)t}.$$

Case 2: *The production function takes the Cobb-Douglas form:* $f(k(t)) = [k(t)]^{\alpha}$ *where* $\alpha \in [0, 1]$.
In this case, the equation for $\dot{k}$ *represents a Bernoulli non-linear differential equation:*

$$\dot{k}(t) + (n + g + \delta)k(t) = s[k(t)]^{\alpha}, \qquad k(0) = k_0$$

Dividing both sides of the latter equation by $[k(t)]^{\alpha}$, *and denoting* $m(t) = [k(t)]^{1-\alpha}$, $m(0) = (k_0)^{1-\alpha}$, *we obtain:*

$$\frac{1}{1 - \alpha}\dot{m}(t) + (n + g + \delta)m(t) = s,$$

or

$$\dot{m}(t) + (1 - \alpha)(n + g + \delta)m(t) = s(1 - \alpha).$$

The solution to this first-order linear differential equation is given by:

$$
\begin{aligned}
m(t) &= e^{-\int(1-\alpha)(n+g+\delta)dt}[A + \int s(1 - \alpha)e^{\int(1-\alpha)(n+g+\delta)dt}dt] \\
&= e^{-(1-\alpha)(n+g+\delta)t}[A + s(1 - \alpha)\int e^{(1-\alpha)(n+g+\delta)t}dt] \\
&= Ae^{-(1-\alpha)(n+g+\delta)t} + \frac{s}{n + g + \delta},
\end{aligned}
$$

where A *is determined from the initial condition* $m(0) = (k_0)^{1-\alpha}$. *It follows that* $A = (k_0)^{1-\alpha} - \frac{s}{n+g+\delta}$ *and*

$$m(t) = [(k_0)^{1-\alpha} - \frac{s}{n + g + \delta}]e^{-(1-\alpha)(n+g+\delta)t} + \frac{s}{n + g + \delta},$$

$$k(t) = \{[(k_0)^{1-\alpha} - \frac{s}{n + g + \delta}]e^{-(1-\alpha)(n+g+\delta)t} + \frac{s}{n + g + \delta}\}^{\frac{1}{1-\alpha}}.$$

Economics Application 21 (Ramsey-Cass-Koopmans Model)

This model takes capital per effective labor $k(t)$ and consumption per effective labor $c(t)$ as endogenous variables. Their behavior is described by the following system of differential equations:

$$\dot{k}(t) = f(k(t)) - c(t) - (n + g)k(t)$$
$$\dot{c}(t) = \frac{f'(k(t)) - \rho - \theta g}{\theta}c(t)$$

with the initial conditions

$$k(0) = k_0, c(0) = c_0,$$

where, in addition to the variables already specified in the previous application,

- *ρ stands for the discount rate;*

- *θ is the coefficient of relative risk aversion.*

Assume again that $f(k(t)) = ak(t)$ where $a > 0$ is a constant. The initial system reduces to a system of linear differential equations:

$$\dot{k}(t) = (a - n - g)k(t) - c(t)$$
$$\dot{c}(t) = \frac{a - \rho - \theta g}{\theta}c(t)$$

The second equation implies that:

$$c(t) = c_0 e^{bt}, \text{ where } b = \frac{a - \rho - \theta g}{\theta}.$$

Substituting for $c(t)$ in the first equation, we obtain:

$$\dot{k}(t) = (a - n - g)k(t) - c_0 e^{bt}.$$

The solution to this linear differential equation takes the form:

$$k(t) = k_c(t) + k_p(t)$$

where k_c and k_p are the complementary solution and the particular integral, respectively.

The homogeneous equation $\dot{k}(t) = (a - n - g)k(t)$ gives the complementary solution:

$$k(t) = k_0 e^{(a-n-g)t}.$$

The particular integral should be of the form $k_p(t) = me^{bt}$; the value of m can be determined by substituting k_p in the non-homogeneous equation for $\dot{k}$:

$$mbe^{bt} = (a - n - g)me^{bt} - c_0 e^{bt}$$

It follows that

$$m = \frac{c_0}{a - n - g - b}, \qquad k_p(t) = \frac{c_0}{a - n - g - b}e^{bt}$$

and

$$k(t) = k_0 e^{(a-n-g)t} + \frac{c_0}{a - n - g - b}e^{bt}.$$

4.1.5 Simultaneous Differential Equations and Types of Equilibria

Let $x = x(t)$, $y = y(t)$, where t is an independent variable. Then consider a two-dimensional autonomous system of simultaneous differential equations

$$\begin{cases} \dfrac{dx}{dt} = f(x, y), \\ \dfrac{dy}{dt} = g(x, y). \end{cases}$$

The point (x^*, y^*) is called an equilibrium of this system if $f(x^*, y^*) = g(x^*, y^*) = 0$.

Let $\mathcal{J}$ be the Jacobian matrix

$$\mathcal{J} = \begin{pmatrix} \dfrac{\partial f}{\partial x} & \dfrac{\partial f}{\partial y} \\ \dfrac{\partial g}{\partial x} & \dfrac{\partial g}{\partial y} \end{pmatrix}$$

evaluated at (x^*, y^*), and let λ_1, λ_2 be the eigenvalues of this Jacobian.

Recipe 21 – Express Test for Stability of Equilibrium *As we already know, eigenvalues of a matrix J can be expressed in terms of its trace and determinant:*

$$\lambda_{1,2} = \frac{Tr(J) \pm \sqrt{[Tr(J)]^2 - 4Det(J)}}{2}.$$

In particular, signs of trace and determinant may provide a shortcut while checking stability of an equilibrium. For instance, if at (x^*, y^*), $Tr(J) < 0$ and $Det(J) > 0$ then the equilibrium is stable.

Then the equilibrium is

1. a stable (unstable) node if λ_1, λ_2 are real, distinct, and both negative (positive);

2. a saddle if the eigenvalues are real and of different signs, i.e. $\lambda_1 \lambda_2 < 0$;

3. a stable (unstable) focus if λ_1, λ_2 are complex and $Re(\lambda_1) < 0$ $(Re(\lambda_1) > 0)$;

4. a center or a focus if λ_1, λ_2 are complex and $Re(\lambda_1) = 0$;

5. a stable (unstable) improper node if λ_1, λ_2 are real, $\lambda_1 = \lambda_2 < 0$ $(\lambda_1 = \lambda_2 > 0)$ and the Jacobian is not a diagonal matrix;

6. a stable (unstable) star node if λ_1, λ_2 are real, $\lambda_1 = \lambda_2 < 0$ $(\lambda_1 = \lambda_2 > 0)$ and the Jacobian is a diagonal matrix.

The figure below illustrates this classification:

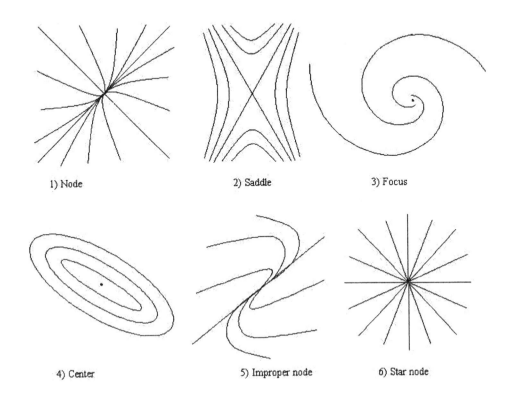

1) Node   2) Saddle   3) Focus

4) Center   5) Improper node   6) Star node

Example 119 *Find the equilibria of the system and classify them*

$$\begin{cases} \dot{x} = 2xy - 4y - 8, \\ \dot{y} = 4y^2 - x^2. \end{cases}$$

Solving the system

$$2xy - 4y - 8 = 0,$$
$$4y^2 - x^2 = 0$$

for x, y, we find two equilibria, $(-2, -1)$ and $(4, 2)$.
    At $(-2, -1)$ the Jacobian

$$J = \begin{pmatrix} 2y & 2x - 4 \\ -2x & 8y \end{pmatrix}_{(-2,-1)} = \begin{pmatrix} -2 & -8 \\ 4 & -8 \end{pmatrix}.$$

The characteristic equation for J is $\lambda^2 - (tr(J))\lambda + \det(J) = 0$.

$(tr(J))^2 - 4\det(J) < 0 \implies \lambda_{1,2}$ *are complex.*

$tr(J) < 0 \implies \lambda_1 + \lambda_2 < 0 \implies Re\lambda_1 < 0 \implies$ *the equilibrium* $(-2, -1)$ *is a stable focus.*

At $(4, 2)$

$$J = \begin{pmatrix} 4 & 4 \\ -8 & 16 \end{pmatrix}.$$

$(tr(J))^2 - 4\det(J)) > 0 \implies \lambda_{1,2}$ *are distinct and real.*

$tr(J) = \lambda_1 + \lambda_2 > 0, \ \det(J) = \lambda_1\lambda_2 > 0 \implies \lambda_1 > 0, \ \lambda_2 > 0 \implies$ *the equilibrium* $(4, 2)$ *is an unstable node.*

Example 120 *Find the equilibria of the system and classify them*

$$\begin{cases} \dot{x} = xy + 4y \\ \dot{y} = x^2 + y^2 - 17. \end{cases}$$

There are four equilibria, $(4, -1), (-4, 1), (1, -4)$ *and* $(-1, 4)$.
The Jacobian

$$J = \begin{pmatrix} y^* & x^* \\ 2x^* & 2y^* \end{pmatrix}$$

with the corresponding characteristic equation $\lambda^2 - 3y^*\lambda + 2((y^*)^2 - (x^*)^2) = 0.$

- *At* $(-1, 4)$ *eigenvalues are real and positive* $\implies$ *unstable node;*

- *At* $(1, -4)$ *eigenvalues are real and negative* $\implies$ *stable node;*

- *At* $(-4, 1)$ *and* $(4, -1)$ *eigenvalues are real and of different signs* $\implies$ *saddles.*

Economics Application 22 (Ramsey–Solow Growth Model)

Consider a growth model in a per capita (Consumption–Capital) plane

$$\begin{cases} \dot{k} &= f(k) - \delta k - c, \\ \dot{c} &= -c(r + \delta - f'(k)), \end{cases}$$

where $f(0) = 0, \ f'(k) > 0, \ f''(k) < 0, \ f'(0) > 0, \ f'(\infty) < 0.$

For simplicity, take $r = \delta = 0$ *and* $f(k) = ak - bk^2$. *Thus the system becomes*

$$\begin{cases} \dot{k} = ak - bk^2 - c, \\ \dot{c} = -c(2bk - a). \end{cases}$$

It has three equilibria $(c = 0, k = a/b)$, $(k = c = 0)$, *and* $(k^* = a/2b, c^* =^2 /4b)$. *The first two equilibria are ruled out due to strict positivity of* k *and* c. *The Jacobian, evaluated at the last equilibrium*

$$J = \begin{pmatrix} a - 2bk^* & -1 \\ -2bc^* & 0 \end{pmatrix}$$

has a negative determinant; therefore, eigenvalues of J *are real and of opposite signs, and the equilibrium is a saddle located at the intersection of the isoclines* $\dot{k} = 0$ $(c(k) = ak - bk^2)$ *and* $\dot{c} = 0$ $(k = a/2b)$.

4.2   Difference Equations

Let y denote a real-valued function defined over the set of natural numbers. Hereafter y_k stands for $y(k)$ — the value of y at k, where $k = 0, 1, 2, \ldots$.

Definition 94

- *The* first difference *of* y *evaluated at* k *is* $\Delta y_k = y_{k+1} - y_k$.

- *The* second difference *of* y *evaluated at* k *is* $\Delta^2 y_k = \Delta(\Delta y_k) = y_{k+2} - 2y_{k+1} + y_k$.

- *In general, the* nth difference *of* y *evaluated at* k *is defined as* $\Delta^n y_k = \Delta(\Delta^{n-1} y_k)$, *for any integer* n, $n > 1$.

Definition 95 *A* difference equation *in the unknown* y *is an equation relating* y *and any of its differences* $\Delta y, \Delta^2 y, \ldots$ *for each value of* k, $k = 0, 1, \ldots$.

Solving a difference equation means finding all functions y which satisfy the relation specified by the equation.

Example 121 (Examples of Difference Equations)

- $2\Delta y_k - y_k = 1$ *(or, equivalently,* $2\Delta y - y = 1$*)*

- $\Delta^2 y_k + 5y_k \Delta y_k + (y_k)^2 = 2$ *(or $\Delta^2 y + 5y\Delta y + (y)^2 = 2$)*

Note that any difference equation can be written as

$$f(y_{k+n}, y_{k+n-1}, \ldots, y_{k+1}, y_k) = g(k), \ n \in \mathbf{N}$$

For example, the above two equations may also be expressed as follows:

- $2y_{k+1} - 3y_k = 1$

- $y_{k+2} - 2y_{k+1} + y_k + 5y_k y_{k+1} - 4(y_k)^2 = 2$

Definition 96 *A difference equation is* linear *if it takes the form*

$$f_0(k)y_{k+n} + f_1(k)y_{k+n-1} + \ldots + f_{n-1}(k)y_{k+1} + f_n(k)y_k = g(k),$$

where $f_0, f_1, \ldots, f_n, g$ are each functions of k (but not of some y_{k+i}, $i \in \mathbf{N}$) defined for all $k = 0, 1, 2, \ldots$.

Definition 97 *A linear difference equation is of order n if both $f_0(k)$ and $f_n(k)$ are different from zero for each $k = 0, 1, 2, \ldots$.*

From now on, we focus on the problem of solving linear difference equations of order n with constant coefficients. The general form of such an equation is:

$$f_0 y_{k+n} + f_1 y_{k+n-1} + \ldots + f_{n-1} y_{k+1} + f_n y_k = g_k,$$

where, this time, $f_0, f_1, \ldots, f_n$ are real constants and f_0, f_n are different from zero.

Dividing this equation by f_0 and denoting $a_i = \frac{f_i}{f_0}$ for $i = 0, \ldots, n$, $r_k = \frac{g_k}{f_0}$, the general form of a linear difference equation of order n with constant coefficients becomes:

$$y_{k+n} + a_1 y_{k+n-1} + \ldots + a_{n-1} y_{k+1} + a_n y_k = r_k, \tag{4.3}$$

where a_n is non-zero.

Recipe 22 – How to Solve a Difference Equation: General Procedure:

The general procedure for solving a linear difference equation of order n with constant coefficients involves three steps:

Step 1. Consider the homogeneous equation

$$y_{k+n} + a_1 y_{k+n-1} + \ldots + a_{n-1} y_{k+1} + a_n y_k = 0,$$

and find its solution Y. (First we will describe and illustrate the solving procedure for equations of order 1 and 2. Following that, the theory for the case of order n is obtained as a natural extension of the theory for second-order equations.)

Step 2. Find a particular solution y^* of the complete equation (4.3). The most useful technique for finding particular solutions is the method of undetermined coefficients which will be illustrated in the examples that follow.

Step 3. The general solution to equation (4.3) is:

$$y_k = Y + y^*.$$

To conclude, let us consider the stability properties of an equilibrium of a first-order non-linear autonomous difference equation $x_{t+1} = f(x_t)$.

Definition 98 *A solution x^* to a non-linear autonomous difference equation $x_{t+1} = f(x_t)$ is called equilibrium, or steady state, if x^* is constant for any t and $x^* = f(x^*)$.*

Proposition 58 *Let x^* be an equilibrium of an autonomous non-linear difference equation $x_{t+1} = f(x_t)$. Assume continuous differentiability of $f(\cdot)$. If $|f'(x^*)| < 1$ then x^* is locally asymptotically stable. If $|f'(x^*)| > 1$ then x^* is unstable.*

4.2.1 First-Order Linear Difference Equations

The general form of a first-order linear differential equation is:

$$y_{k+1} + a y_k = r_k.$$

The corresponding homogeneous equation is:

$$y_{k+1} + a y_k = 0,$$

and has the solution $Y = C(-a)^k$, where C is an arbitrary constant.

Recipe 23 – How to Solve a First-Order Non-Homogeneous Linear Difference Equation:

First, consider the case in which r_k is constant over time, i.e. $r_k \equiv r$ for all k.

Much as in the case of linear differential equations, the general solution to the non-homogeneous difference equation y_t is the sum of the general solution to the homogeneous equation and a particular solution to the non-homogeneous equation. Therefore

$$y_k = A \cdot (-a)^k + \frac{r}{1+a}, \quad a \neq -1,$$

$$y_k = A \cdot (-a)^k + rk = A + rk, \quad a = -1,$$

where A is an arbitrary constant.

If we have the initial condition $y_k = y_0$ when $k = 0$, the general solution to the non-homogeneous equation takes the form

$$y_k = \left(y_0 - \frac{r}{1+a} \right) \times (-a)^k + \frac{r}{1+a}, \quad a \neq -1,$$

$$y_k = y_0 + rk, \quad a = -1.$$

If r now depends on k, $r = r_k$, then the solution takes the form

$$y_k = (-a)^k y_0 + \sum_{i=0}^{k-1} (-a)^{k-1-i} r_i, \quad k = 1, 2, \ldots$$

A particular solution to a non-homogeneous difference equation can be found using the method of undetermined coefficients.

Example 122 $y_{k+1} - 3y_k = 2$.

The solution to the homogeneous equation $y_{k+1} - 3y_k = 0$ is: $Y = C \cdot 3^k$. The right-hand term suggests looking for a constant function as a particular solution. Assuming $y^* = A$ and substituting into the initial equation, we obtain $A = -1$; hence $y^* = -1$ and the general solution is $y_k = Y + y^* = C3^k - 1$.

Example 123 $y_{k+1} - 3y_k = k^2 + k + 2$.

The homogeneous equation is the same as in the preceding example. But the right-hand term suggests finding a particular solution of the form:

$$y^* = Ak^2 + Bk + D.$$

Substituting y^* into a non-homogeneous equation, we have:

$$A(k + 1)^2 + B(k + 1) + D - 3Ak^2 - 3Bk - 3D = k^2 + k + 2, \text{ or}$$

$$-2Ak^2 + 2(A - B)k + A + B - 2D = k^2 + k + 2.$$

In order to satisfy this equality for each k, we must have:

$$\begin{cases} -2A = 1 \\ 2(A - B) = 1 \\ A + B - 2D = 2. \end{cases}$$

It follows that $A = -\frac{1}{2}, B = -1, D = -\frac{3}{4}$ and $y^* = -\frac{1}{2}k^2 - k - \frac{3}{4}$. The general solution is $y_k = Y + y^* = C3^k - \frac{1}{2}k^2 - k - \frac{3}{4}$.

Example 124 $y_{k+1} - 3y_k = 4e^k$.

This time we try a particular solution of the form: $y^* = Ae^k$. After substitution, we find $A = \frac{4}{e-3}$. The general solution is $y_k = Y + y^* = C3^k + \frac{4e^k}{e-3}$.

Two remarks:

- The method of undetermined coefficients illustrated by these examples applies to the general case of order n as well. The trial solutions corresponding to some simple functions r_k are given in the following table:

Right-hand term r_k	Trial solution y^*
a^k	Aa^k
k^n	$A_0 + A_1 k + A_2 k^2 + \ldots + A_n k^n$
$\sin ak$ or $\cos ak$	$A \sin ak + B \cos ak$

- If the function r is a combination of simple functions for which we know the trial solutions, then a trial solution for r can be obtained by combining the trial solutions for the simple functions. For example, if $r_k = 5^k k^3$, then the trial solution would be $y^* = 5^k(A_0 + A_1 k + A_2 k^2 + A_3 k^3)$.

Economics Application 23 (Simple and Compound Interest)

Let S_0 denote an initial sum of money. There are two basic methods for computing the interest earned in a period, for example, one year.

S_0 earns simple interest at rate r if in each period the interest equals a fraction r of S_0. If S_k denotes the sum accumulated after k periods, S_k is computed in the following way:

$$S_{k+1} = S_k + rS_0.$$

This is a first-order difference equation which has the solution $S_k = S_0(1 + kr)$.

S_0 earns compound interest at rate r if in each period the interest equals a fraction r of the sum accumulated at the beginning of that period. In this case, the computation formula is:

$$S_{k+1} = S_k + rS_k \ or \ S_{k+1} = S_k(1+r).$$

The solution to this homogeneous first-order difference equation is $S_k = (1+r)^k S_0$.

Economics Application 24 (Dynamic Model of Economic Growth)

Let Y_t, C_t, I_t denote national income, consumption, and investment in period t, respectively. A simple dynamic model of economic growth is described by the equations:

$$\begin{aligned} Y_t &= C_t + I_t \\ C_t &= c + mY_t \\ Y_{t+1} &= Y_t + rI_t \end{aligned}$$

where

- c is a positive constant;

- m is the marginal propensity to consume, $0 < m < 1$;

- r is the growth factor.

From the above system of three equations, we can express the dynamics of national income in the form of a first-order difference equation:

$$\begin{aligned} Y_{t+1} - Y_t &= rI_t \\ &= r(Y_t - C_t) \\ &= rY_t - r(c + mY_t), \end{aligned}$$

or

$$Y_{t+1} - [1 + r(1 - m)]Y_t = -rc.$$

The solution to this equation is

$$Y_t = (Y_0 - \frac{c}{1 - m})[1 + r(1 - m)]^t + \frac{c}{1 - m}.$$

The behavior of investment can be described in the following way:

$$
\begin{aligned}
I_{t+1} - I_t &= (Y_{t+1} - C_{t+1}) - (Y_t - C_t) \\
&= (Y_{t+1} - Y_t) - (C_{t+1}) - C_t) \\
&= rI_t - m(Y_{t+1} - Y_t) \\
&= rI_t - mrI_t,
\end{aligned}
$$

or

$$I_{t+1} = [1 + r(1 - m)]I_t.$$

This homogeneous difference equation has the solution $I_t = [1 + r(1 - m)]^t I_0$.

4.2.2 Second-Order Linear Difference Equations

The general form is:

$$y_{k+2} + a_1 y_{k+1} + a_2 y_k = r_k. \tag{4.4}$$

The solution to the corresponding homogeneous equation

$$y_{k+2} + a_1 y_{k+1} + a_2 y_k = 0 \tag{4.5}$$

depends on the solutions to the quadratic equation:

$$m^2 + a_1 m + a_2 = 0,$$

which is called the auxiliary (or characteristic) equation of equation (4.4). Let m_1, m_2 denote the roots of the characteristic equation. Then, both m_1 and m_2 will be non-zero because a_2 is non-zero for (4.4) to be of second order.

Case 1: m_1 and m_2 are real and distinct.
The solution to (4.5) is $y_k = C_1 m_1^k + C_2 m_2^k$, where C_1, C_2 are arbitrary constants.

Case 2: m_1 and m_2 are real and equal.

The solution to (4.5) is $y_k = C_1 m_1^k + C_2 k m_1^k = (C_1 + C_2 k) m_1^k$.

Case 3: m_1 and m_2 are complex with the polar forms $r(\cos\theta \pm i\sin\theta)$, $r > 0$, $\theta \in (-\pi, \pi]$.

The solution to (4.5) is $y_k = C_1 r^k \cos(k\theta + C_2)$.

Any complex number $a + bi$ admits a representation in the polar (or trigonometric) form $r(\cos\theta \pm i\sin\theta)$. The transformation is given by

- $r = \sqrt{a^2 + b^2}$

- θ is the unique angle (in radians) such that $\theta \in (-\pi, \pi]$ and

$$\cos\theta = \frac{a}{\sqrt{a^2 + b^2}} \ , \quad \sin\theta = \frac{b}{\sqrt{a^2 + b^2}}$$

Example 125 *Solve* $y_{k+2} - 5y_{k+1} + 6y_k = 0$.

The characteristic equation is $m^2 - 5m + 6 = 0$ *and has the roots* $m_1 = 2, m_2 = 3$. *The difference equation has the solution* $y_k = C_1 2^k + C_2 3^k$.

Example 126 *Solve* $y_{k+2} - 8y_{k+1} + 16y_k = 2^k + 3$.

The characteristic equation $m^2 - 8m + 16 = 0$ *has the roots* $m_1 = m_2 = 4$; *therefore, the solution to the corresponding homogeneous equation is* $Y = (C_1 + C_2 k)4^k$.

To find a particular solution, we consider the trial solution $y^* = A2^k + B$. *By substituting* y^* *into the non-homogeneous equation and performing the computations we obtain* $A = \frac{1}{4}$, $B = \frac{1}{3}$. *Thus, the solution to the non-homogeneous equation is* $y_k = (C_1 + C_2 k)4^k + \frac{1}{4}2^k + \frac{1}{3}$.

Example 127 *Solve* $y_{k+2} + y_{k+1} + y_k = 0$.

The equation $m^2 + m + 1 = 0$ *gives the roots* $m_1 = \frac{-1 - i\sqrt{3}}{2}$, $m_2 = \frac{-1 + i\sqrt{3}}{2}$ *or, written in the polar form,* $m_{1,2} = \cos\frac{2\pi}{3} \pm i\sin\frac{2\pi}{3}$. *Therefore, the given difference equation has the solution:* $y^* = C_1 \cos(\frac{2\pi}{3}k + C_2)$.

Example 128 *Consider a non-homogeneous equation* $y_{t+2} + ay_{t+1} + by_t = C$. *The functional form of a particular solution depends on parameters* a *and* b:

- *If* $1 + a + b \neq 0$ *then the particular solution* $x_t^* \equiv \dfrac{C}{1 + a + b}$;

- *If* $1 + a + b = 0$ *and* $a \neq 2$ *then we search for a particular solution in the form* $x_t^* = At$, *where the constant* A *can be uniquely determined, once* x_t^* *is substituted into the original equation;*

- *If $1 + a + b = 0$ and $a = 2$ then a particular solution is of the form $x_t^* = At^2$, where the constant A again is to be determined from the original equation.*

4.2.3 The General Case of Order n

In general, any difference equation of order n can be written as:

$$y_{k+n} + a_1 y_{k+n-1} + \ldots + a_{n-1} y_{k+1} + a_n y_k = r_k. \qquad (14)$$

The corresponding characteristic equation becomes

$$m^n + a_1 m^{n-1} + \ldots + a_{n-1} m + a_n = 0$$

and has exactly n roots denoted as $m_1, \ldots, m_n$.

The general solution to the homogeneous equation is a sum of terms which are produced by the roots $m_1, \ldots, m_n$ in the following way:

- a real simple root m generates the term $C_1 m^k$.

- a real and repeated p times root m generates the term

$$(C_1 + C_2 k + C_3 k^2 + \ldots + C_p k^{p-1}) m^k.$$

- each pair of simple complex conjugate roots $r(\cos \theta \pm i \sin \theta)$ generates the term

$$C_1 r^k \cos(k\theta + C_2).$$

- each repeated p times pair of complex conjugate roots $r(\cos \theta \pm i \sin \theta)$ generates the term

$$r^k [C_{1,1} \cos(k\theta + C_{1,2}) +$$

$$+ C_{2,1} k \cos(k\theta + C_{2,2}) + \ldots + C_{p,1} k^{p-1} \cos(k\theta + C_{p,2})].$$

The sum of terms will contain n arbitrary constants.

A particular solution y^* to the non-homogeneous equation again can be obtained by means of the method of undetermined coefficients (as was illustrated for the cases of first-order and second-order difference equations).

Example 129 $y_{k+4} - 5y_{k+3} + 9y_{k+2} - 7y_{k+1} + 2y_k = 0$

   The characteristic equation

$$m^4 - 5m^3 + 9m^2 - 7m + 2 = 0$$

has the roots $m_1 = 1$, which is repeated 3 times, and $m_2 = 2$. The solution is $y^* = (C_1 + C_2 k + C_3 k^2) + C_4 2^k$.

4.2.4 Systems of Simultaneous First-Order Difference Equations with Constant Coefficients

A typical first-order linear non-homogeneous system written in vector form

$$x_t = Ax_{t-1} + b, \quad x(0) = x_0$$

with x_t, x_{t-1}, b, x_0 being $[n \times 1]$ real column vectors and A $[n \times n]$ real matrix (b, x_0 and A are constant and given), has the solution

$$x_t = A^t(x_0 - x_e) + x_e,$$

where $x_e = (I - A)^{-1}b$ is the equilibrium (or particular) solution to x, assuming $(I - A)$ is a non-singular matrix.

   By iteration,

$$x_1 = Ax_0 + b$$

$$x_2 = Ax_1 + b = A(Ax_0 + b) + b = A^2 x_0 + Ab + b$$

$$x_3 = A^3 x_0 + (I + A + A^2)b, \text{ etc.,}$$

and, therefore, for any t,

$$x_t = A^t x_0 + (I + A + \dots + A^{t-1})b.$$

Since $(I + A + A^2 + \dots + A^{t-1}) \equiv (I - A^t)(I - A)^{-1}$, by substitution

$$\begin{aligned} x_t &= A^t x_0 + (I - A^t)(I - A)^{-1}b \\ &= A^t[x_0 - (I - A)^{-1}b] + (I - A)^{-1}b \equiv A^t(x_0 - x_e) + x_e. \end{aligned}$$

If A is diagonalizable (i.e. the matrix containing eigenvectors $P \equiv [v_1, v_2, ..., v_n]$ is non-singular), and $P^{-1}AP = \Lambda \equiv diag(\lambda_i)$, then $A = P\Lambda P^{-1}$, $A^2 = (P\Lambda P^{-1})(P\Lambda P^{-1}) = P\Lambda^2 P^{-1}$, $...$, $A^t = P\Lambda^t P^{-1}$. Thus

$$x_t = P\Lambda^t P^{-1}(x_0 - x_e) + x_e \equiv c_1 v_1 \lambda_1^t + c_2 v_2 \lambda_2^t + ... + c_n v_n \lambda_n^t + x_e,$$

where c_i is the ith element of a constant vector $c \equiv P^{-1}(x_0 - x_e)$ that has to be determined by the initial conditions and equilibrium solution; v_i denotes an eigenvector associated with the eigenvalue λ_i.

In the case of a homogeneous system (that is, $b = 0$), the solution becomes

$$x_t = A^t x_0 = P\Lambda^t P^{-1} x_0.$$

In the case of a planar system of difference equations, similarly to a differential equation, we can determine the type of equilibrium, and sketch a corresponding phase diagram.

Indeed, for a system of difference equations $x_t = Ax_{t-1}$ with A being $[2 \times 2]$ square matrix, its phase diagrams resemble the one constructed for a planar system of ordinary differential equations. The only difference between the two phase spaces is that in the former, solution points are discrete (but if connected to a continuous curve for the sake of visual simplicity, the analogy will become obvious).

The characteristic equation $c(\lambda) = |A - \lambda I| = \lambda^2 - \tau\lambda + \sigma = 0$ has the roots $\lambda = \tau/2 \pm \sqrt{\Delta}/2$ where $\Delta \equiv \tau^2 - 4\sigma$, and $\sigma = Tr(A)$, $\tau = Det(A)$. Three cases according to the sign of Δ can be distinguished.

Case 1:
$\Delta > 0$: two real distinct roots λ_1, λ_2. The type of equilibrium is:

- *stable node if $|\lambda_i| < 1$*

- *unstable node if $|\lambda_i| > 1$*

- *saddle point if $|\lambda_i| < 1 < |\lambda_j|$*

for $i, j = 1, 2$ and $i \neq j$.

Case 2:
$\Delta = 0$: equal roots $\lambda_1 = \lambda_2 = \lambda$, $J = \begin{pmatrix} \lambda & 1 \\ 0 & \lambda \end{pmatrix}$. The equilibrium is:

- *stable improper node* if $|\lambda| < 1$

- *unstable improper node* if $|\lambda| > 1$

Case 3.
$\Delta < 0$: $\lambda = \alpha \pm i\beta$, where $\alpha = \tau/2$, $\beta = \sqrt{-\Delta}/2$.
$$J = \begin{pmatrix} \alpha & \beta \\ -\beta & \alpha \end{pmatrix} = r \begin{pmatrix} \cos\phi & \sin\phi \\ -\sin\phi & \cos\phi \end{pmatrix}.$$ The type of equilibrium
is classified as:

- *stable focus* (SF) if $r = \sqrt{\alpha^2 + \beta^2} < 1$

- *unstable focus* (UF) if $r = \sqrt{\alpha^2 + \beta^2} > 1$

For all three cases the fundamental condition for stability is that both eigenvalues lie inside the unit circle. For complex eigenvalues $|\lambda|^2 \equiv \lambda\tilde{\lambda} = (\alpha + i\beta)(\alpha - i\beta) = \alpha^2 + \beta^2 = r^2$, $|\lambda| = |r| < 1$, i.e. the modulus of each eigenvalue must be smaller than one.

4.3   Introduction to Dynamic Optimization

4.3.1   First-Order Conditions

A typical dynamic optimization problem takes the following form:

$$
\begin{array}{rcl}
\max_{c(t)} \quad V & = & \int_0^T f(k(t), c(t), t)\, dt \\
\text{subject to} \quad \dot{k}(t) & = & g(k(t), c(t), t), \\
k(0) & = & k_0 > 0,
\end{array}
$$

where

- $[0, T]$ is the horizon over which the problem is considered (T can be finite or infinite);

- V is the value of the objective function as seen from the initial moment $t_0 = 0$;

- $c(t)$ is the *control variable* and the objective function V is maximized with respect to this variable;

- $k(t)$ is the *state variable* and the first constraint (called the *equation of motion*) describes the evolution of this state variable over time ($\dot{k}$ denotes $\frac{dk}{dt}$).

Similar to static optimization, in order to solve the optimization program we need first to find the first-order conditions:[4]

Recipe 24 – How to Obtain First-Order Conditions:

- *Step 1. Construct the Hamiltonian function*

$$H = f(k, c, t) + \lambda(t)g(k, c, t)$$

where $\lambda(t)$ is a Lagrange multiplier.

- *Step 2. Take the derivative of the Hamiltonian with respect to the control variable and set it to 0:*

$$\frac{\partial H}{\partial c} = \frac{\partial f}{\partial c} + \lambda \frac{\partial g}{\partial c} = 0.$$

- *Step 3. Take the derivative of the Hamiltonian with respect to the state variable and set it to equal the negative of the derivative of the Lagrange multiplier with respect to time:*

$$\frac{\partial H}{\partial k} = \frac{\partial f}{\partial k} + \lambda \frac{\partial g}{\partial k} = -\dot{\lambda}.$$

- *Step 4.* Transversality condition

  – **Case 1:** *Finite horizons:*

$$\lambda(T)k(T) = 0.$$

---

[4] An accessible derivation of necessary and sufficient conditions for a typical optimal control problem can be found in Kamien, M. I. and N. I. Schwartz, *Dynamic Optimization: The Calculus of Variations and Optimal Control in Economics and Management.*

-   **Case 2:** *Infinite horizons with f of the form*
    $f[k(t), c(t), t] = e^{-pt}u[k(t), c(t)]$:

$$\lim_{t \to \infty} [\lambda(t)k(t)] = 0.$$

-   **Case 3:** *Infinite horizons with f in a form different from that specified in case 2:*

$$\lim_{t \to \infty} [H(t)] = 0.$$

Example 130 *Solve*

$$
\begin{aligned}
\max_{u(t)} \quad V &= \int_0^2 (2x + 3u - 0.5u^2)dt \\
\text{subject to} \quad \dot{x}(t) &= x + u, \\
x(0) &= 5, \quad x(2) \text{ free.}
\end{aligned}
$$

Solution:
The Hamiltonian is $H = 2x + 3u - 0.5u + \lambda(x + u)$. The F.O.C. (maximum principle) are

$$
\begin{aligned}
H_u &= 3 - u + \lambda = 0, & (4.6)\\
H_x &= 2 + \lambda = -\dot{\lambda}. & (4.7)
\end{aligned}
$$

Equation (4.6) implies $u^{opt} = 3 + \lambda$. Equation (4.7), that is $\dot{\lambda} + \lambda + 2 = 0$, solves to $\lambda(t) = C_1 e^{-t} - 2$, and from the transversality condition $\lambda(2) = 0$ we find $C_1 = 2e^2$. Therefore,

$$u^{opt}(t) = 3 - 2 + 2e^2 e^{-t} = 1 + 2e^2 e^{-t}.$$

Thus the equation of motion becomes $\dot{x} = x + 1 + 2e^2 e^{-t}$ and has a solution

$$x(t) = C_2 e^t - 1 - e^2 e^{-t}.$$

Plugging in the initial condition $x(0) = 5$, we find C_2 from the equation $5 = C_2 - 1 - e^2$ as $C_2 = 6 + e^2$. Collecting the pieces together, the optimal x is

$$x^{opt}(t) = (6 + e^2)e^t - e^2 e^{-1} - 1.$$

Recipe 25 – F.O.C. with More than One State and/or Control Variable:
It may happen that the dynamic optimization problem contains more than one control variable and more than one state variable. In that case we need an equation of motion for each state variable. To write the first-order conditions, the algorithm specified above should be modified in the following way:

- **Step 1a.** The Hamiltonian includes the right-hand side of each equation of motion times the corresponding multiplier.

- **Step 2a.** Applies for each control variable.

- **Step 3a.** Applies for each state variable.

Recipe 26 – F.O.C. and the Calculus of Variations
In the calculus of variations, the control variable $c(t)$ is not present, and the objective functional V is optimized by finding the optimal time path for the state variable $k(t)$. A typical problem of the calculus of variations takes the following form:

$$\max_{k(t)} \quad V = \int_0^T f(k(t), \dot{k}(t), t)dt$$
$$\text{subject to} \qquad k(0) = k_0, \ k(T) = k_T,$$
$$k_0, \ k_T, \ T \text{ given.}$$

Now the first-order conditions are derived in the form of the so-called Euler equation:

$$\frac{\partial f}{\partial k} - \frac{d}{dt}\left(\frac{\partial f}{\partial \dot{k}}\right) = 0,$$

or

$$\frac{\partial^2 f}{\partial k \partial \dot{k}} \cdot \ddot{k} + \frac{\partial^2 f}{\partial k \partial \dot{k}} \cdot \dot{k} + \frac{\partial^2 f}{\partial t \partial \dot{k}} - \frac{\partial f}{\partial k} = 0.$$

If the terminal state k_T or the terminal time T is not given then, in addition to the Euler equation, transversality conditions are imposed:

- k_T *is free* $\implies$ $\dfrac{\partial f}{\partial \dot{k}} = 0$ *at* $t = T$

- T *is free* $\implies$ $\left(f - \dot{k}\dfrac{\partial f}{\partial \dot{k}}\right) = 0$ *at* $t = T$

Example 131 *Find such $k(t)$ that optimizes $\int_0^1 (tk + \dot{k}^2)dt$. The initial condition is $k(0) = 0$, and the terminal point $k(1)$ is free.*

Solution:

The Euler equation reads as $2\ddot{k} - t = 0$. Integrating it twice with respect to t we obtain $\dot{k}(t) = \frac{1}{4}t^2 + c_1$, and $k(t) = \frac{1}{12}t^3 + c_1 t + c_2$. The initial condition $k(0) = 1$ yields $c_2 = 0$. Since the terminal condition is not given, the other constant c_1 should be determined with the help of the transversality condition $\dfrac{\partial f}{\partial \dot{k}} = 0$ at $t = 1$, that is, $2\dot{k}(1) = 0$. At time $t = 1$, the equation $\dot{k}(t) = \frac{1}{4}t^2 + c_1$ becomes $\dot{k}(1) = 1/4 + c_1$, thus $c_1 = -1/4$.

4.3.2 Present-Value and Current-Value Hamiltonians

In economic problems, the objective function is usually of the form

$$f[k(t), c(t), t] = e^{-\rho t} u[k(t), c(t)],$$

where ρ is a constant discount rate and $e^{-\rho t}$ is a discount factor. The Hamiltonian

$$H = e^{-\rho t} u(k, c) + \lambda(t) g(k, c, t)$$

is called the *present-value Hamiltonian* (since it represents a value at the time $t_0 = 0$).

Sometimes it is more convenient to work with the *current-value Hamiltonian*, which represents a value at time t:

$$\hat{H} = He^{\rho t} = u(k, c) + \mu(t) g(k, c, t)$$

where $\mu(t) = \lambda(t) e^{\rho t}$. The first-order conditions written in terms of the current-value Hamiltonian appear to be a slight modification of the present-value type:

- $\frac{\partial \hat{H}}{\partial c} = 0$;

- $\frac{\partial \hat{H}}{\partial k} = \rho\mu - \dot{\mu}$;

- $\mu(T)k(T) = 0$ (the transversality condition in the case of finite horizons), or
  $\lim_{t \to \infty} [\mu(t)k(t)e^{-\rho t}] = 0$ (the transversality condition in the case of infinite horizon).

4.3.3 Dynamic Problems with Inequality Constraints

Let us add an inequality constraint to the dynamic optimization problem considered above.

$$
\begin{array}{rcl}
\max_{c(t)} \ V & = & \int_0^T f[k(t), c(t), t]dt \\
\text{subject to} \ \ \dot{k}(t) & = & g[k(t), c(t), t], \\
h(t, k, c) & \geq & 0, \\
k(0) & = & k_0 > 0.
\end{array}
$$

The following algorithm allows us to write the first-order conditions:

Recipe 27 – F.O.C. with Inequality Constraints:

- *Step 1. Construct the Hamiltonian*

$$
H = f(k, c, t) + \mu(t)g(k, c, t) + w(t)h(t, k, c)
$$

  where $\mu(t)$ and $w(t)$ are Lagrange multipliers.

- *Step 2.*

$$
\frac{\partial H}{\partial c} = 0.
$$

- *Step 3.*

$$
\frac{\partial H}{\partial k} = -\dot{\mu}.
$$

- *Step 4.*

$$
w(t) \geq 0, \qquad w(t)h(t, k, c) = 0.
$$

- *Step 5. Transversality condition if T is finite:*

$$
\mu(T)k(T) = 0.
$$

Economics Application 25 (Ramsey Model)

This example illustrates how the tools of dynamic optimization can be applied to solve a model of economic growth, namely the Ramsey model (for more details on the economic illustrations presented in this section see, for instance, R. J. Barro, X. Sala-i-Martin, Economic Growth*).*

The model assumes that the economy consists of households and firms but here we further assume, for simplicity, that households carry out production directly. Each household chooses, at each moment t, the level of consumption that maximizes the present value of lifetime utility

$$U = \int_0^\infty u[A(t)c(t)]e^{nt}e^{-\rho t}dt$$

subject to the dynamic budget constraint

$$\dot{k}(t) = f[k(t)] - c(t) - (x + n + \delta)k(t)$$

and the initial level of capital $k(0) = k_0$.

Throughout the example we use the following notation:

- *$A(t)$ is the level of technology;*

- *c denotes consumption per unit of effective labor (effective labor is defined as the product of labor force $L(t)$ and the level of technology $A(t)$);*

- *k denotes the amount of capital per unit of effective labor;*

- *$f(\cdot)$ is the production function written in intensive form ($f(k) = F(K/AL, 1)$);*

- *n is the rate of population growth;*

- *ρ is the rate at which households discount future utility;*

- *x is the growth rate of technological progress;*

- *δ is the depreciation rate.*

The expression $A(t)c(t)$ in the utility function represents consumption per person and the term e^{nt} captures the effect of the growth in family size at rate n. The dynamic constraint says that the change in the stock of capital (expressed in units of effective labor) is given by the difference between the level of output (in units of effective labor) and the sum of consumption (per unit of effective labor), depreciation, and the additional effect $(x + n)k(t)$ from expressing all variables in terms of effective labor.

As can be observed, households face a dynamic optimization problem with c as a control variable and k as a state variable. The present-value Hamiltonian is

$$H = u(A(t)c(t))e^{nt}e^{-\rho t} + \mu(t)[f[k(t)] - c(t) - (x + n + \delta)k(t)]$$

and the first-order conditions state that

$$\frac{\partial H}{\partial c} = \frac{\partial u(Ac)}{\partial c}e^{nt}e^{-\rho t} = \mu,$$
$$\frac{\partial H}{\partial k} = \mu[f'(k) - (x + n + \delta)] = -\dot{\mu}.$$

In addition the transversality condition requires:

$$\lim_{t \to \infty} [\mu(t)k(t)] = 0.$$

Eliminating μ from the first-order conditions and assuming that the utility function takes the functional form $u(Ac) = \frac{(Ac)^{(1-\theta)}}{1-\theta}$ leads us to an expression[5] for the growth rate of c:

$$\frac{\dot{c}}{c} = \frac{1}{\theta}[f'(k) - \delta - \rho - \theta x].$$

---

[5] To reconstruct these calculations, note first that

$$\frac{\partial u(Ac)}{\partial c} = A^{1-\theta}c^{\theta}$$

and

$$\frac{d}{dt}\frac{\partial u(Ac)}{\partial c} = \left((1-\theta)\frac{\dot{A}}{A} - \theta\frac{\dot{c}}{c}\right) \cdot \frac{\partial u(Ac)}{\partial c}$$

or, assuming that $\frac{\dot{A}}{A} = x$,

$$\frac{d}{dt}\frac{\partial u(Ac)}{\partial c} = \left((1-\theta)x - \theta\frac{\dot{c}}{c}\right) \cdot \frac{\partial u(Ac)}{\partial c}.$$

This equation, together with the dynamic budget constraint forms a system of differential equations which completely describe the dynamics of the Ramsey model. The boundary conditions are given by the initial condition k_0 and the transversality condition.

Economics Application 26 (Model of Economic Growth with Physical and Human Capital)

This model illustrates the case in which the optimization problem has two dynamic constraints and two control variables.

As was the case in the previous example, we assume that households perform production directly. The model also assumes that physical capital K and human capital H enter the production function Y in a Cobb-Douglas manner:

$$Y = AK^\alpha H^{1-\alpha},$$

where A denotes a constant level of technology and $0 \le \alpha \le 1$. Output is divided among consumption C, investment in physical capital I_K and investment in human capital I_H:

$$Y = C + I_K + I_H.$$

The two capital stocks change according to the dynamic equations:

$$\begin{aligned} \dot{K} &= I_K - \delta K, \\ \dot{H} &= I_H - \delta H, \end{aligned}$$

where δ denotes the depreciation rate.

Therefore,

$$\dot{\mu} = \frac{\partial u(Ac)}{\partial c} e^{nt} e^{-\rho t} \cdot \left(n - \rho + (1-\theta)x - \theta \frac{\dot{c}}{c} \right).$$

Eliminating μ from the second equation in the F.O.C. we get

$$\theta \frac{\dot{c}}{c} - (1-\theta)x - n + \rho = f'(k) - x - n - \delta,$$

which after rearranging terms gives an expression for the growth rate of consumption c.

The households' problem is to choose C, I_K, I_H such that they maximize the present-value of lifetime utility

$$U = \int_0^\infty u(C)e^{-\rho t}dt$$

subject to the above four constraints. We again assume that $u(C) = \frac{C^{(1-\theta)}}{1-\theta}$. By eliminating Y and I_K, the households' problem becomes

$$\begin{aligned}
\max \quad &U \\
\text{subject to} \quad &\dot{K} = AK^\alpha H^{1-\alpha} - C - \delta K - I_H, \\
&\dot{H} = I_H - \delta H,
\end{aligned}$$

and it contains two control variables (C, I_H) and two state variables (K, H).

The Hamiltonian associated with this dynamic problem is:

$$\mathcal{H} = u(C) + \mu_1(AK^\alpha H^{1-\alpha} - C - \delta K - I_H) + \mu_2(I_H - \delta H)$$

and the first-order conditions require

$$\frac{\partial \mathcal{H}}{\partial C} = u'(C)e^{-\rho t} - \mu_1 = 0, \tag{4.8}$$

$$\frac{\partial \mathcal{H}}{\partial I_H} = -\mu_1 + \mu_2 = 0, \tag{4.9}$$

$$\frac{\partial \mathcal{H}}{\partial K} = -\mu_1(\alpha AK^{\alpha-1}H^{1-\alpha} - \delta) = -\dot{\mu}_1, \tag{4.10}$$

$$\frac{\partial \mathcal{H}}{\partial H} = -\mu_1(1-\alpha)AK^\alpha H^{-\alpha} - \mu_2\delta = -\dot{\mu}_2. \tag{4.11}$$

Eliminating μ_1 from (4.8) and (4.10) we obtain a result for the growth rate of consumption:

$$\frac{\dot{C}}{C} = \frac{1}{\theta}\left[\alpha A\left(\frac{K}{H}\right)^{-(1-\alpha)} - \delta - \rho\right].$$

Since $\mu_1 = \mu_2$ (from (4.9)), conditions (4.10) and (4.11) imply

$$\alpha A\left(\frac{K}{H}\right)^{-(1-\alpha)} - \delta = (1-\alpha)A\left(\frac{K}{H}\right)^\alpha - \delta,$$

which enables us to find the ratio:

$$\frac{K}{H} = \frac{\alpha}{1-\alpha}.$$

After substituting this ratio in the result for $\frac{\dot{C}}{C}$, it turns out that consumption grows at a constant rate given by:

$$\frac{\dot{C}}{C} = \frac{1}{\theta}[A\alpha^\alpha(1-\alpha)^{(1-\alpha)} - \delta - \rho].$$

The substitution of $\frac{K}{H}$ into the production function implies that Y is a linear function of K:

$$Y = AK\left(\frac{1-\alpha}{\alpha}\right)^{(1-\alpha)}$$

(Therefore, such a model is also called an AK model.)
    Formulas for Y and the K/H-ratio suggest that K, H and Y grow at a constant rate. Moreover, it can be shown that this rate is the same as the growth rate of C found above.

Economics Application 27 (Investment-Selling Decisions)
    Finally, let us exemplify the case of a dynamic problem with inequality constraints.
    Assume that a firm produces, at each moment t, a good which can either be sold or reinvested to expand the productive capacity $y(t)$ of the firm. The firm's problem is to choose at each moment the proportion $u(t)$ of the output $y(t)$ which should be reinvested such that it maximizes total sales over the period $[0, T]$.
    Expressed in mathematical terms, the firm's problem is:

$$\begin{aligned}
\max \quad & \int_0^T [1-u(t)]y(t)dt \\
\text{subject to} \quad & \dot{y}(t) = u(t)y(t), \\
& y(0) = y_0 > 0, \\
& 0 \le u(t) \le 1,
\end{aligned}$$

where u is the control and y is the state variable.
    The Hamiltonian takes the form:

$$H = (1-u)y + \lambda uy + w_1(1-u) + w_2 u.$$

The optimal solution satisfies the conditions:

$$H_u = (\lambda - 1)y + w_2 - w_1 = 0,$$
$$\dot{\lambda} = u - 1 - u\lambda,$$
$$w_1 \geq 0, \qquad w_1(1 - u) = 0,$$
$$w_2 \geq 0, \qquad w_2 u = 0,$$
$$\lambda(T) = 0.$$

These conditions imply that

$$\lambda > 1, \quad u = 1 \text{ and } \dot{\lambda} = -\lambda \qquad\qquad (4.12)$$

or

$$\lambda < 1, \quad u = 0 \text{ and } \dot{\lambda} = -1; \qquad\qquad (4.13)$$

thus, $\lambda(t)$ is a decreasing function over $[0, T]$. Since $\lambda(t)$ is also a continuous function, there exists t^* such that (4.13) holds for any t in $[t^*, T]$. Hence

$$u(t) = 0,$$
$$\lambda(t) = T - t,$$
$$x(t) = x(t^*),$$

for any t in $[t^*, T]$.
  t^* must satisfy $\lambda(t^*) = 1$, which implies that $t^* = T - 1$ if $T \geq 1$. If $T \leq 1$, we have $t^* = 0$.
  If $T > 1$, then for any t in $[0, T - 1]$ (4.12) must hold. Using the fact that y is continuous over $[0, T]$ (and, therefore, continuous at $t^* = T - 1$), we obtain

$$u(t) = 1,$$
$$\lambda(t) = e^{T-t-1},$$
$$x(t) = y_0 e^t,$$

for any t in $[0, T - 1]$.
  In conclusion, if $T > 1$ then it is optimal for the firm to invest all the output until $t = T - 1$ and to sell all the output after that date. If $T < 1$ then it is optimal to sell all the output.

Further Reading:

- Arrowsmith, D. K. and Place, C. M. (1990). *An Introduction to Dynamical Systems.* Cambridge: Cambridge University Press.

- Barro, R. J. and Sala-i-Martin, X. (2003). *Economic Growth* (2nd ed.). MIT Press.

- Chiang, A. C. (1992). *Elements of Dynamic Optimization.* New York: McGraw Hill.

- Goldberg, S. (1986). *Introduction to Difference Equations.* New York: Dover.

- Hartman, P. (2002). *Ordinary Differential Equations* (2nd ed.). Society for Industrial and Applied Math.

- Kamien, M. I. and Schwartz, N. I. (1991). *Dynamic Optimization: The Calculus of Variations and Optimal Control in Economics and Management* (2nd ed.). New York: North Holland.

- Lorenz, H.W. (1989). *Nonlinear Dynamical Economics and Chaotic Motion.* Berlin: Springer-Verlag.

- Pontryagin, L. S. (1974). *Ordinary Differential Equations* (4th ed.). Moscow: Nauka.

- Seierstad, A. and Sydsæter, K. (1987). *Optimal Control Theory with Economic Applications* Amsterdam; New York: Elsevier Science Pub. Co.

- Simonovits, A. (2000). *Mathematical Methods in Dynamic Economics.* New York: Macmillan.

- Turkington, D.A. (2007). *Mathematical Tools for Economics.* New York: Blackwell-Wiley.

- Turnovsky, S. (2000). *Macroeconomic Dynamics* (2nd ed.). Cambridge MA: MIT Press.

Chapter 5

Exercises

This chapter serves the purpose of a testing device. Its objective is to give the reader a chance to master mathematical skills and to enhance his or her command of mathematical tools and methods by solving various problems of a practical nature. As has been mentioned in the introduction, in general the problems presented throughout the text are not 'brand new', because similar – if not nearly identical – setups can be found in other moderately advanced and advanced textbooks on mathematical economics (here I am making a group reference to the sources already listed in this book, such as Chiang, Goldberg, Ostaszewski, Simon and Blume, Sydsæter et al., Turkington, etc.). (*Editor's note*: additional mostly Russian-language sources are listed below). This is unavoidable while writing a textbook: On the one hand, once you have learned how to solve a class of problems (for example, the linear differential equation $y''(x)+ay'(x)+by(x) = 0$) then any particular case of the form, say, $y'' + 3y' + 2y = 0$ becomes redundant from the perspective of pure mathematics. On the other hand, while teaching an introductory graduate course of mathematics for economists at CERGE, I repeatedly received a request for more *solved* problems for self study from my students. For that reason I have decided to include a separate chapter solely dedicated to exercises.

The structure of the chapter is as follows. The 'Practice Problems' section contains basic problems with answers only. In the 'Solved Problems' section, problems, roughly sorted by topic, are provided with a detailed solution. The 'Economics Applications' section is a collection of solved problems explicitly related to various fields of economics. The 'Written Assignments' section can be viewed as a pacemaker for a distance learning scholar who attempts to cover the course of mathematics for economists in one semester. The 'Sample Problem Sets' section again is designed for a distance learning student and exemplifies written exams that cover the content of the whole book. And finally the 'Unsolved Problems' section gives the reader an opportunity for further practice.

Additional Problem Books

- Demidovich, B. P. (1997). *Problems in Mathematical Analysis* (in Russian)(13th ed.). Moscow: MIR Publishers.

- Filippov, A. F. (1973). *Problems of Differential Equations* (in Russian) (4th ed.). Moscow: Nauka.

- Filippov, A.F. (1988). *Differential Equations: Problems and Examples.* (in Russian) Moscow University.

- Krutickaya, N.C. and Shishkin, A.A. (1985). *Linear Algebra: Questions and Problems* (in Russian) Moscow: Vysshaya Shkola.

- Zorich, V.A. (1984) *Mathematical Analysis* (in Russian) (Vols. 1–2). Moscow: Nauka.

5.1   Practice Problems

5.1.1   Problems

1. Evaluate the Vandermonde determinant

$$\det \begin{vmatrix} 1 & x_1 & x_1^2 \\ 1 & x_2 & x_2^2 \\ 1 & x_3 & x_3^2 \end{vmatrix}.$$

2. Find the rank of matrix A,

$$A = \begin{pmatrix} 0 & 2 & 2 \\ 1 & -3 & -1 \\ -2 & 0 & -4 \\ 4 & 6 & 14 \end{pmatrix}.$$

3. Solve the system $Ax = b$, given

$$A = \begin{pmatrix} 1 & 2 & 3 \\ 2 & -1 & -1 \\ 1 & 3 & 4 \end{pmatrix}, \qquad b = \begin{pmatrix} 5 \\ 1 \\ 6 \end{pmatrix}.$$

4. Find the unknown matrix X from the equation

$$\begin{pmatrix} 1 & 2 \\ 2 & 5 \end{pmatrix} X = \begin{pmatrix} 4 & -6 \\ 2 & 1 \end{pmatrix}.$$

5. Find all solutions to the system of linear equations

$$\begin{cases} 2x + y - z & = & 3, \\ 3x + 3y + 2z & = & 7, \\ 7x + 5y & = & 13. \end{cases}$$

6. (a) Given $v = \begin{pmatrix} x \\ y \\ z \end{pmatrix}$ find the quadratic form $v'Av$ if

$$A = \begin{pmatrix} 3 & 4 & 6 \\ 4 & -2 & 0 \\ 6 & 0 & 1 \end{pmatrix}.$$

   (b) Given the quadratic form $x^2 + 2y^2 + 3z^2 + 4xy - 6yz + 8xz$ find matrix A of $v'Av$.

7. Find the eigenvalues and corresponding eigenvectors of the following matrix:

$$\begin{pmatrix} 1 & 0 & 1 \\ 0 & 1 & 1 \\ 1 & 1 & 2 \end{pmatrix}.$$

   Is this matrix positive definite?

8. Show that the matrix

$$\begin{pmatrix} a & d & e \\ d & b & f \\ e & f & c \end{pmatrix}$$

is positive definite if and only if

$$a > 0, \quad ab - d^2 > 0, \quad abc + 2edf > af^2 + be^2 + cd^2.$$

9. Find the following limits:

a) $\displaystyle\lim_{x \to -2} \frac{x^2 + x - 2}{x^2 + 2x}$

b) $\displaystyle\lim_{x \to 1} \frac{x - 1}{\ln x}$

c) $\displaystyle\lim_{x \to +\infty} x^n \cdot e^{-x}$

d) $\displaystyle\lim_{x \to 0} x^x$

10. Show that a parabola $f(x) = x^2/2e$ is tangent to the curve $g(x) = \ln x$ and find the tangent point.

(HINT: Check that at the point of tangency x_0, $f(x_0) = g(x_0)$ and $f'(x_0) = g'(x_0)$.)

11. Find $y'(x)$ if

a) $y = \ln \sqrt{\dfrac{e^{4x}}{e^{4x} + 1}}$

b) $y = \ln(\sin x + \sqrt{1 + \sin^2 x})$

c) $y = x^{1/x}$

12. Show that the function $y(x) = x(\ln x - 1)$ is a solution to the differential equation $y''(x) x \ln x = y'(x)$.

13. Given $f(x) = 1/(1 - x^2)$, show that $f^{(n)}(0) = \begin{cases} n!, & n \text{ is even,} \\ 0, & n \text{ is odd.} \end{cases}$

14. Find $\dfrac{d^2y}{dx^2}$, given

    a) $\begin{cases} x = t^2, \\ y = t + t^3 \end{cases}$

    b) $\begin{cases} x = e^{2t}, \\ y = e^{3t} \end{cases}$

15. Can we apply Rolle's Theorem to the function $f(x) = 1 - \sqrt[3]{x^2}$, where $x \in [-1, 1]$? Explain your answer and show it graphically.

16. Given $y(x) = x^2|x|$, find $y'(x)$ and draw a graph of the function.

17. Find the approximate value of $\sqrt[5]{33}$.

18. Find $\dfrac{\partial z}{\partial x}$ and $\dfrac{\partial z}{\partial y}$, given

    a) $z = x^3 + 3x^2y - y^3$

    b) $z = \dfrac{xy}{x - y}$

    c) $z = xe^{-xy}$

19. Show that the function $z(x, y,) = xe^{-y/x}$ is a solution to the differential equation

$$x\frac{\partial^2 z}{\partial x \partial y} + 2\left(\frac{\partial z}{\partial x} + \frac{\partial z}{\partial y}\right) = y\frac{\partial^2 z}{\partial y^2}.$$

20. Find the extrema of the following functions:

    a) $z = y\sqrt{x} - y^2 - x + 6y$

    b) $z = e^{x/2}(x + y^2)$

21. Evaluate the following indefinite integrals:

a) $\int \dfrac{(x^2+1)^2}{x^3} dx$

b) $\int a^x(1+a^{-x}/\sqrt{x^3})dx$

c) $\int e^{x^3} x^2 dx$

d) $\int x\sqrt{x^2+1}dx$

e) $\int x \ln(x-1)dx$

22. Evaluate the following definite integrals:

a) $\int_1^2 (x^2 + 1/x^4)dx$

b) $\int_4^9 \dfrac{dx}{\sqrt{x}-1}$ (HINT: substitution $x = t^2$)

23. Find the area bounded by the curves:

a) $y = x^3,\ y = 8,\ x = 0$

b) $4y = x^2,\ y^2 = 4x$

24. Check whether the improper integral

$$\int_0^{+\infty} \dfrac{dx}{(x-3)^2}$$

converges or diverges.

25. Find extrema of the functions

    a) $z = 3x + 6y - x^2 - xy - y^2$

    b) $z = 3x^2 - 2x\sqrt{y} + y - 8x + 8$

26. Find the maximum and minimum values of the function $f(x, y) = x^3 + y^3 - 3xy$ in the domain $\{0 \le x \le 2, \ -1 \le y \le 2\}$.

    (HINT: Don't forget to test the function on the boundary of the domain)

27. Find dy/dx, given

    a) $x^2 + y^2 - 4x + 6y = 0$

    b) $xe^{2y} - ye^{2x} = 0$

28. Given the equation $2\sin(x + 2y - 3z) = x + 2y - 3z$, show that $z_x + z_y = 1$.

29. Check that the Cobb-Douglas production function $z = x^a y^b$ for $0 < a, b < 1, \ a + b \le 1$ is concave for $x, y > 0$.

30. Show that $z = (x^{1/2} + y^{1/2})^2$ is a concave function.

31. For which values of x, y is the function $f(x, y) = x^3 + xy + y^2$ convex?

32. Find dx/dz and dy/dz given

    a) $x + y + z = 0, \ x^2 + y^2 + z^2 = 1$

    b) $x^2 + y^2 - 2z = 1, \ x + xy + y + z = 1$

33. Find the conditional extrema of the following functions and classify them:

    a) $u = x + y + z^2, \quad$ s.t. $z - x = 1, \ y - xz = 1$

    b) $u = x + y, \quad$ s.t. $1/x^2 + 1/y^2 = 1/a^2$

c) $u = (x + y)z,$ s.t. $1/x^2 + 1/y^2 + 2/z^2 = 4$

d) $u = xyz,$ s.t. $x^2 + y^2 + z^2 = 3$

e) $u = xyz,$ s.t. $x^2 + y^2 + z^2 = 1, \ x + y + z = 0$

f) $u = x - 2y + z,$ s.t. $x + y^2 - z^2 = 1$

34. A consumer is known to have a Cobb-Douglas utility of the form $u(x, y) = x^\alpha y^{1-\alpha}$, where the parameter α is unknown. However, it is known that when faced with the utility maximization problem

$$\max u(x, y) \text{ subject to } x + y = 3,$$

the consumer chooses $x = 1, y = 2$. Find the value of α.

35. Solve the following differential equations:

a) $x(1 + y^2) + y(1 + x^2)y' = 0$

b) $y' = xy(y + 2)$

c) $y' + y = 2x + 1$

d) $y^2 + x^2 y' = xyy'$

e) $2x^3 y' = y(2x^2 - y^2)$

f) $x^2 y' + xy + 1 = 0$

g) $y = x(y' - x \cos x)$

h) $xy' - 2y = 2x^4$

i) $2y'' - 5y' + 2y = 0$

j) $y'' - 2y' = 0$

k) $y''' - 3y' + 2y = 0$

l) $y'' - 2y' + y = 6xe^x$

m) $y'' + 2y' + y = 3e^{-x}\sqrt{x + 1}$

36. Solve the following systems of differential equations:

    a) $\dot{x} = 5x + 2y, \quad \dot{y} = -4x - y$

    b) $\dot{x} = 2y, \quad \dot{y} = 2x$

    c) $\dot{x} = 2x + y, \quad \dot{y} = x + 2y$

37. Find $f'(x)$ if

    (a) $f(x) = (1 + nx^m)(1 + mx^n)$, m and n are constants

    (b) $f(x) = \frac{x^p(1-x)^q}{1+x}$, $x \neq -1$, p and q are constants

    (c) $f(x) = ((1-x)^m \cdot (1-x)^n)^{\frac{1}{m+n}}$, m and n are constants

    (d) $f(x) = e^{-x^2}(x^2 - 2x + 2)$

    (e) $f(x) = a^{x-1}x^{1-a}$, $x > 0$ and a is a positive constant

    (f) $f(x) = [\ln(x^2)]^3$

    (g) $f(x) = (e^{ax} - e^{-ax})^2$, a is a constant

    (h) $f(x) = \ln\left(\sqrt{\frac{1+2x}{1-2x}}\right)$

    (i) $f(x) = \frac{1+e^x}{1-e^x}$

    (j) $f(x) = \sqrt[x]{x} \left(= x^{\frac{1}{x}}\right)$

    (k) $f(x) = \log_x(e)$

    (l) $f(x) = \frac{1+x-x^2}{1-x+x^2}$

38. Let $f(x) = (x - a_1)^{\alpha_1} \cdot (x - a_2)^{\alpha_2} \cdot \ldots \cdot (x - a_n)^{\alpha_n}$, a_i, α_i are constant for $i = 1, \ldots, n$

    Find the log-derivative $\frac{d}{dx}[\ln(f(x))]$ (that is, essentially, the growth rate: $\frac{d}{dx}[\ln(f(x))] = \frac{f'(x)}{f(x)}$)

39. Find the second derivative $f''(x)$ if

    (a) $f(x) = \frac{x}{\sqrt{1-x^2}}$

    (b) $f(x) = e^{-x^2}$

    (c) $f(x) = x \ln(x)$, $x > 0$

40. Find a linear approximation (tangent line) of

    (a) $f(x) = 4x - x^2$ at $x = 0$ and $x = 4$

    (b) $f(x) = \ln(x)$ at $x = 1$

41. Use second-order approximation of $f(x)$ around x_0 $f(x) \approx f(x_0) + f'(x_0)(x - x_0) + \frac{f''(x_0)}{2!}(x - x_0)^2$ to compute approximately $\sqrt{101}$. (HINT: Take $f(x) = \sqrt{x}$, $x_0 = 100$, $x = 101$)

42. Find extrema of the function $f(x)$ and classify them if

    (a) $f(x) = 3x - x^3$; also sketch a graph of $f(x)$

    (b) $f(x) = x^3 - 12x + 1$

    (c) $f(x) = x^2 e^{-x}$

    (d) $f(x) = x \ln(x)$

    (e) $f(x) = \frac{x^2 - 2x + 2}{x - 1}$

43. Sketch a graph of $f(x)$ if

    (a) $f(x) = \frac{1}{3}x^3 - \frac{5}{2}x^2 + 6x$

    (b) $f(x) = x^2 e^{-x^2}$ (HINT: $\lim\limits_{x \to \pm\infty} x^2 e^{-x^2} = 0$)

44. Find maximum and minimum values of

    (a) $f(x) = x^3 - 3x^2 + 3x$ on $[-1, 2]$

    (b) $f(x) = \frac{x}{1 + x^2}$ on $[0, 10]$

    (c) $f(x) = x \cdot \ln(x)$ on $[1, e]$

45. Find the following indefinite integrals:

    (a) $\int (x^2 + 2x + \frac{1}{x}) \, dx$

    (b) $\int \frac{x^2 - 2x}{x^4} \, dx$

    (c) $\int e^x \left(1 - \frac{e^{-x}}{x^2}\right) dx$

    (d) $\int e^{-3x} \, dx$

    (e) $\int (3 - 2x)^4 \, dx$ (HINT: Introduce a new variable $u = 3 - 2x$, $\frac{du}{dx} = -2$, thus $-\frac{1}{2} \int (3 - 2x)^4 \cdot (-2) \, dx = -\frac{1}{2} \int u^4 \, du$)

    (f) $\int \frac{dx}{\sqrt{3 - 2x}}$

(g) $\int (e^x + e^{-x})^2 \, dx$

46. Use integration by substitution to find

    (a) $\int e^{\sqrt{x}} \cdot \frac{1}{\sqrt{x}} \, dx$ (HINT: $u = \sqrt{x}$)

    (b) $\int \sqrt{x^2 + 1} \cdot x \, dx$ (HINT: $u = x^2 + 1$)

    (c) $\int (1 + \ln(x))^{\frac{1}{2}} \cdot \frac{1}{x} \, dx$ (HINT: $u = 1 + \ln(x)$)

47. Use integration by parts to find

    (a) $\int \ln(x) \, dx$ (HINT: In $\int f'g \, dx = fg - \int fg' \, dx$, take $f'(x) = 1$)

    (b) $\int xe^{2x} \, dx$ (HINT: $g(x) = x$)

    (c) $\int x \ln(x - 1) \, dx$ (HINT: $g(x) = x$; $\frac{x^2}{x-1} = \frac{x^2-1+1}{x-1} = \frac{x^2-1}{x-1} + \frac{1}{x-1}$)

48. Find the definite integrals:

    (a) $\int\limits_{1}^{4} \sqrt{x} \, dx$

    (b) $\int\limits_{-2}^{0} xe^{-x} \, dx$

    (c) $\int\limits_{-5}^{5} x^3 \sqrt{x^2 + 1} \, dx$

49. Find the area bounded by the curves $y = 4 - x^2$ and $y = 0$
    (HINT: Plot the graphs first)

50. Find partial derivatives of the function f if

    (a) $f(x, y) = x^3 + 3x^2y - y^3$

    (b) $f(x, y) = \frac{y}{x}$

    (c) $f(x, y) = \ln(x^2 + y^2)$

    (d) $f(x, y) = xe^{-xy}$

51. Find df if $f(x, y, z) = \sqrt{x^2 + y^2 + z^2}$
    Further show that $(f_x)^2 + (f_y)^2 + (f_z)^2 = 1$

52. Prove Euler's theorem:

    If $f(x, y)$ is homogeneous of degree k then $k \cdot f(x, y) = x \cdot f_x(x, y) + y \cdot f_y(x, y)$.

    Find the degree of homogeneity of $f(x, y)$ and verify Euler's theorem if

    (a) $f(x, y) = x^3 + xy^2 - 2y^3$

    (b) $f(x, y) = e^{-\frac{x}{y}}$

53. Find $\frac{df(x,y)}{dt}$ if $f(x, y) = \frac{y}{x}$ and $x = e^t$, $y = 1 - e^{-2t}$.

54. Find $d^2 f$ if

    (a) $f(x, y) = \frac{y^2}{x^2}$

    (b) $f(x, y) = x \ln\left(\frac{y}{x}\right)$

55. Show that $f(x, y) = Ax^\alpha y^\beta$ is concave if $\alpha + \beta < 1$, A, α, β are positive constants and x, $y > 0$

56. Given $f(x, y) = 3x + 6y - x^2 - xy - y^2$, use F.O.C. to find extremum of $f(x, y)$ and S.O.C. to check whether this extremum point is minimum or maximum.

57. Find critical points of the following functions and classify them ((local) min, (local) max, saddle point, or can't tell):

    (a) $f(x, y) = x^2 - xy + y^2 + 9x - 6y + 20$

    (b) $f(x, y) = y\sqrt{x} - y^2 - x + 6y$

    (c) $f(x, y) = 2xy - 4x - 2y$

    (d) $f(x, y) = e^{\frac{x}{2}} \cdot (x + y^2)$

    (e) $f(x, y) = 3x^4 + 3x^2 y - y^3$

    (f) $f(x, y) = x^4 + x^2 - 6xy + 3y^2$

    (g) $f(x, y) = xyz(4a - x - y - z)$, a is a parameter

    (h) $f(x, y) = (x^2 + 2y^2 + 3z^2)e^{-(x^2 + y^2 + z^2)}$

58. What functions in the preceding problem are globally convex or globally concave?

    (HINT: Such functions can only have a unique point of extremum, thus you only need to check (a), (b) and (d)).

59. Find conditional extrema of the following functions and try to check whether these extrema are (local) minimum or (local) maximum:

    (a) $f(x, y) = x + y$, s.t. $\frac{1}{x^2} + \frac{1}{y^2} = \frac{1}{2}$

    (b) $f(x, y) = xy$, s.t. $x^2 + y^2 = 2$

60. Solve the following equality-constrained optimization problems and check the concavity of the Lagrangian function to verify that you indeed obtain maximum:

    (a) $\max f(x, y, z) = 100 - x^2 - y^2 - z^2$, s.t. $x + 2y + z = a$, a is a parameter

    (b) $\max f(x, y, z) = x + 4y + z$, s.t. $x^2 + y^2 + z^2 = 216$, and $x + 2y + 3z = 0$

61. Find $\max f(x, y) = -y$ subject to $y^3 - x^2 = 0$.

62. Extremize $f(x, y) = x^2 - y^2$ subject to $1 - x - y = 0$

63. Find $\max f(x, y) = 2x^3 - 3x^2$ subject to $(3 - x)^3 - y^2 = 0$.

64. Extremize $f(x, y) = x^2 - y^2$ subject to $1 - x^2 - y^2 = 0$

65. Extremize $f(x, y) = \frac{1}{3}x^3 - \frac{3}{2}y^2 + 2x$ subject to $x - y = 0$

66. Find $\max f(x, y) = 2x + 3y$ subject to $g(x, y) = \sqrt{x} + \sqrt{y} = 5$

67. Extremize $f(x, y) = x^2 + y^2$ subject to $g(x, y) = x^2 + xy + y^2 = 3$

68. Find $\min f(x, y, z) = (x - 4)^2 + (y - 4)^2 + (z - \frac{1}{2})^2$ subject to $x^2 + y^2 = z$

69. Extremize $x^2 + y^2 + z^2$ subject to $\begin{cases} x + 2y + z = 30 \\ 2x - y - 3z = 10 \end{cases}$

70. Extremize $f(x) = x^3$ subject to $x \geq 0$

71. Find $\max f(x, y) = -(x^2 + y^2)$ subject to $(x - 1)^3 - y^2 \geq 0$

72. Find $\max f(x) = x^2 - x$ subject to $x \geq 0$

73. Find $\max f(x, y) = x^2 - y$ subject to $1 - x^2 - y^2 \geq 0$

74. Extremize $f(x, y) = x^2 + y^2 + y - 1$ s.t. $g(x, y) = x^2 + y^2 \leq 1$

75. Find max $f(x, y) = x^2 + 2y^2 - x$ subject to $g(x, y) = x^2 + y^2 \leq 1$

76. A firm has a total of L units of labor to allocate to the production of two goods. These can be sold at fixed positive prices a and b respectively. Producing x units of the first good requires αx^2 units of labor, whereas producing y units of the second good requires βy^2 units of labor, where α and β are positive constants. Find what output levels of the two goods maximize the revenue that the firm can earn by using this fixed amount of labor.

5.1.2   Answers

1. $(x_2 - x_1)(x_3 - x_1)(x_3 - x_2)$

2. 2

3. $$x = \begin{pmatrix} 1 \\ -1 \\ 2 \end{pmatrix}$$

4. $$X = \begin{pmatrix} 16 & -32 \\ -6 & 13 \end{pmatrix}$$

5. The third line in the matrix is a linear combination of the first two, therefore we have to solve the system of two equations with three variables. Taking one of the variables, say z, as a parameter, we find the solution to be

$$x = \frac{2 + 5z}{3}, \quad y = \frac{5 - 7z}{3}$$

6. a) $3x^2 - 2y^2 + z^2 + 8xy + 12xz$

   b)

$$A = \begin{pmatrix} 1 & 2 & 4 \\ 2 & 2 & -3 \\ 4 & -3 & 3 \end{pmatrix}$$

7. Eigenvalues are $\lambda = 0,\ 1,\ 3$

   Corresponding eigenvectors are $(1, -1, -1)'$, $(1, -1, 0)'$, $(1, 1, 2)'$

   The matrix is non-negative definite

9. a) $3/2$ b) 1 c) 0 d) 1

10. $x_0 = \sqrt{e}$, $y_0 = 1/2$

11. a) $2/(e^{4x} + 1)$ b) $\cos x/\sqrt{1 + \sin^2 x}$ c) $x^{1/x}\frac{1 - \ln x}{x^2}$.

14. a) $(3t^2 - 1)/4t^3$ b) $3/4e^t$

17. 2.0125

18. a) $3x(x + 2y)$, $3(x^2 - y^2)$ b) $-y^2/(x - y)^2$, $x^2/(x - y)^2$ c)
    $e - xy(1 - xy)$, $-x^2 e - xy$

20. a) $z_{max} = 12$ at $x = y = 4$ b) $z_{min} = -2/e$ at $x = -2$, $y = 0$.

21. a) $x^2/2 + 2\ln|x| - 1/2x^2 + C$ b) $a^x/\ln a - 2/\sqrt{x} + C$ c) $e^{x^3}/3 + C$
    d) $\sqrt{(x^2 + 1)^3}/3 + C$
    e) $x^2 \ln|x - 1|/2 - (x^2/2 + x + \ln|x - 1|)/2 + C$

22. a) $10/8$ b) $2(1 + \ln 2)$

23. a) 12 b) $16/3$

24. Diverges

25. a) $z_{max} = 9$ at $x = 0$, $y = 3$ b) $z_{min} = 0$ at $x = 2$, $y = 4$

26. $f_{min} = -1$ at $x = y = 1$, $f_{max} = 13$ at $x = 2$, $y = -1$

27. a) $\frac{2 - x}{y + 3}$ b) $\frac{2ye^{2x} - e^{2y}}{2xe^{2y} - e^{2x}}$

31. For any y and $x \geq 1/12$

32. a) $x' = (y - z)/(x - y)$, $y' = (z - x)/(x - y)$.
    b) $y' = -x' = 1/(y - x)$

33. a) minimum at $x = -1$, $y = 1$, $z = 0$

    b) minimum at $x = y = -\sqrt{2}a$, maximum at $x = y = \sqrt{2}a$

    c) maximum at $x = y = z = 1$ and $x = y = z = -1$; minimum at $x = y = 1$, $z = -1$ and $x = y = -1$, $z = 1$

    d) minimum at points $(-1, 1, 1)$, $(1, -1, 1)$, $(1, 1, -1)$, $(-1, -1, -1)$,

    maximum at points $(1, 1, 1)$, $(-1, -1, 1)$, $(-1, 1, -1)$, $(1, -1, -1)$

    e) minimum at points $(a, a, -2a)$, $(a, -2a, a)$, $(-2a, a, a)$, maximum at points $(-a, -a, 2a)$, $(-a, 2a, -a)$, $(2a, -a, -a)$, where $a = 1/\sqrt{6}$

    f) no extrema

34. $\alpha = 1/3$

35. a) $(1 + x^2)(1 + y^2) = C$

    b) $y = \frac{2Ce^{x^2}}{1 - Ce^{x^2}}$

    c) $y = 2x - 1 + Ce^{-x}$

    d) $y = Ce^{y/x}$

    e) $x = \pm y\sqrt{\ln Cx}$

    f) $xy = C - \ln|x|$

    g) $y = x(C + \sin x)$

    h) $y = Cx^2 + x^4$

    i) $y = C_1 e^{2x} + C_2 e^{x/2}$

    j) $y = C_1 + C_2 e^{2x}$

    k) $y = e^x(C_1 + C_2 x) + C_3 e^{-2x}$

    l) $y = (C_1 + C_2 x)e^x + x^3 e^x$

    m) $y = e^{-x}(\frac{4}{5}(x + 1)^{5/2} + C_1 + C_2 x)$

36. a) $x = C_1 e^t + C_2 e^{3t}$, $y = -2C_1 e^t - C_2 e^{3t}$

    b) $x = C_1 e^{2t} + C_2 e^{-2t}$, $y = C_1 e^{2t} - C_2 e^{-2t}$

    c) $x = C_1 e^t + C_2 e^{3t}$, $y = -C_1 e^t + C_2 e^{3t}$

37. (a) $mn(x^{m-1} + x^{n-1} + (m + n)x^{m+n-1})$

    (b) $\frac{x^{p-1}(1-x)^{q-1}}{(1+x)^2} \cdot (p - (q + 1)x - (p + q - 1)x^2)$

    (c) $\dfrac{(n-m)-(n+m)x}{(n+m)((1-x)^m(1+x)^n)^{\frac{1}{n+m}}}$

    (d) $e^{-x^2}(-2x^3 + 4x^2 - 2x - 2)$

    (e) $a^{x-1}x^{1-a}\left(\ln(a) + \frac{1-a}{x}\right)$

(f) $\frac{24[\ln(x)]^2}{x}$

(g) $2a\left(e^{2ax} - e^{-2ax}\right)$

(h) $\frac{2}{1-4x^2}$

(i) $\frac{2e^x}{(1-e^x)^2}$

(j) $x^{\frac{1}{x}-2}$

(k) $-\frac{[\ln(x)]^{-2}}{x}\left(= -\frac{[\log_x(e)]^2}{x}\right)$

(l) $\frac{2(1-2x)}{(1-x+x^2)^2}$

38. $\sum_{i=1}^{n} \frac{\alpha_i}{x-\alpha_i}$

39. (a) $\frac{3x}{(1-x^2)^{\frac{5}{2}}}$

   (b) $2xe^{-x^2}(2x^2-1)$

   (c) $\frac{1}{x}$

40. (a) $y = 4x, \ y = -4x + 16$

   (b) $y = x - 1$

41. $\sqrt{101} \approx \sqrt{100} + \frac{1}{2\sqrt{100}}\cdot(101-100) - \frac{1}{8\sqrt{100}^3}\cdot(101-100)^2 = 10.049875$

   (The exact value is $\sqrt{101} = 10.04987562112\ldots$)

42. (a) $x = -1$ – local min   and $x = 1$ – local max

   (b) $x = -2$ – min   and $x = 2$ – max

   (c) $x = 0$ – min   and $x = 2$ – max

   (d) $x = \frac{1}{e}$ – min

   (e) $x = 0$ – max   and $x = 2$ – min

43. (a) $f_{\min} = f(3) = \frac{9}{2}, \quad f_{\max} = f(2) = \frac{14}{3}$

   (b) $f_{\min} = f(0) = 0, \quad f_{\max} = f(\pm 1) = \frac{1}{e}$

44. (a) $f_{\min} = f(-1) = -7, \quad f_{\max} = f(2) = 2$

   (b) $f_{\min} = f(0) = 0, \quad f_{\max} = f(1) = \frac{1}{2}$

   (c) $f_{\min} = f(1) = 0, \quad f_{\max} = f(e) = e$

45. (a) $\frac{x^3}{3} + x^2 + \ln(x) + const$

    (b) $\frac{1-x}{x^2}$ + const

    (c) $e^x + \frac{1}{x}$ + const

    (d) $-\frac{1}{3}e^{-3x}$ + const

    (e) $-\frac{1}{10}(3 - 2x)^5$ + const

    (f) $-\sqrt{3 - 2x}$ + const

    (g) $\frac{1}{2}(e^{2x} - e^{-2x}) + 2x$ + const

46. (a) $2e^{\sqrt{x}}$ + const

    (b) $\frac{1}{3}\sqrt{(x^2 + 1)^3}$ + const

    (c) $\frac{2}{3}(1 + \ln(x))^{\frac{3}{2}}$ + const

47. (a) $x\ln(x) - x$ + const

    (b) $\frac{1}{2}e^{2x}\left(x - \frac{1}{2}\right)$ + const

    (c) $\frac{x^2}{2}\ln(x - 1) - \frac{1}{2}\left(\frac{x^2}{2} + x - \ln(x - 1)\right)$ + const

48. (a) $\frac{14}{3}$

    (b) $-1 - 2^{-2}$

    (c) 0 (Because $x^3\sqrt{x^2 + 1}$ is symmetric with respect to the origin)

49. $f(-2) = f(2) = 0$ and $f(0) = 4$

    The area $= \int\limits_{-2}^{2}(4 - x^2)\,dx = \frac{32}{5}$

50. (a) $f_x = 3x(x + 2y),\ f_y = 3(x^2 - y^2)$

    (b) $f_x = -\frac{y}{x^2},\ f_y = \frac{1}{y}$

    (c) $f_x = \frac{2x}{x^2+y^2}, f_y = \frac{2y}{x^2+y^2}$

    (d) $f_x = (1 - xy)e^{-xy},\ f_y = -x^2e^{-xy}$

51. $df = \frac{x}{\sqrt{x^2+y^2+z^2}}dx + \frac{y}{\sqrt{x^2+y^2+z^2}}dy + \frac{z}{\sqrt{x^2+y^2+z^2}}dz$

52. (a) Homogeneous of degree 3

    (b) Homogeneous of degree 1

53. $\frac{df}{dt} = -(e^t + e^{-t})$

54. (a) $d^2 f = \frac{6y^2}{x^4} dx^2 - \frac{8y}{x^3} dx dy + \frac{2}{x^2} dy^2$

    (b) $d^2 f = -\frac{1}{x} dx^2 + \frac{1}{y} dx dy - \frac{x}{y^2} dy^2$

55. Compute the Hessian matrix:

$$\mathcal{H} = \begin{pmatrix} A\alpha(\alpha-1)x^{\alpha-2}y^\beta & A\alpha\beta x^{\alpha-1}y^{\beta-1} \\ A\alpha\beta x^{\alpha-1}y^{\beta-1} & A\alpha(\beta-1)x^\alpha y^{\beta-2} \end{pmatrix}$$

$f_{xx} = A\alpha(\alpha-1)x^{\alpha-2}y^\beta < 0$ because $0 < \alpha < 1$

$f_{xx}f_{yy} - (f_{xy})^2 = A\alpha\beta x^{2\alpha-2}y^{2\beta-2} \cdot ((\alpha-1)(\beta-1) - \alpha\beta) > 0$ because $(\alpha-1)(\beta-1) - \alpha\beta = 1 - \alpha - \beta > 0$.

56. $(x^* = 0,\ y^* = 3)$ is the minimum of $f(x,y)$.

57. (a) min at $x = -4,\ y = 1$ (for short, at $(-4,1)$)

    (b) Local max at $(4,4)$

    (c) Saddle point at $(1,2)$

    (d) Local max at $(-2,0)$

    (e) Local min at $(0.5, -0.5)$

    Local max at $(-0.5, -0.5)$

    The point $(0,0)$ cannot be classified (Hessian $\equiv 0$ at $(0,0)$)

    (f) Saddle points at $(0,0), (0,1), (1,0), (-1,0)$

    Local min at $(\frac{1}{\sqrt{5}}, 0.4)$

    (g) Local max at (a,a,a)

    The critical point $(0,0,0)$ cannot be classified because at this point all elements of the Hessian matrix are zeroes.

    (h) Local max at $(0,0,1)$ and $(0,0,-1)$

    Local min at $(0,0,0)$

    Saddle points at $(1,0,0), (-1,0,0), (0,1,0), (0,-1,0)$

58. The function in 1(a) is globally convex

59. (a) Local min at $(2,2)$

    Local max at $(-2,-2)$

    (b) Local min at $(1,-1)$ and $(-1,1)$

    Local max at $(1,1)$ and $(-1,-1)$

60. (a) $x_{opt} = \frac{a}{6}$, $y_{opt} = \frac{a}{3}$, $z_{opt} = \frac{a}{6}$, $\lambda = -\frac{a}{3}$

(b) $\mathcal{L} = x + 4y + z - \lambda_1(x^2 + y^2 + z^2 - 216) - \lambda_2(x + 2y + 3z)$

$x_{opt} = \frac{2}{\sqrt{7}}$, $y_{opt} = \frac{32}{\sqrt{7}}$, $z_{opt} = -\frac{22}{\sqrt{7}}$, $\lambda_1 = \frac{1}{4\sqrt{7}}$, $\lambda_2 = \frac{6}{7}$

61. It follows from the constraint that $y \geq 0$. Moreover, $y = 0$ if and only if $x = 0$. Therefore, f attains a global maximum at $(0,0)$.

62. There is no maxima and minima (neither globally nor locally).

63. It follows from the constraint that $x \leq 3$. The function itself is increasing in the interval $(-\infty, 0] \cup [1, 3]$. Moreover, the function is positive only in the interval $\left[\frac{3}{2}, 3\right]$. Therefore, f attains a global maximum at $(3, 0)$. In addition, first-order conditions yield that $(0, \sqrt{27})$ is a local maximum.

64. $(-1, 0), (1, 0)$ are global maxima and $(0, 1), (0, -1)$ are global minima.

65. Direct substitution yields $f(x, y) = \frac{1}{3}x^3 - \frac{3}{2}x^2 + 2x$. Thus $(1,1)$ is a local maximum, $(2,2)$ is a local minimum, and there are no global extrema.

66. Since in the objective function more weight is given to y relative to x, and in the constraint equal weights are attached to x and y, we have a corner solution $(0, 25)$.

67. $(1, 1)$ and $(-1, -1)$ are local minima while local maxima are $(\sqrt{3}, -\sqrt{3})$ and $(-\sqrt{3}, \sqrt{3})$.

68. $(1, 1, 2)$ is a global minimum.

69. $(10, 10, 0)$ is a local minimum.

70. $x^* = 0$ is a local minimum of f.

71. It follows from the constraint that $x \geq 1$. As a result, we have a corner solution $(1, 0)$, which is a global maximum.

72. Bounded solution does not exist since the objective function is monotonically increasing for $x \geq 0$.

73. $\left(\frac{\sqrt{3}}{2}, -\frac{1}{2}\right)$ and $\left(-\frac{\sqrt{3}}{2}, -\frac{1}{2}\right)$.

74. $\left(0, -\frac{1}{2}\right)$ is minimum and $(0, 1)$ maximum.

75. $\left(-\frac{1}{2}, \frac{\sqrt{3}}{2}\right)$ and $\left(-\frac{1}{2}, -\frac{\sqrt{3}}{2}\right)$ are maxima.

76. The firm's revenue maximization problem is

$$\max_{x,y} ax + by \text{ s.t. } \alpha x^2 + \beta y^2 \leq L \Rightarrow x^* = \frac{a\sqrt{L}}{\alpha\mu} \text{ and } y^* = \frac{b\sqrt{L}}{\beta\mu}.$$

5.2 Solved Problems

5.2.1 Linear Algebra

1. Given the system of equations

$$\begin{cases} x_1 + 2x_2 + 4x_3 = -2 \\ x_1 + 3x_2 + 6x_3 = 1 \\ x_1 + 3x_2 + \alpha x_3 = -3 - \beta \end{cases}$$

how many solutions may it have, depending on different values of α and β? Next, solve the system given $\alpha = 7$ and $\beta = 0$.

Solution: Subtracting from the second equation the first equation we will get $x_2 + 2x_3 = 3$. Subtracting from the second equation the third equation we will get $(6 - \alpha)x_3 = 4 + \beta$. Therefore, we can rewrite the given system of equations in the following way:

$$\begin{cases} x_1 = -8 \\ x_2 = -2x_3 + 3 \\ (6 - \alpha)x_3 = 4 + \beta \end{cases}$$

If $\alpha = 6$ and $\beta = -4$, then the system of equations has infinitely many solutions.
If $\alpha = 6$ and $\beta \neq -4$, then the system of equations has no solution.
If $\alpha \neq 6$ and $\beta = -4$, then the system of equations has a unique solution: $x_1 = -8$, $x_2 = 3$, and $x_3 = 0$.
For $\alpha = 7$ and $\beta = 0$: $x_1 = -8$, $x_2 = 11$, and $x_3 = -4$.

2. Find all eigenvalues and corresponding eigenvectors of the matrix

$$A = \begin{pmatrix} 0 & 0 & 0 & 1 \\ 0 & 0 & 1 & 0 \\ 0 & 1 & 0 & 0 \\ 1 & 0 & 0 & 0 \end{pmatrix}$$

Can you find a transformation that diagonalizes A?

<u>Solution:</u> Let us find the eigenvalues of matrix A :

$$\begin{vmatrix} 0 - \lambda_{1,1} & 0 & 0 & 1_{1,4} \\ 0 & 0 - \lambda & 1 & 0 \\ 0 & 1 & 0 - \lambda & 0 \\ 1 & 0 & 0 & 0 - \lambda \end{vmatrix} = 0$$

$$\Longrightarrow (-1)^{1+1}(-\lambda)\begin{vmatrix} -\lambda & 1 & 0 \\ 1 & -\lambda & 0 \\ 0 & 0 & -\lambda \end{vmatrix} + (-1)^{1+4}\begin{vmatrix} 0 & -\lambda & 1 \\ 0 & 1 & -\lambda \\ 1 & 0 & 0 \end{vmatrix} = 0$$

$$\Longrightarrow -\lambda\left[-\lambda^3 + \lambda\right] - \left[\lambda^2 - 1\right] = 0 \Rightarrow (\lambda^2 - 1)^2 = 0.$$

Therefore, the eigenvalues are $\lambda = \pm 1$ each of multiplicity 2.
Now let us find the eigenvectors corresponding to each eigenvalue.
For $\lambda_1 = -1$,

$$v_1 = \begin{pmatrix} \alpha_1 \\ \beta_1 \\ \gamma_1 \\ \delta_1 \end{pmatrix} \Longrightarrow \begin{pmatrix} 1 & 0 & 0 & 1 \\ 0 & 1 & 1 & 0 \\ 0 & 1 & 1 & 0 \\ 1 & 0 & 0 & 1 \end{pmatrix} \cdot \begin{pmatrix} \alpha_1 \\ \beta_1 \\ \gamma_1 \\ \delta_1 \end{pmatrix} = \begin{pmatrix} 0 \\ 0 \\ 0 \\ 0 \end{pmatrix}$$

$$\Longrightarrow \begin{cases} \alpha_1 + \delta_1 = 0 \\ \beta_1 + \gamma_1 = 0 \end{cases}.$$

Similarly, for $\lambda_2 = -1$,

$$\begin{cases} \alpha_2 + \delta_2 = 0 \\ \beta_2 + \gamma_2 = 0 \end{cases}$$

We also need to impose an orthogonality condition between eigenvectors v_1 and v_2, i.e. $\alpha_1\alpha_2 + \beta_1\beta_2 + \gamma_1\gamma_2 + \delta_1\delta_2 = 0$. So we have

5 equations and 8 unknowns. From $\begin{cases} \alpha_1 = -\delta_1 & \& \quad \alpha_2 = -\delta_2 \\ \beta_1 = -\gamma_1 & \& \quad \beta_2 = -\gamma_2 \end{cases}$

it follows that $\delta_1\delta_2 + \gamma_1\gamma_2 = 0$. We can choose values for δ_1, δ_2, γ_1, γ_2 such that $\delta_1\delta_2 + \gamma_1\gamma_2 = 0$ holds:

$$\begin{array}{cccc} \delta_1 = 1 & \gamma_1 = 0 & \gamma_2 = 1 & \delta_2 = 0 \\ \downarrow & \downarrow & \downarrow & \downarrow \\ \alpha_1 = -1 & \beta_1 = 0 & \beta_2 = -1 & \alpha_2 = 0 \end{array}$$

Therefore, $v_1 = \begin{pmatrix} -1 \\ 0 \\ 0 \\ 1 \end{pmatrix}$ and $v_2 = \begin{pmatrix} 0 \\ -1 \\ 1 \\ 0 \end{pmatrix}$. The corresponding

normalized eigenvectors are $e_1 = \begin{pmatrix} -\frac{1}{\sqrt{2}} \\ 0 \\ 0 \\ \frac{1}{\sqrt{2}} \end{pmatrix}$ and $e_2 = \begin{pmatrix} 0 \\ -\frac{1}{\sqrt{2}} \\ \frac{1}{\sqrt{2}} \\ 0 \end{pmatrix}$.

In a similar way, for $\lambda_3 = 1$,

$$v_3 = \begin{pmatrix} \alpha_3 \\ \beta_3 \\ \gamma_3 \\ \delta_3 \end{pmatrix} \implies \begin{pmatrix} -1 & 0 & 0 & 1 \\ 0 & -1 & 1 & 0 \\ 0 & 1 & -1 & 0 \\ 1 & 0 & 0 & -1 \end{pmatrix} \cdot \begin{pmatrix} \alpha_3 \\ \beta_3 \\ \gamma_3 \\ \delta_3 \end{pmatrix} = \begin{pmatrix} 0 \\ 0 \\ 0 \\ 0 \end{pmatrix}$$

$$\implies \begin{cases} -\alpha_3 + \delta_3 = 0 \\ -\beta_3 + \gamma_3 = 0 \end{cases}.$$

Similarly, for $\lambda_4 = 1$,

$$\begin{cases} -\alpha_4 + \delta_4 = 0 \\ -\beta_4 + \gamma_4 = 0 \end{cases}$$

We also need to impose an orthogonality condition between eigenvectors v_3 and v_4, i.e. $\alpha_3\alpha_4 + \beta_3\beta_4 + \gamma_3\gamma_4 + \delta_3\delta_4 = 0$. So we have 5 equations and 8 unknowns. From $\begin{cases} \alpha_3 = \delta_3 & \& \quad \alpha_4 = \delta_4 \\ \beta_3 = \gamma_3 & \& \quad \beta_4 = \gamma_4 \end{cases}$ it follows that $\delta_3\delta_4 + \gamma_3\gamma_4 = 0$. Assuming

$$\begin{array}{cccc} \delta_3 = 1 & \gamma_3 = 0 & \gamma_4 = 1 & \delta_4 = 0 \\ \downarrow & \downarrow & \downarrow & \downarrow \\ \alpha_3 = 1 & \beta_3 = 0 & \beta_4 = 1 & \alpha_4 = 0 \end{array}$$

Therefore, $v_3 = \begin{pmatrix} 1 \\ 0 \\ 0 \\ 1 \end{pmatrix}$ and $v_4 = \begin{pmatrix} 0 \\ 1 \\ 1 \\ 0 \end{pmatrix}$. Finally, the corre-

sponding normalized eigenvectors are $e_3 = \begin{pmatrix} \frac{1}{\sqrt{2}} \\ 0 \\ 0 \\ \frac{1}{\sqrt{2}} \end{pmatrix}$ and $e_4 =$

$\begin{pmatrix} 0 \\ \frac{1}{\sqrt{2}} \\ \frac{1}{\sqrt{2}} \\ 0 \end{pmatrix}$.

Using the derived eigenvalues and corresponding eigenvectors we can write down a spectral decomposition of matrix A: $A = C \Lambda C^{-1}$, where C is an orthogonal matrix whose columns are represented by the normalized eigenvectors and Λ is a diago-

nal matrix. That is, $C = \begin{pmatrix} -\frac{1}{\sqrt{2}} & 0 & \frac{1}{\sqrt{2}} & 0 \\ 0 & -\frac{1}{\sqrt{2}} & 0 & \frac{1}{\sqrt{2}} \\ 0 & \frac{1}{\sqrt{2}} & 0 & \frac{1}{\sqrt{2}} \\ \frac{1}{\sqrt{2}} & 0 & \frac{1}{\sqrt{2}} & 0 \end{pmatrix}$ and $\Lambda =$

$\begin{pmatrix} -1 & 0 & 0 & 0 \\ 0 & -1 & 0 & 0 \\ 0 & 0 & 1 & 0 \\ 0 & 0 & 0 & 1 \end{pmatrix}$. The transformation that diagonalizes ma-

trix A consists of pre-multiplying it by C^T (since matrix C is orthogonal we are using the fact that $C^{-1} = C^T$) and post-multiplying by C, that is, $C^T A C = \Lambda$.

3. Check whether the following vectors are linearly independent:

$$v_1 = (1, 2, 3), \quad v_2 = (4, 5, 6), \quad v_3 = (7, 8, 9)$$

Discuss the tools (methods) you might apply.

Solution: There are several options. For example, one can compute a determinant and if it is zero, the vectors are linearly dependent. A non-zero determinant implies linear independence.

In this case, the determinant is

$$\begin{vmatrix} 1 & 2 & 3 \\ 4 & 5 & 6 \\ 7 & 8 & 9 \end{vmatrix} = \begin{vmatrix} 1 & 2 & 3 \\ 0 & -3 & -6 \\ 0 & -3 & -6 \end{vmatrix} = 0$$

Other possibilities include the upper diagonal form (using elementary row or column operations), finding the rank, or solving a system of linear equations

$$av_1 + bv_2 + cv_3 = 0$$

If this system has a non-zero solution, the vectors are linearly independent.[1]

4. Diagonalize the matrix $\begin{pmatrix} 3 & -1 & 0 \\ -1 & 3 & 0 \\ 0 & 0 & 5 \end{pmatrix}$.

Solution: First, we will compute eigenvalues and eigenvectors

$$\begin{vmatrix} 3-\lambda & -1 & 0 \\ -1 & 3-\lambda & 0 \\ 0 & 0 & 5-\lambda \end{vmatrix} = \left((3-\lambda)^2 - 1\right)(5-\lambda) = 0$$

The latter has a solution $\lambda_1 = 5$, along with the roots of $(3-\lambda)^2 - 1 = 0$ which are $\lambda_2 = 2$, $\lambda_3 = 4$. For $\lambda = 5$, we find eigenvectors as

$$\begin{pmatrix} -2 & -1 & 0 \\ -1 & -2 & 0 \\ 0 & 0 & 0 \end{pmatrix} \begin{pmatrix} x_1 \\ x_2 \\ x_3 \end{pmatrix} = \begin{pmatrix} 0 \\ 0 \\ 0 \end{pmatrix}$$

which has infinitely many solutions $x_3 = t$, $x_2 = 0$, $x_1 = 0$. For $\lambda = 2$, eigenvectors are solutions to the equation

$$\begin{pmatrix} 1 & -1 & 0 \\ -1 & 1 & 0 \\ 0 & 0 & 3 \end{pmatrix} \begin{pmatrix} x_1 \\ x_2 \\ x_3 \end{pmatrix} = \begin{pmatrix} 0 \\ 0 \\ 0 \end{pmatrix}$$

---

[1] You could, in principle, use the eigenvalues to check the linear independence. If at least one of the eigenvalues is 0, then the vectors are linearly independent. This, however, typically requires more work than just computing a determinant, for example.

which are $x_3 = 0$, $x_2 = q$, $x_3 = q$. For the remaining eigenvalue $\lambda = 4$, corresponding eigenvectors

$$
\begin{pmatrix} -1 & -1 & 0 \\ -1 & -1 & 0 \\ 0 & 0 & 1 \end{pmatrix} \begin{pmatrix} x_1 \\ x_2 \\ x_3 \end{pmatrix} = \begin{pmatrix} 0 \\ 0 \\ 0 \end{pmatrix}
$$

are $x_3 = 0$, $x_2 = s$, $x_3 = -s$. We have to normalize eigenvectors to obtain orthonormal vectors

$$
\begin{pmatrix} 0 \\ 0 \\ 1 \end{pmatrix}, \quad \begin{pmatrix} \frac{1}{\sqrt{2}} \\ \frac{1}{\sqrt{2}} \\ 0 \end{pmatrix}, \quad \begin{pmatrix} -\frac{1}{\sqrt{2}} \\ \frac{1}{\sqrt{2}} \\ 0 \end{pmatrix}
$$

The spectral theorem implies that the original matrix can be decomposed as follows:

$$
\begin{pmatrix} 3 & -1 & 0 \\ -1 & 3 & 0 \\ 0 & 0 & 5 \end{pmatrix} =
$$

$$
= \begin{pmatrix} 0 & \frac{1}{2}\sqrt{2} & -\frac{1}{2}\sqrt{2} \\ 0 & \frac{1}{2}\sqrt{2} & \frac{1}{2}\sqrt{2} \\ 1 & 0 & 0 \end{pmatrix} \begin{pmatrix} 5 & 0 & 0 \\ 0 & 2 & 0 \\ 0 & 0 & 4 \end{pmatrix} \begin{pmatrix} 0 & 0 & 1 \\ \frac{1}{\sqrt{2}} & \frac{1}{\sqrt{2}} & 0 \\ -\frac{1}{\sqrt{2}} & \frac{1}{\sqrt{2}} & 0 \end{pmatrix}
$$

5. Let matrix $A = \begin{pmatrix} 25 & 7 \\ 7 & 13 \end{pmatrix}$. Find the vector x that minimizes $y = x'Ax + 2x_1 + 3x_2 - 10$. What is the value of the objective function at the minimum? Now, minimize y s.t. $x_1 + x_2 = 1$. Further, how do y and optimal x_1 and x_2 change if the constraint marginally changes?

<u>Solution:</u>   If unsure, you can just plug in for $x = (x_1, x_2)$, multiply and proceed:

$$
\begin{pmatrix} 25x_1 + 7x_2, & 7x_1 + 13x_2 \end{pmatrix} \begin{pmatrix} x_1 \\ x_2 \end{pmatrix} =
$$

$$
= (25x_1 + 7x_2)\, x_1 + (7x_1 + 13x_2)\, x_2,
$$

therefore

$$y = 25x_1^2 + 13x_2^2 + 14x_2x_1 + 2x_1 + 3x_2 - 10$$

which leads to the first order conditions[2,3]

$$50x_1 + 14x_2 + 2 = 0$$
$$14x_1 + 26x_2 + 3 = 0$$

The solution is evaluated as $\left[x_1 = -\dfrac{5}{552}, x_2 = -\dfrac{61}{552} \right]$. The value of the objective function is

$$\left(-\frac{5}{552}, -\frac{61}{552} \right) \left(\begin{array}{cc} 25 & 7 \\ 7 & 13 \end{array} \right) \left(\begin{array}{c} -\frac{5}{552} \\ -\frac{61}{552} \end{array} \right) +$$

$$+ \left(-\frac{5}{552}, -\frac{61}{552} \right) \left(\begin{array}{c} 2 \\ 3 \end{array} \right) - 10$$

Solving the problem subject to the constraint $x_1 + x_2 = a$ using a Lagrangian function

$$L = 25x_1^2 + 13x_2^2 + 14x_2x_1 + 2x_1 + 3x_2 - 10 + \lambda(a - x_1 - x_2)$$

---

[2] Note that A is positive definite, therefore the second-order conditions for minimum are satisfied indeed.

[3] Alternatively, we can differentiate according to the vector to get

$$2 \left(\begin{array}{cc} 25 & 7 \\ 7 & 13 \end{array} \right) \left(\begin{array}{c} x_1 \\ x_2 \end{array} \right) + \left(\begin{array}{c} 2 \\ 3 \end{array} \right) = \left(\begin{array}{c} 0 \\ 0 \end{array} \right)$$

or

$$\left(\begin{array}{c} x_1 \\ x_2 \end{array} \right) = \left(\begin{array}{cc} 25 & 7 \\ 7 & 13 \end{array} \right)^{-1} \left(\begin{array}{c} -1 \\ -3/2 \end{array} \right)$$

Thus

$$\left(\begin{array}{c} x_1 \\ x_2 \end{array} \right) = \frac{1}{25*13 - 49} \left(\begin{array}{cc} 13 & -7 \\ -7 & 25 \end{array} \right) \left(\begin{array}{c} -1 \\ -3/2 \end{array} \right) = \left(\begin{array}{c} -\frac{5}{552} \\ -\frac{61}{552} \end{array} \right)$$

leads to the following first-order conditions

$$50x_1 + 14x_2 + 2 - \lambda = 0$$
$$26x_2 + 14x_1 + 3 - \lambda = 0$$
$$x_1 + x_2 - a = 0$$

which has the solution[4]

$$\left[\lambda = 23a + \frac{11}{4}, x_1 = \frac{1}{4}a + \frac{1}{48}, x_2 = \frac{3}{4}a - \frac{1}{48} \right]$$

Taking $a = 1$, we obtain $x_1 = \frac{13}{48}$, $x_2 = \frac{35}{48}$. Moreover, comparative statics on a gives us

$$\frac{\partial x_1}{\partial a} = \frac{1}{4}, \quad \frac{\partial x_2}{\partial a} = \frac{3}{4}$$

Therefore, if we marginally relax the constraint by increasing a, the first component increases by $1/4$ of the amount by which we relaxed the constraint and the second one by $3/4$ of that amount.

Furthermore, from the envelope theorem we know that the marginal change of the optimal value function $y(x_1^*, x_2^*, a)$ with respect to a marginal change in a is equal to λ (the shadow price of a).

We can also find how the value of objective function changes by computing

$$\frac{\partial L}{\partial a} = 50x_1\frac{\partial x_1}{\partial a} + 26x_2\frac{\partial x_2}{\partial a} + 14(x_2\frac{\partial x_1}{\partial a} + x_1\frac{\partial x_2}{\partial a}) + 2\frac{\partial x_1}{\partial a} + 3\frac{\partial x_2}{\partial a}$$

and plugging in for x_1, x_2, $\frac{\partial x_1}{\partial a}$, $\frac{\partial x_2}{\partial a}$, and $a = 1$.

6. Prove that $(I_m + AB)^{-1} = I_m - A(I_n + BA)^{-1}B$, where I_m is $m \times m$ unit matrix, I_n is $n \times n$ unit matrix, A is $m \times n$, B is $n \times m$, and $|I_m + AB| \neq 0$.

---

[4] Note that we are not interested in λ per se, but its expression can be useful to check the solution.

Solution: Let us multiply $(I_m + AB)^{-1} = I_m - A(I_n + BA)^{-1}B$ by $(I_m + AB)$ from the left. Then $LHS = I_m$, and

$$
\begin{aligned}
\text{RHS} &= (I_m + AB)(I_m - A(I_n + BA)^{-1}B) \\
&= I_m + AB - A(I_n + BA)^{-1}B - ABA(I_n + BA)^{-1}B \\
&= I_m + AB - A((I_n + BA)^{-1} - BA(I_n + BA)^{-1})B \\
&= I_m + AB - A((I_n + BA)^{-1}(I_n + BA))B \\
&= I_m + AB - AB = I_m,
\end{aligned}
$$

i.e. $LHS = RHS$.

5.2.2   Calculus

1. Show that for any sufficiently differentiable functions u and v the function $\phi(x, y) = u[x + v(y)]$ solves the following equation in partial derivatives: $\frac{\partial \phi}{\partial x} \frac{\partial^2 \phi}{\partial x \partial y} = \frac{\partial \phi}{\partial y} \frac{\partial^2 \phi}{\partial x^2}$.

   Solution: First we will find expressions for the first partial derivative with respect to x $\left(\frac{\partial \phi}{\partial x}\right)$, first partial derivative with respect to y $\left(\frac{\partial \phi}{\partial y}\right)$, second partial derivative with respect to x $\left(\frac{\partial^2 \phi}{\partial x^2}\right)$, and mixed partial derivative $\left(\frac{\partial^2 \phi}{\partial x \partial y}\right)$:

$$
\begin{aligned}
\frac{\partial \phi}{\partial x} &= u' \cdot 1, & \frac{\partial^2 \phi}{\partial x^2} &= u'', \\
\frac{\partial \phi}{\partial y} &= u' \cdot v' \cdot 1, & \frac{\partial^2 \phi}{\partial x \partial y} &= u'' \cdot v'.
\end{aligned}
$$

   Direct substitution into $\dfrac{\partial \phi}{\partial x} \dfrac{\partial^2 \phi}{\partial x \partial y} = \dfrac{\partial \phi}{\partial y} \dfrac{\partial^2 \phi}{\partial x^2}$ leads to $u' \cdot u'' v' = u' v' \cdot u''$.

2. Evaluate $\int_0^1 \ln(1 + \sqrt{x}) \, dx$. (HINT: Substitute $y = 1 + \sqrt{x}$ and integrate by parts.)

   Solution:

$$
\int\limits_0^1 \ln(1 + \sqrt{x}) \, dx \;=\; \begin{vmatrix} y = 1 + \sqrt{x} \\ x = 0 \Rightarrow y = 1 \\ x = 1 \Rightarrow y = 2 \\ x = (y-1)^2 \end{vmatrix} \;=\; \int\limits_1^2 \ln y \, d(y-1)^2
$$

[Integrating by parts:

$$\int u\,dv = uv - \int v\,du]$$

$$= \ln y \cdot (y-1)^2\big|_1^2 - \int_1^2 (y-1)^2 \cdot \frac{1}{y}\,dy$$

$$= \ln 2 - \int_1^2 (y-2+\frac{1}{y})\,dy$$

$$= \ln 2 - (\frac{1}{2}y^2 - 2y + \ln y)\Big|_1^2$$

$$= \ln 2 - \left(\frac{1}{2}\cdot 2^2 - 2\cdot 2 + \ln 2\right) +$$

$$+ \left(\frac{1}{2}\cdot 1^2 - 2\cdot 1 + \ln 1\right)$$

$$= \frac{1}{2}.$$

3. Consider a production function Y that depends on three inputs, labor L, physical capital K, and human capital H: $Y(L, K, H) = L^\alpha K^\beta H^\gamma$. Is Y homogeneous? If yes, of what degree? What are the conditions to make Y strictly concave?

Solution: Yes, the function is homogeneous of degree $\alpha + \beta + \gamma$. Let us derive this result: $Y(tL, tK, tH) = (tL)^\alpha (tK)^\beta (tH)^\gamma = t^{\alpha+\beta+\gamma} L^\alpha K^\beta H^\gamma$. That is, it follows that

$$Y(tL, tK, tH) = t^{\alpha+\beta+\gamma} Y(L, K, H).$$

In order to derive conditions making Y strictly concave we will construct a Hessian. First, compute

$$\frac{\partial Y}{\partial L} = \alpha L^{\alpha-1} K^\beta H^\gamma,$$

$$\frac{\partial Y}{\partial K} = \beta L^\alpha K^{\beta-1} H^\gamma,$$

$$\frac{\partial Y}{\partial H} = \gamma L^\alpha K^\beta H^{\gamma-1},$$

$$\frac{\partial^2 Y}{\partial L^2} = \alpha(\alpha - 1)\frac{Y}{L^2},$$

$$\frac{\partial^2 Y}{\partial K^2} = \beta(\beta - 1)\frac{Y}{K^2},$$

$$\frac{\partial^2 Y}{\partial H^2} = \gamma(\gamma - 1)\frac{Y}{H^2},$$

$$\frac{\partial^2 Y}{\partial L \partial K} = \alpha\beta\frac{Y}{LK},$$

$$\frac{\partial^2 Y}{\partial L \partial H} = \alpha\gamma\frac{Y}{K^2},$$

$$\frac{\partial^2 Y}{\partial K \partial H} = \beta\gamma\frac{Y}{KH}.$$

Let us construct the corresponding Hessian matrix:

$$\begin{pmatrix} \alpha(\alpha-1)\frac{Y}{L^2} & \alpha\beta\frac{Y}{LK} & \alpha\gamma\frac{Y}{K^2} \\ \alpha\beta\frac{Y}{LK} & \beta(\beta-1)\frac{Y}{K^2} & \beta\gamma\frac{Y}{KH} \\ \alpha\gamma\frac{Y}{K^2} & \beta\gamma\frac{Y}{KH} & \gamma(\gamma-1)\frac{Y}{H^2} \end{pmatrix} =$$

$$= \frac{Y}{L^2 K^2 H^2}\begin{pmatrix} \alpha(\alpha-1)\cdot K^2 H^2 & \alpha\beta\cdot LKH^2 & \alpha\gamma\cdot LK^2 H \\ \alpha\beta\cdot LKH^2 & \beta(\beta-1)\cdot L^2 H^2 & \beta\gamma\cdot L^2 KH \\ \alpha\gamma\cdot LK^2 H & \beta\gamma\cdot L^2 KH & \gamma(\gamma-1)\cdot L^2 K^2 \end{pmatrix}$$

Now we need to impose conditions on positive α, β, and γ to have principal minors alternating in sign starting from minus (negative definite Hessian → strict concavity condition).

$$\alpha(\alpha - 1) < 0 \Leftrightarrow \alpha \in (0, 1);$$

$$\alpha\beta(\alpha - 1)(\beta - 1) - (\alpha\beta)^2 > 0 \Leftrightarrow \alpha + \beta \in (0, 1);$$

$$\alpha\beta\gamma(\alpha - 1)(\beta - 1)(\gamma - 1) + \alpha^2\beta^2\gamma^2 + \alpha^2\beta^2\gamma^2 -$$

$$- \left[\alpha^2\beta(\beta - 1)\gamma^2 + \alpha(\alpha - 1)\beta^2\gamma^2 + \alpha^2\beta^2\gamma(\gamma - 1)\right] < 0 \Leftrightarrow$$

$$\Leftrightarrow \alpha + \beta + \gamma \in (0, 1).$$

So, the conditions for positive α, β, and γ which make Y strictly concave are $\alpha + \beta + \gamma \in (0, 1)$. Economically, this means that Y has to exhibit decreasing returns to scale.

4. What is the maximum value of a function $f(x) = 3x^3 - 5x^2 + x$, subject to $x \in [a, b]$, where a and b are given parameters? To answer, find and classify all the critical points of unrestricted $f(x)$, plot the graph of $f(x)$, and apply Weierstrass's theorem.[5]

Solution: To find the critical points, we equalize the first derivative to zero:

$$9x^2 - 10x + 1 = 0$$

which has a solution

$$x_1 = \frac{1}{9}, \quad x_2 = 1$$

Using second-order condition

$$18x - 10 \quad < \quad 0 \text{ for } x_1 = \frac{1}{9},$$
$$18x - 10 \quad > \quad 0 \text{ for } x_2 = 1,$$

we can conclude that the function has a local maximum in x_1 and a local minimum in x_2. The function is concave up to $\frac{5}{9}$ and concave from there on. Its limits are

$$\lim_{x \to -\infty} f(x) = -\infty, \quad \lim_{x \to \infty} f(x) = \infty$$

and it does not have asymptotes, because

$$\lim_{x \to \pm\infty} \frac{f(x)}{x} = \lim_{x \to \pm\infty} x^2 = \infty$$

---

[5] Weierstrass's theorem says that a continuous function on a closed and bounded subset of R^1 reaches there its maximum and minimum.

Since a and b are unknown parameters, we have to distinguish several possibilities

Location of local extrema $x_1 = \frac{1}{9}, x_2 = 1$	Maximum value
$a < b < x_1$	b
$a < x_1 < b < x_2$	x_1
$x_2 < a < b < x_1$	a
$x_1 < a < x_2 < b$	$\arg\max\{f(a), f(b)\}$
$x_2 < a < b$	b
$a < x_1 < x_2 < b$	$\arg\max\{f(b), f(x_1)\}$

In all cases except $x_1 < a < x_2 < b$ and $a < x_1 < x_2 < b$ we can tell where the maximum lies. In the exceptional cases we have to compare the value in two corners a, b and x_1.

5. Compute $\int\int (x^2 + y^2 + 1)^{-3/2} dx dy$ over R^2.

Solution: The presence of $x^2 + y^2$ hints that it may be helpful to switch to polar coordinates by using the substitution

$$x = r\sin\theta$$
$$y = r\cos\theta$$

Using the formula for substitution in integrals[6]

$$\int \cdots \int f(x_1, \ldots, x_n) dx_1 \cdots dx_n =$$

$$= \int \cdots \int f(\varphi_1(\mathbf{z}), \ldots, \varphi_n(\mathbf{z})) |Det(D\varphi(\mathbf{z}))| dz_1 \cdots dz_n$$

where $\mathbf{z} = (z_1, \ldots, z_n)$, $|Det(D\varphi(\mathbf{z}))|$ is the absolute value for a determinant of a Jacobian matrix $\left(\dfrac{\partial \varphi_i}{\partial z_j}\right)$, and

$$x_i = \varphi_i(z_1, \ldots, z_n), \quad i = 1, \ldots, n$$

---

[6] If you are computing the definite integral, do not forget to also transform the boundaries of integration!

are the substitutions. In our case, the determinant is

$$\begin{vmatrix} \sin\theta & r\cos\theta \\ \cos\theta & -r\sin\theta \end{vmatrix} = -r$$

and its absolute value is r. Thus, the integral after the substitution becomes

$$
\begin{aligned}
I &= \int_{-\infty}^{\infty}\int_{-\infty}^{\infty}(x^2+y^2+1)^{-3/2}dxdy \\
&= \lim_{A\to\infty}\int_{-A}^{A}\int_{-A}^{A}(x^2+y^2+1)^{-3/2}dxdy \\
&= \lim_{A\to\infty}\int_{0}^{2\pi}\int_{0}^{A}(r^2+1)^{-3/2}rdrd\theta.
\end{aligned}
$$

Using the substitution

$$
\begin{aligned}
t &= r^2+1 \\
dt &= 2rdr
\end{aligned}
$$

we get[7]

$$
\begin{aligned}
2\pi\frac{1}{2}\int_{1}^{\infty}t^{-3/2}dt &= \lim_{A\to\infty}2\pi\frac{1}{2}\int_{1}^{A}t^{-3/2}dt \\
&= \lim_{A\to\infty}-2\pi[t^{-1/2}]_{1}^{A} \\
&= 2\pi
\end{aligned}
$$

6. Let $f(x)$ be strictly concave. Find optimum of $f(x_1)+\ldots+f(x_n)$, s.t. $x_1+\ldots+x_n=1$.

---

[7] If you want to compute an indefinite integral, simply substitute back. Expressing r and θ as functions of x and y

$$
\begin{aligned}
r^2 &= x^2+y^2 \\
\theta &= \arctan\frac{x}{y}
\end{aligned}
$$

we get $I = \dfrac{-\arctan\frac{x}{y}}{\sqrt{x^2+y^2+1}}$

<u>Solution:</u> Intuitively, this problem can be answered as follows: Since the objective function $F(x_1, \ldots, x_n) = f(x_1) + \ldots + f(x_n)$ and the constraint are both symmetric in arguments $x_1, \ldots, x_n$, the optimal solution should be symmetric as well, i.e. $x_1^* = \ldots = x_n^* = \frac{1}{n}$.

More formally, let use the Langrangian function

$$L(x_1, x_2, \ldots, x_n, \lambda) = f(x_1) + \ldots + f(x_n) +$$

$$+ \lambda(1 - x_1 - \ldots - x_n).$$

The first-order conditions are

$$\frac{\partial f(x_i)}{\partial x_i} = \lambda = \frac{\partial f(x_j)}{\partial x_j}$$

for all $i \neq j$. Moreover

$$x_1 + \ldots + x_n = 1.$$

Since $f(x)$ is strictly concave,[8] we know that $x_i = x_j$ which implies

$$x_1 = x_2 = \ldots = x_n = \frac{1}{n}$$

The optimum is then $nf\left(\dfrac{1}{n}\right)$

7. Given the implicit function $y = x + \ln y$, find $\dfrac{d^2 y}{dx^2}$.

---

[8] In fact, strict concavity does not guarantee that $\dfrac{\partial^2 f(x)}{\partial x} < 0$ but only $\dfrac{\partial^2 f(x)}{\partial x} \leq 0$ with equality in isolated points. However, if there would be an (small) interval in which $\dfrac{\partial^2 f(x)}{\partial x} = 0$ the function f would not be strictly concave on this interval. Thus, the equation $\dfrac{\partial f(x)}{\partial x} = c$ has at most one solution for any c.

Solution: By the Implicit Function Theorem, $\dfrac{dy}{dx} = \dfrac{y}{y-1} =$

$1 + \dfrac{1}{y-1}$, for all $y \neq 1$.

$$\dfrac{d^2y}{dx^2} = \dfrac{d}{dx}\dfrac{dy}{dx} = -\dfrac{1}{(y-1)^2}\dfrac{dy}{dx} = \dfrac{y}{(y-1)^3}, \ y \neq 1.$$

8. Prove or disprove: $\dfrac{e^x + e^{-x}}{2} > 1 + \dfrac{x^2}{2}$ for all $x \neq 0$.

Solution: TRUE. Just note that Taylor expansion of $\dfrac{e^x + e^{-x}}{2}$

is $1 + \dfrac{x^2}{2} + \dfrac{x^4}{4} + \dfrac{x^6}{6} + \ldots > 1 + \dfrac{x^2}{2} \ \forall x \neq 0$. (Alternatively, you can check that the global minimum of the function $e^x + e^{-x} - x^2 - 2$ is 0 and reached at $x = 0$.

9. Find extrema (if any) of the function $z(x,y) = 1 - \sqrt{x^2 + y^2}$ and classify
   them.

   Solution: F.O.C. have no solution. However, $z(x,y)$ is not differentiable at the origin, therefore the point $(0,0)$ has to be checked separately. It is clear that $\sqrt{x^2 + y^2} > 0 \ \forall (x,y) \neq (0,0)$. Therefore $z(0,0) > z(x,y) \ \forall (x,y) \neq (0,0)$, and $(0,0)$ is a global maximum.

10. Evaluate the following integrals:

a) $\int (-x)\sqrt{1-x^2}dx = |^{1-x^2=z}_{-2xdx=dz}| = \int \dfrac{\sqrt{z}}{2}dz = \dfrac{\sqrt{z^3}}{3} + c = \dfrac{\sqrt{(1-x^2)^3}}{3} +$
   c

b) $\int (2x+1)\sqrt{x^2+x}dx = |^{x^2+x=z}_{(2x+1)dx=dz}| = \int \sqrt{z}dz = \dfrac{2}{3}z^{\frac{3}{2}} + c =$
   $= \dfrac{2}{3}(x^2+x)^{\frac{3}{2}} + c$

c) $\int (x^3+1)^2 dx = \int x^6 + 2x^3 + 1 dx = \dfrac{x^7}{7} + \dfrac{x^4}{2} + x + c$

d) $\int \dfrac{x+1}{\sqrt{x^2+2x+3}}dx = |^{x^2+2x+3=z}_{2(x+1)dx=dz}| = \int \dfrac{1}{2\sqrt{z}}dz = \sqrt{z} + c =$
   $= \sqrt{x^2 + 2x + 3} + c$

e) $\int \dfrac{1}{(x-3)^3}dx = |^{x-3=z}_{dx=dz}| = \int z^{-3}dz = -\dfrac{z^{-2}}{2} + c = \dfrac{1}{-2(x-3)^2} + c$

f) $\int (2x-3)^{1/3}dx = |^{2x-3=z}_{2dx=dz}| = \int \dfrac{z^{\frac{1}{3}}}{2}dz = \dfrac{3}{8}z^{\frac{4}{3}} + c = \dfrac{3}{8}(2x-3)^{\frac{4}{3}} + c$

g) $\int\limits_{1}^{3}(x^3+2)\mathrm{d}x = \left[\frac{x^4}{4}+2x\right]_{1}^{3} = \frac{81}{4}+6-\frac{1}{4}-2 = 24$

h) $\int\limits_{-3}^{2}(3x^2+2x+1)\mathrm{d}x = \left[x^3+x^2+x\right]_{-3}^{2} = 8+4+2-(-27+9-3) =$
35

i) $\int\limits_{0}^{1}5(x^2+1)^4 2x\mathrm{d}x = |_{2x\mathrm{d}x=\mathrm{d}z}^{1+x^2=z}| = \int\limits_{1}^{2}5z^4\mathrm{d}z = \left[z^5\right]_{1}^{2} = 31$

j) $\int\limits_{2}^{5}\frac{1}{\sqrt{x-1}}\mathrm{d}x = |_{\mathrm{d}x=\mathrm{d}z}^{x-1=z}| = \int\limits_{1}^{4}z^{-\frac{1}{2}}\mathrm{d}z = \left[2\sqrt{z}\right]_{1}^{4} = 2$

k) $\int\limits_{0}^{1}x\sqrt{1-x^2}\mathrm{d}x = |_{-2x\mathrm{d}x=\mathrm{d}z}^{1-x^2=z}| = \int\limits_{1}^{0}-\frac{1}{2}z^{\frac{1}{2}}\mathrm{d}z = \int\limits_{0}^{1}\frac{1}{2}z^{\frac{1}{2}}\mathrm{d}z =$
$= \left[\frac{1}{3}z^{\frac{3}{2}}\right]_{0}^{1} = \frac{1}{3}$

l) $\int\limits_{0}^{1}(2x-1)\sqrt{x-x^2}\mathrm{d}x = |_{(1-2x)\mathrm{d}x=\mathrm{d}z}^{x-x^2=z}| = \int\limits_{0}^{0}-z^{\frac{1}{2}}\mathrm{d}z = 0$

m) $\int\limits_{a}^{b}\ln x\mathrm{d}x = |_{v'=1}^{u=\ln x}| = \left[x\ln x\right]_{a}^{b} - \int\limits_{a}^{b}1\mathrm{d}x = b\ln b - a\ln a - b + a$

n) $\int\limits_{0}^{2}e^x x^2\mathrm{d}x = |_{v'=e^x}^{u=x^2}| = \left[x^2 e^x\right]_{0}^{2} - 2\int\limits_{0}^{2}xe^x\mathrm{d}x = |_{v'=e^x}^{u=x}| =$
$= \left[x^2 e^x\right]_{0}^{2} - 2\left\{\left[xe^x\right]_{0}^{2} - \int\limits_{0}^{2}e^x\mathrm{d}x\right\} = 2e^2 - 2$

11. Find partial derivatives of the function $f(\cdot)$ with respect to all variables:

a) $f(x,y) = x^3 + 5x^2 y + 3xy^2 + 2y^3$
$\dfrac{\partial f}{\partial x} = 3x^2 + 10xy + 3y^2$
$\dfrac{\partial f}{\partial y} = 5x^2 + 6xy + 6y^2$

b) $f(x,y) = \dfrac{x+y}{x-y}$
$\dfrac{\partial f}{\partial x} = \dfrac{x-y-x-y}{(x-y)^2} = \dfrac{-2y}{(x-y)^2}$
$\dfrac{\partial f}{\partial y} = \dfrac{x-y+x+y}{(x-y)^2} = \dfrac{2x}{(x-y)^2}$

c) $f(x, y, z) = x^3 y z^2$

$$\frac{\partial f}{\partial x} = 3x^2 y z^2$$

$$\frac{\partial f}{\partial y} = x^3 z^2$$

$$\frac{\partial f}{\partial z} = 2x^3 y z$$

d) $f(x, y, z) = e^{xyz} \sin x \cos y$

$$\frac{\partial f}{\partial x} = yz e^{xyz} \sin x \cos y + e^{xyz} \cos x \cos y$$

$$\frac{\partial f}{\partial y} = xz e^{xyz} \sin x \cos y - e^{xyz} \sin x \sin y$$

$$\frac{\partial f}{\partial z} = xy e^{xyz} \sin x \cos y$$

12. Find the first- and second-order total differentials of the function $f(\cdot)$:

a) $f(x, y) = x^2 y^2$
   $\mathrm{d}f = 2xy^2 \mathrm{d}x + 2x^2 y \mathrm{d}y$
   $\mathrm{d}^2 f = 2y^2 \mathrm{d}x^2 + 8xy \mathrm{d}x \mathrm{d}y + 2x^2 \mathrm{d}y^2$

b) $f(x, y) = y^x$
   $\mathrm{d}f = y^x \ln y \mathrm{d}x + xy^{x-1} \mathrm{d}y$
   $\mathrm{d}^2 f = y^x (\ln y)^2 \mathrm{d}x^2 + 2(xy^{x-1} \ln y + y^{x-1}) \mathrm{d}x \mathrm{d}y + x(x-1)y^{x-2} \mathrm{d}y^2$

13. Find $\dfrac{\partial u}{\partial s}$ and $\dfrac{\partial u}{\partial t}$:

a) $u = x^2 - y^2, \quad x = s + 2t, \quad y = 3s - t$
   $$\frac{\partial u}{\partial s} = 2x - 6y$$
   $$\frac{\partial u}{\partial t} = 4x + 2y$$

b) $u = \dfrac{x}{y}, \quad x = s^2 + t, \quad y = t^2 + s$
   $$\frac{\partial u}{\partial x} = \frac{1}{y}$$
   $$\frac{\partial u}{\partial y} = -\frac{x}{y^2}$$
   $$\frac{\partial u}{\partial s} = \frac{1}{y} 2s - \frac{x}{y^2} = \frac{2sy - x}{y^2}$$
   $$\frac{\partial u}{\partial t} = \frac{1}{y} - \frac{x}{y^2} 2t = \frac{y - 2tx}{y^2}$$

14. Find all stationary points of the function $f(x,y)$ and characterize them:

a) $f(x,y) = x^3 + y^3 - 3xy$

Solution: The first-order necessary conditions (F.O.C.) are

$$\frac{\partial f}{\partial x} = 3x^2 - 3y = 0,$$

$$\frac{\partial f}{\partial y} = 3y^2 - 3x = 0.$$

From the F.O.C. we find two stationary points $(0,0)$ and $(1,1)$. To classify them let us compute the second-order patrials

$$\frac{\partial^2 f}{\partial^2 x} = 6x, \qquad \frac{\partial^2 f}{\partial^2 y} = 6y, \qquad \frac{\partial^2 f}{\partial x \partial y} = -3$$

to form the Hessian matrix: $H = \begin{pmatrix} 6x & -3 \\ -3 & 6y \end{pmatrix}$.

At the stationary point $(0,0)$,

$$H = \begin{pmatrix} 6x & -3 \\ -3 & 6y \end{pmatrix} \Big|_{(0,0)} = \begin{pmatrix} 0 & -3 \\ -3 & 0 \end{pmatrix}.$$

Leading principle minors of the Hessian matrix at $(0,0)$ are $|H_1| = 0$ and $|H_2| = -9$. Therefore, $(0,0)$ is a saddle point. At $(1,1)$,

$$H = \begin{pmatrix} 6x & -3 \\ -3 & 6y \end{pmatrix} \Big|_{(1,1)} = \begin{pmatrix} 6 & -3 \\ -3 & 6 \end{pmatrix} \Rightarrow$$

$\Rightarrow |H_1| = 6$ and $|H_2| = 27 \Rightarrow (1,1)$ is a local minimum.

b) $f(x,y) = xy + \frac{2}{x} + \frac{4}{y}$

Solution: The F.O.C.

$$\frac{\partial f}{\partial x} = y - 2x^{-2} = 0$$

$$\frac{\partial f}{\partial y} = x - 4y^{-2} = 0$$

yield a unique stationary point $(1,2)$.

Given the S.O.C.

$$\frac{\partial^2 f}{\partial^2 x} = 4x^{-3}, \qquad \frac{\partial^2 f}{\partial^2 y} = 8y^{-3}, \qquad \frac{\partial^2 f}{\partial x \partial y} = 1,$$

we construct the Hessian matrix as $\quad H = \begin{pmatrix} 4x^{-3} & 1 \\ 1 & 8y^{-3} \end{pmatrix}.$

At the stationary point,

$$H = \begin{pmatrix} 4x^{-3} & 1 \\ 1 & 8y^{-3} \end{pmatrix}\Big|_{(1,2)} = \begin{pmatrix} 4 & 1 \\ 1 & 1 \end{pmatrix} \Rightarrow$$

$\Rightarrow |H_1| = 1 \quad$ and $\quad |H_2| = 3 \Rightarrow (1,2)$ is a local minimum.

c) $f(x, y) = xye^{-(3x+2y)}$

Solution: The F.O.C.:

$$\frac{\partial f}{\partial x} = ye^{-(3x+2y)} - 3xye^{-(3x+2y)} = 0,$$

$$\frac{\partial f}{\partial y} = xe^{-(3x+2y)} - 2xye^{-(3x+2y)} = 0.$$

From the F.O.C. we derive two stationary points $(0,0)$ and $(\frac{1}{3}, \frac{1}{2})$. The second-order necessary conditions written directly into the Hessian matrix with $Q = e^{-(3x+2y)}$ are

$$\begin{aligned} H &= \begin{pmatrix} (-3y - 3y + 3xy)Q & (1 - 3x - 2y + 6xy)Q \\ (1 - 3x - 2y + 6xy)Q & (-2x - 2x + 4xy)Q \end{pmatrix} \\ &= Q \begin{pmatrix} -6y - 9xy & 1 - 3x - 2y + 6xy \\ 1 - 3x - 2y + 6xy & -4x + 4xy \end{pmatrix} \end{aligned}$$

At $(0,0)$, $H|_{(0,0)} = \begin{pmatrix} 0 & 1 \\ 1 & 0 \end{pmatrix} \Rightarrow$

$\Rightarrow |H_1| = 0 \quad$ and $\quad |H_2| = -1 \neq 0 \Rightarrow (0,0)$ is a saddle point;

at $(1/3, 1/2)$, $H|_{(\frac{1}{3}, \frac{1}{2})} = e^{-2} \begin{pmatrix} -\frac{3}{2} & 0 \\ 0 & -\frac{2}{3} \end{pmatrix} \Rightarrow$

$\Rightarrow \quad |H_1| = -\frac{3}{2}e^{-2} \quad$ and $\quad |H_2| = e^{-2} > 0$

$\Rightarrow (\frac{1}{3}, \frac{1}{2})$ is a local maximum.

15. Find whether the following functions are convex or concave:

a) $f(x, y) = x^2 + y^2$

Solution: We need to write the Hessian matrix of the second-order partial derivatives:

$$H = \begin{pmatrix} 2 & 0 \\ 0 & 2 \end{pmatrix}$$

Since this matrix is positive definite ($|H_1| = 2$ and $|H_2| = 4$), $f(x, y)$ is strictly convex.

b) $f(x, y) = -x^2 - y^2 + xy$

Solution: The Hessian matrix is

$$H = \begin{pmatrix} -2 & 1 \\ 1 & -2 \end{pmatrix}$$

H is negative definite because $|H_1| = -2$ and $|H_2| = 3$. Therefore, $f(x, y)$ is strictly concave.

16. Apply the implicit function theorem to derive $\dfrac{dx}{dy}$:

a) $x^2 + y^2 = 1$

Solution:

$$\left. \begin{array}{l} F'_y = 2y \\ F'_x = 2x \end{array} \right\} \Rightarrow \frac{dy}{dx} = -\frac{x}{y}$$

b) $xe^x = y^2 + xy$

Solution:

$$\left. \begin{array}{l} F'_y = -2y - x \\ F'_x = e^x + xe^x - y \end{array} \right\} \Rightarrow \frac{dy}{dx} = -\frac{e^x(1+x) - y}{x + 2y}$$

17. Apply implicit function theorem to derive $\frac{dx}{dy}$ at a given point:

(a) $e^{xy} + 2 = x + y;$ $\quad (3, 0)$

Solution:

$$\left. \begin{array}{l} F'_y = xe^{xy} - 1 \\ F'_x = ye^{xy} - 1 \end{array} \right\} \Rightarrow \frac{dy}{dx}\Big|_{(3,0)} = -\frac{ye^{xy} - 1}{xe^{xy} - 1}\Big|_{(3,0)} = \frac{1}{2} > 0$$

(b) $\quad x^{\frac{2}{3}} + y^{\frac{2}{3}} = 5; \quad (1,8)$

Solution:

$$\left. \begin{array}{l} F'_y = \frac{2}{3}y^{-\frac{1}{3}} \\ F'_x = \frac{2}{3}x^{-\frac{1}{3}} \end{array} \right\} \Rightarrow \frac{dy}{dx}\Big|_{(1,8)} = -\frac{x^{-\frac{1}{3}}}{y^{-\frac{1}{3}}}\Big|_{(1,8)} = -2 < 0$$

18. Find the limit of $\ln(x)\ln(1-x)$ as x approaches 1 from the left.

    Solution: Writing the product as a ratio, we get a limit of the type $\dfrac{0}{0}$, so we can use L'Hôpital's rule:

    $$\lim_{x \to 1^-} \ln(x)\ln(1-x) = \lim_{x \to 1^-} \frac{\ln(1-x)}{1/\ln x}$$

    $$= \lim_{x \to 1^-} \frac{\frac{-1}{1-x}}{\frac{1}{x}\frac{1}{\ln^2 x}}$$

    $$= \lim_{x \to 1^-} \frac{\ln^2 x}{(x-1)x}$$

    This is limit of the type $\frac{0}{0}$, so we again use L'Hôpital's rule:

    $$\lim_{x \to 1^-} \frac{\ln^2 x}{(x-1)x} = \lim_{x \to 1^-} \frac{2\ln x \cdot \frac{1}{x}}{2x-1} = 0$$

    Thus we obtain $\lim_{x \to 1^-} \ln(x)\ln(1-x) = 0$.

19. Plot the graph of the function $f(x) = (1+x^2)e^{-x^2}$.

    Solution: First note that the function is symmetric (also called even), i.e., $f(x) = f(-x)$ for all $x \in \mathbb{R}$. Therefore, we can restrict our analysis to only non-negative numbers, i.e., interval $[0, +\infty)$. Using the result of problem 26 (see below) we obtain that $f(x) \to 0$, for $x \to +\infty$. Computing the first derivative we get $f'(x) = -2x^3 e^{-x^2}$, which means that the function is decreasing on $[0, +\infty)$. Because of the symmetricity, it will be increasing on $(-\infty, 0]$, which also implies that at $x = 0$, the function attains its maximum $f(0) = 1$. Checking the second-order derivative, can also find intervals of convexity and concavity.

20. Find dy/dx and d^2y/dx^2 if $x = t\ln t$ and $y = \frac{\ln t}{t}$.

    Solution: To have clear notation, denote $y = f(x)$. So computing dy/dx and d^2y/dx^2 we actually compute f' and f''. When we write $y(t) = f(x(t))$. Differentiating this equality with respect to (further abbreviated as w.r.t.) t and using the chain rule we obtain

$$\frac{dy}{dt} = f'(x(t))\frac{dx}{dt},$$

which gives the formula for f'. When evaluating f'', we differentiate the last formula w.r.t. t and obtain:

$$\frac{d^2y}{dt^2} = f''(x(t))\frac{dx}{dt} + f'(x(t))\frac{d^2x}{dt^2}. \tag{5.1}$$

From this we can again express f''.

Having $x(t) = t\ln t$ and $y(t) = \frac{1}{t}\ln t$, we obtain[9]

$$\frac{dx}{dt} = 1 + \ln t, \quad \frac{dy}{dt} = \frac{1}{t^2}(1 - \ln t),$$

and so

$$f'(x(t)) = \frac{1}{t^2} \cdot \frac{1 - \ln t}{1 + \ln t}.$$

Now evaluate second derivations:

$$\frac{d^2x}{dt^2} = \frac{1}{t}, \quad \frac{d^2y}{dt^2} = -\frac{1}{t^3}(3 - 2\ln t).$$

Plugging this into (5.1) and using the result for f' we get

$$f''(x(t)) = -\frac{2}{t^3} \cdot \frac{2 - \ln^2 t}{(1 + \ln t)^2}.$$

---

[9] Be aware of the difference between $f'(x(t))$ and $f'(x)$.

21. Evaluate the following definite integrals: $\int_{1/e}^{e} |\ln(x)| dx$.

Solution: Using the definition of the absolute value, we have $|\ln x|$ equal to either $\ln x$ or $-\ln x$ depending on whether $\ln x \geq 0$ or $\ln x < 0$, i.e., $x \geq 1$ or $x < 1$. So we split the interval $[1/e, e]$ into two parts and compute the integral for each part separately:

$$\int_{1/e}^{e} |\ln x| \, dx = \int_{1/e}^{1} -\ln x \, dx + \int_{1}^{e} \ln x \, dx$$

(integrating by parts)

$$= -x(\ln x - 1)|_{1/e}^{1} + x(\ln x - 1)|_{1}^{e} = 2 - \frac{2}{e}.$$

22. Find du and d^2u given (a) $u = \ln(\sqrt{x^2 + y^2})$, (b) $u = e^{xy}$, (c) $u = \frac{z}{x^2 + y^2}$.

Solution: To evaluate the differential we just need to compute the derivations and write them into the appropriate formulas:

$$du = \frac{\partial u}{\partial x} dx + \frac{\partial u}{\partial y} dy,$$

$$d^2u = \frac{\partial^2 u}{\partial x^2} dx^2 + 2\frac{\partial^2 u}{\partial x \partial y} dx dy + \frac{\partial^2 u}{\partial y^2} dy^2,$$

where $u = u(x, y)$; similar formulas can be written for u being the function of three variables.

(a) We have

$$du = \frac{x}{\sqrt{x^2 + y^2}} dx + \frac{y}{\sqrt{x^2 + y^2}} dy,$$

$$d^2u = \frac{y^2 - x^2}{\sqrt{x^2 + y^2}} dx^2 - 4\frac{xy}{\sqrt{x^2 + y^2}} dx dy +$$

$$+ \frac{x^2 - y^2}{\sqrt{x^2 + y^2}} dy^2.$$

(b) We obtain

$$du = ye^{xy} dx + xe^{xy} dy,$$
$$d^2u = y^2 e^{xy} dx^2 + 2(1 + xy)e^{xy} dx dy + x^2 e^{xy} dy^2.$$

(c) Now we have three variables and we get

$$
\begin{aligned}
du &= -\frac{2xz}{(x^2 + y^2)^2}dx - \frac{2yz}{(x^2 + y^2)^2}dy + \frac{1}{x^2 + y^2}dz, \\
d^2u &= \frac{2z(3x^2 - y^2)}{(x^2 + y^2)^3}dx^2 + \frac{2z(3y^2 - x^2)}{(x^2 + y^2)^3}dy^2 + \\
&\quad + \frac{8xyz}{(x^2 + y^2)^3}dxdy - \\
&\quad - \frac{2x}{(x^2 + y^2)^2}dxdz - \frac{2y}{(x^2 + y^2)^2}dydz.
\end{aligned}
$$

23. Let $z(x, y) = \frac{y^2}{3x} + \phi(xy)$, where $\phi(\cdot)$ is an arbitrary continuously differentiable function. Show that $x^2\dfrac{\partial z}{\partial x} - xy\dfrac{\partial z}{\partial y} + y^2 = 0$.

    Solution: Differentiating the given equality wrt x and wrt y we obtain

$$
\begin{aligned}
\frac{\partial z}{\partial x} &= -\frac{y^2}{3x^2} + y\phi'(xy), \\
\frac{\partial z}{\partial y} &= \frac{2y}{3x} + x\phi'(xy).
\end{aligned}
$$

    Then, clearly

$$
x^2\frac{\partial z}{\partial x} - xy\frac{\partial z}{\partial y} + y^2 = 0.
$$

24. Find $\frac{\partial z}{\partial x}$ and $\frac{\partial z}{\partial y}$ at the point $(u = 1, v = 1)$, given that $x = u + \ln v$, $y = v - \ln u$, $z = 2u + v$.

    Solution: Similarly as in part (c), we denote $z = F(x, y)$ and compute $F_x = \partial z/\partial x$, $F_y = \partial z/\partial y$. Having x, y, z expressed as functions of u and v, we can write $z(u, v) = F(x(u, v), y(u, v))$. Differentiating this equality w.r.t. u and w.r.t. v, we obtain:

$$
\begin{aligned}
\frac{\partial z}{\partial u} &= F_x\frac{\partial x}{\partial u} + F_y\frac{\partial y}{\partial u}, \\
\frac{\partial z}{\partial v} &= F_x\frac{\partial x}{\partial v} + F_y\frac{\partial y}{\partial v}.
\end{aligned}
$$

When computing all the partial derivatives w.r.t. u and v, we obtain a system of two equations with two unknowns F_x and F_y:

$$2 \;=\; F_x - \frac{1}{u}F_y,$$

$$1 \;=\; \frac{1}{v}F_x + F_y.$$

Solving the system we get

$$F_x = \frac{v(2u+1)}{uv+1}, \qquad F_y = \frac{u(v-2)}{uv+1},$$

which after substitution $u = v = 1$ gives $F_x = \frac{3}{2}$ and $F_y = -\frac{1}{2}$.

25. Find extrema of the function $u = xy^2z^3(a - x - 2y - 3z)$ and classify them (a is a positive parameter).

Solution: First note that if any of the arguments x, y, z is equal to zero, then also $f(x,y,z) = 0$. By the definition of maximum, t is a local maximum of function F, if there is some neighborhood $U(t)$ of t such that $F(t) > F(s)$ for any $s \in U(t)$, $s \neq t$. In our case, a point (x,y,z) with at least one coordinate being equal zero (for example, $x = 0$) cannot be maximum, because in its neighborhood we can find another point for which the value of f is the same or greater. For this reason, in what follows we will consider only $x, y, z \neq 0$.

Differentiating our function w.r.t. x, y, z, we obtain the first-order conditions:

$$y^2 z^3 (a - 2x - 2y - 3z) \;=\; 0,$$
$$2xyz(a - x - 3y - 3z) \;=\; 0,$$
$$3xyz(a - x - y - 4z) \;=\; 0.$$

Recalling that $x, y, z \neq 0$, we can solve the system of three equations with three unknowns and parameter a to obtain $x = y = z = a/7$. Now, we need to classify this point. To do so, we compute the Hessian matrix:

$$\left(\frac{a}{7}\right)^5 \begin{pmatrix} -2 & -2 & -3 \\ -2 & -6 & -6 \\ -3 & -6 & -12 \end{pmatrix},$$

with leading principal minors $D_1 = -2(a/7)^5 < 0$, $D_2 = 8(a/7)^{10} > 0$, $D_3 = -48(a/7)^{15} < 0$, which implies that the point $(x, y, z) = (a/7, a/7, a/7)$ is a local maximum.

26. Show that for large x an exponential function e^{ax}, $a > 0$, increases faster than a polynomial function x^n, $n > 0$. In other words, in the limit $x^n e^{-ax} \to 0$ for any positive a and n as $x \to \infty$. (This statement proves to be helpful when discussing intertemporal optimization over an infinite time horizon.)

    Solution: When $x \to +\infty$, then $x^n \to +\infty$ and $e^{-ax} \to 0$ (because $a > 0$). So we have a limit of the type $\infty \times 0$. Rewriting it as a ratio x^n/e^{ax}, we can use L'Hôpital's rule n-times:

$$\lim_{x \to +\infty} \frac{x^n}{e^{ax}} = \lim_{x \to +\infty} \frac{n x^{n-1}}{a e^{ax}} = \lim_{x \to +\infty} \frac{n(n-1)x^{n-2}}{a^2 e^{ax}} = \dots =$$

$$= \lim_{x \to +\infty} \frac{n(n-1)\dots 2x}{a^{n-1} e^{ax}} = \lim_{x \to +\infty} \frac{n(n-1)\dots 2 \cdot 1}{a^n e^{ax}} = 0.$$

27. Let $\int f(x)dx = F(x) + C$. Show that for any $a \neq 0$ and b, $\int f(ax+b)dx = \frac{1}{a}F(ax+b)+C$. (This exercise relates to location-scale invariance in distribution theory.)

    Solution: The equality $\int f(x)\, dx = F(x) + C$ is equivalent to $f(x) = F'(x)$. Then

$$\frac{d}{dx}\left(\frac{1}{a}F(ax+b)\right) = F'(ax+b) = f(ax+b),$$

and so $\int f(ax+b)\, dx = \frac{1}{a}F(ax+b) + C$, which we wanted to prove.

28. Show that $\int_o^x e^{x^2} dx \sim \frac{1}{2x}e^{x^2}$ as $x \to \infty$.

    Solution: One (of more) usual interpretation of approximation[10] $f(x) \approx g(x)$ for $x \to +\infty$ is that

$$\lim_{x \to +\infty} \frac{f(x)}{g(x)} = 1$$

---

[10] Used in particular when some variable is approaching infinity.

So, we will prove that

$$\lim_{x \to +\infty} \frac{\int_0^x e^{t^2} \, dt}{\frac{1}{2x} e^{x^2}} = 1.$$

Note that we have used t instead of x in the integral (to have clear notation). Obviously, $e^{t^2} \geq 1$ for $t > 0$, so $\int_0^x e^{t^2} \, dt \geq \int_0^x dt = x$. Therefore, the numerator approaches $+\infty$, as $x \to +\infty$. Using the result of part (a), we know that the denominator also approaches $+\infty$. Thus we have $\frac{\infty}{\infty}$ limit, and we can use L'Hôpital's rule. Derivation of the numerator w.r.t. x is e^{x^2} (use the Newton-Leibniz formula); derivation of the denominator is $(1 - \frac{1}{2x^2})e^{x^2}$. Then

$$\lim_{x \to +\infty} \frac{\int_0^x e^{t^2} \, dt}{\frac{1}{2x} e^{x^2}} = \lim_{x \to +\infty} \frac{e^{x^2}}{(1 - \frac{1}{2x^2})e^{x^2}} = \lim_{x \to +\infty} \frac{1}{1 - \frac{1}{2x^2}} = 1.$$

The proof is complete.

29. Find the minima and maxima of the function $f(x_1, x_2, \ldots, x_n) = \ln(1 + x_1^2 + x_2^2 + \ldots + x_n^2) - a(x_1^2 + x_2^2 + \ldots + x_n^2)$, where a is a positive parameter, and $x_i \in [-1, 1] \; \forall i = 1, 2, \ldots, n$.

    Solution: In this problem we want to find global extrema. Looking carefully at the function, we can see some kind of symmetricity; in particular, if $f(x_1, \ldots, x_n)$ is the same for any $(x_1, \ldots, x_n)$ such that $x_1^2 + \ldots + x_n^2 = t$ (where $t \in [0, n]$, because $x_i \in [-1, 1]$ for $i = 1, 2, \ldots, n$). For that reason, let us consider the function $g(t) = \ln(1+t) - at$. Now, we find extrema of the function g. Taking a first derivative, we get the F.O.C.: $g'(t) = 1/(1+t) - a = 0$, which yields $t = \frac{1}{a} - 1$. This quantity belongs to $[0, n]$, iff $a \in [\frac{1}{n+1}, 1]$. Moreover, we can see that $g'(t) > 0$ for $t < \frac{1}{a} - 1$ and $g'(t) < 0$ for $t > \frac{1}{a} - 1$.

    Now, we discuss several cases:

    - If $a \in (0, \frac{1}{n+1}]$, then $\frac{1}{a} - 1 \geq n$, and g is increasing on $[0, n]$. Therefore it attains minimum 0 in $t = 0$ and maximum $\ln(1 + n) - an$ in $t = n$. Then f attains the same minimum for $x_1 = \ldots = x_n = 0$ and the same maximum for any $(x_1, \ldots, x_n)$ such that $|x_1| = \ldots = |x_n| = 1$.

- If $a \geq 1$, then $\frac{1}{a} - 1 \leq 0$, and g is decreasing on $[0, n]$. Using a similar argument as in the previous case, we obtain that f attains maximum 0 for $x_1 = \ldots = x_n = 0$ and minimum $\ln(1 + n) - an$ for any $(x_1, \ldots, x_n)$ such that $|x_1| = \ldots = |x_n| = 1$.

- If $a \in (\frac{1}{n+1}, 1)$, then $\frac{1}{a} - 1 \in (0, n)$. Our analysis of g implies that g attains its maximum $a - \ln a - 1$ in $t = \frac{1}{a} - 1$. Therefore, f attains the same maximum for any $(x_1, \ldots, x_n)$ such that $x_1^2 + \ldots + x_n^2 = \frac{1}{a} - 1$.

  To find the minimum, we just need to compare the values on the boundary: $f(0) = 0$ and $f(n) = \ln(1 + n) - an$. So, we have two cases:

  - If, in addition $a < \frac{1}{n} \ln(n + 1)$, then g attains its minimum 0 in $t = 0$, and f attains the same minimum for $x_1 = \ldots = x_n = 0$.
  - If, in addition $a < \frac{1}{n} \ln(n + 1)$, then g attains its minimum $\ln(1 + n) - an$ in $t = n$, and attains minimum $\ln(1+n) - an$ for any $(x_1, \ldots, x_n)$ such that $|x_1| = \ldots = |x_n| = 1$.

30. A twice continuously differentiable function $f(x_1, \ldots, x_n)$ is defined to be convex (concave) if $d^2 f$ is positive (negative) semi-definite and strictly convex (strictly concave) if $d^2 f$ is positive (negative) definite. Which values of p and q will make the linear additive utility function $f(x, y) = x^{-p} + y^{-q}$ (strictly) concave/convex?

    Solution: We can easily compute the Hessian matrix:

$$H = \begin{pmatrix} p(p + 1)x^{-p-2} & 0 \\ 0 & q(q + 1)y^{-q-2} \end{pmatrix}.$$

To have f strictly concave, we must have H positive definite, i.e., $p(p+1) > 0$ and $p(p+1)q(q+1) > 0$, which is equivalent to $p, q \in (-\infty, -1) \cup (0, +\infty)$. Similarly, for f to be strictly concave, we have $p, q \in (-1, 0)$. For f to be convex, H must be positive semi-definite, i.e., $p(p + 1) \geq 0$, $q(q + 1) \geq 0$, and $p(p + 1)q(q + 1) \geq 0$, which is equivalent to $p, q \in (-\infty, -1] \cup [0, +\infty)$. Similarly, for f to be concave, we must have $p, q \in [-1, 0]$.

5.2.3   Constrained Optimization

1. Find an extremum of an additive utility function $U(x_1, \ldots, x_n) = x_1^\alpha + \ldots + x_n^\alpha$, with $\alpha > 0$, subject to a linear constraint $x_1 + \ldots + x_n = M > 0$.

(a) Show that the extremum is unique.

(b) Check the second-order sufficient conditions (S.O.S.C.) to find whether the extremum is maximum or minimum, depending on α.

(c) Can you determine the type of extremum point without checking the S.O.S.C.?

(d) What will happen to the shadow price λ (Lagrange multiplier) if M marginally decreases?

(e) What will happen to the optimal solution if α marginally changes?

Solution:

(a) The Lagrangian function is

$$L = x_1^\alpha + \ldots + x_n^\alpha + \lambda(M - x_1 - \ldots - x_n).$$

F.O.C.:

$$L_{x_i} = \alpha x_i^{\alpha-1} - \lambda = 0, \text{ or } x_i = \left(\frac{\lambda^*}{\alpha}\right)^{1/(\alpha-1)}.$$

$M - x_1 - \ldots - x_n = 0$, thus $M - n\left(\frac{\lambda^*}{\alpha}\right)^{1/(\alpha-1)} = 0$, or

$$\lambda^* = \left(\frac{M}{n}\right)^{\alpha-1} \cdot \alpha.$$

Finally, the unique stationary point is $x_1^* = \ldots = x_n^* = \dfrac{M}{n}$.

(b) $d^2L = \sum \alpha \cdot (\alpha - 1) x_i^{\alpha-2} d^2 x_i$. $d^2L < 0$ when $\alpha \in (0, 1)$ and, therefore, $x_1^* = \ldots = x_n^* = \dfrac{M}{n}$ is maximum and $u_{max} = \dfrac{M^\alpha}{n^{\alpha-1}}$. $d^2L > 0$ when $\alpha > 1$ and, therefore, $x_1^* = \ldots = x_n^* = \dfrac{M}{n}$ is minimum and $u_{min} = \dfrac{M^\alpha}{n^{\alpha-1}}$.

(c) When $\alpha \in (0,1)$ it is sufficient to notice the concavity of the utility function and (weak) concavity of the budget constraint, which imply that the Lagrangian function is concave and, therefore, the stationary point is a global maximum. Similarly, when $\alpha > 1$ it is sufficient to notice the convexity of the utility function and (weak) convexity of the budget constraint, which imply that the Lagrangian function is convex and, therefore, the stationary point is a global minimum.

(d) $\dfrac{\partial \lambda^*}{\partial M} = (\alpha-1)M^{\alpha-2} \cdot \dfrac{\alpha}{n^{\alpha-1}}$. When $\alpha \in (0,1)$ $\dfrac{\partial \lambda^*}{\partial M} < 0$, i.e. λ^* increases as M decreases. Similarly, when $\alpha > 1$ $\dfrac{\partial \lambda^*}{\partial M} > 0$, i.e. λ^* increases as M increases.

(e) $\dfrac{\partial x_i}{\partial \alpha} = \dfrac{\partial}{\partial \alpha}\left(\dfrac{M}{n}\right) = 0$, i.e. the optimal solution does not change as α marginally changes.

2. Consider a constrained optimization problem

$$\text{extremize } u(x_1,\ldots,x_n) = x_1^{a_1} x_2^{a_2} \ldots x_n^{a_n} \qquad (a_i > 0 \ \forall i)$$

$$\text{s.t.} \quad x_1 + \ldots + x_n = \alpha > 0, \quad x_i > 0, \ i = 1,\ldots,n$$

(a) Find the extremum (or extrema) of $u(x_1,\ldots,x_n)$ and check whether it is minimum or maximum.

(b) Let $(x_1^*,\ldots,x_n^*)$ be a point of optimality. What will happen to $(x_1^*,\ldots,x_n^*)$ if α marginally increases?

Solution:

(a) Using monotonic transformation $v = \ln u = \sum_{i=1}^{n} a_i \ln x_i$, the Lagrangian function takes the form

$$L = \sum_{i=1}^{n} a_i \ln x_i + \lambda\left(\sum_{i=1}^{n} x_i - \alpha\right),$$

and the F.O.C. become

$$\frac{a_i}{x_i} + \lambda = 0, \ i = 1,\ldots,n, \qquad \sum_{i=1}^{n} x_i = \alpha.$$

Therefore,

$$\lambda = -\frac{\sum_{i=1}^{n} a_i}{\alpha},$$

and

$$x_j^* = \frac{\alpha a_j}{\sum_{i=1}^{n} a_i}, \quad j = 1, \ldots, n.$$

Since at x^*

$$d^2 L = -\sum_{i=1}^{n} \left(\frac{a_i}{x_i^2}\right) dx_i^2$$

$$= -\left(\sum_{i=1}^{n} a_i\right)^2 \frac{1}{\alpha^2} \cdot \sum_{i=1}^{n} \frac{dx_i^2}{a_i} < 0,$$

$x^* = (x_1^*, \ldots, x_n^*)$ is a point of maximum.

(b) Clearly, $\dfrac{\partial x_j^*}{\partial \alpha} = \dfrac{a_j}{\sum_{i=1}^{n} a_i} > 0$ for any $j = 1, \ldots, n$.

3. Consider the following non-linear optimization problem

$$\max_{x,y} -8x^2 - 10y^2 + 12xy - 50x + 80y$$

s.t. $\quad x + y \leq 1, \quad 12x^2 + y^2 \leq 3, \quad x \geq 0, \ y \geq 0.$

(a) Do Kuhn-Tucker conditions provide a unique point of optimality? If so, find it and illustrate your solution graphically.

(b) What will change in Kuhn-Tucker conditions and the optimal solution if the non-negativity constraints on x and y are removed?

Solution:

(a) The objective function is strictly concave and the constraint functions are convex. Therefore the Kuhn-Tucker conditions provide a unique solution that is a global maximum. The

Lagrangian function is $L = -8x^2 - 10y^2 + 12xy - 50x + 80y + \lambda_1(1 - x - y) + \lambda_2(3 - 12x^2 - y^2)$, and the F.O.C. are

$$\frac{\partial L}{\partial x} \leq 0, \quad x\frac{\partial L}{\partial x} = 0, \quad \frac{\partial L}{\partial y} \leq 0, \quad y\frac{\partial L}{\partial y} = 0,$$

$$\frac{\partial L}{\partial \lambda_1} \geq 0, \quad \lambda_1\frac{\partial L}{\partial \lambda_1} = 0, \quad \frac{\partial L}{\partial \lambda_2} \geq 0, \quad \lambda_2\frac{\partial L}{\partial \lambda_2} = 0.$$

Since the objective function is increasing upward-leftward, it reaches its maximum in the corner of the feasible set at $(x = 0, y = 1)$.

(b) Again, the solution is corner, but in the absence of non-negativity constraints it shifts to the point of intersection of $x + y = 1$ and $12x^2 + y^2 = 3$ in the negative quadrant at $(x \approx -1/3, y \approx 4/3)$.

4. Consider the Kuhn-Tucker conditions for the standard maximization problem

$$\max f(x_1, \ldots, x_n)$$
$$\text{subject to } g^i(x_1, \ldots, x_n) \leq b_i, \quad i = 1, 2, \ldots, m.$$
$$x_1 \geq 0, \ldots, x_n \geq 0.$$

How will these Kuhn-Tucker conditions change if the non-negativity restriction on $x_1, \ldots, x_n$ is replaced with the constraint $0 \leq x_1 \leq 10$?

Solution: The K.-T. conditions are

$$\frac{\partial L}{\partial x_i} = 0, \quad i = 2, \ldots, n$$

$$\frac{\partial L}{\partial x_1} \leq 0, \, x_1 \geq 0 \quad \text{and} \quad x_1\frac{\partial L}{\partial x_1} = 0,$$

$$\frac{\partial L}{\partial \lambda_j} \geq 0, \, \lambda_j \geq 0 \quad \text{and} \quad \lambda_j\frac{\partial L}{\partial \lambda_j} = 0, \quad j = 1, \ldots, m$$

$$x_1 \leq 10, \quad \mu \geq 0 \quad \text{and} \quad \mu(10 - x_1) = 0,$$

where

$$L(x_1, \ldots, x_n, \lambda_1, \ldots, \lambda_m, \mu) =$$

$$= f(x_1, \ldots, x_n) + \sum_{j=1}^{m} \lambda_j(b_j - g^j(x_1, \ldots, x_n)) + \mu(10 - x_1)$$

is the Lagrangian function of a non-linear program.

5. Find minimum and maximum of the function $u(x, y, z) = x^2 + 2y^2 + 3z^2$ in the domain $x^2 + y^2 + z^2 \leq 100$.

   Solution: By Weierstrass's theorem, a continuously differentiable function in a closed compact domain reaches its maximum and minimum either at a stationary interior point, or at the boundary of its domain.

   The only interior stationary point in $x^2 + y^2 + z^2 \leq 100$ is the origin.

   To check the boundary, we need to solve the equally constrained optimization problem

$$\text{extremize } u(x, y, z) = x^2 + 2y^2 + 3z^2$$

$$\text{s.t. } x^2 + y^2 + z^2 = 100$$

   Solving for the F.O.C., we find six boundary stationary points: $(\pm 10, 0, 0)$, $(0, \pm 10, 0)$, and $(0, 0, \pm 10)$.

   A direct check yields that $u_{min} = u(0, 0, 0) = 0$, and $u_{max} = u(0, 0, \pm 10) = 300$.

6. Find the conditional extrema of the function and classify them:

$$u(x, y, z) = xy^2 z^3,$$

$$\text{s.t. } x + 2y + 3z = a, \qquad x > 0, y > 0, z > 0, a > 0$$

<u>Solution:</u>   Consider a monotonic transformation $v(x, y, z) = \ln u(x, y, z) = \ln x + 2 \ln y + 3 \ln z$. The equivalent optimization problem is

$$\text{extremize } v(x, y, z) = \ln x + 2 \ln y + 3 \ln z,$$

$$\text{s.t. } x + 2y + 3z = a, \qquad x > 0, y > 0, z > 0, a > 0.$$

The F.O.C. imply that the only stationary point is $x = y = z = a/6$. Since the Lagrangian function is concave, the stationary point should be maximum.

7. Maximize $x_1 + x_2$, subject to $x_1^2 + x_2 \leq 1$, $2x_2 - x_1 \geq 1$, and $x_1 \geq 0$, $x_2 \geq 0$.

   Are the K.-T. conditions sufficient?

   <u>Solution:</u>   Concavity of the Lagrangian function implies that the K.-T. conditions are indeed sufficient for a maximum.

   The K.-T. conditions yield the unique optimal solution
   $x_1 = 1/2, x_2 = 3/4, \lambda_1 = 1, \lambda_2 = 0$.

8. Consider the following maximization problem:

$$
\begin{array}{rlrl}
\max_{x,y} xy & \quad \text{s.t.} & 5x + 4y & \leq 50 \\
 & & 3x + 6y & \leq 40 \\
 & & x & \geq 0 \\
 & & y & \geq 0
\end{array}
$$

   a) Are the Kuhn-Tucker conditions necessary for a solution to this problem?

   <u>Solution:</u> Yes, they are, because each constraint is a linear function.

   b) Are the Kuhn-Tucker conditions sufficient for a solution of this problem?

   <u>Solution:</u> Yes, they are, because the objective function is differentiable and concave, and constraints are differentiable and linear.

c) If possible, use the Kuhn-Tucker conditions to find the solution(s) of this problem.

Solution: The Lagrangian function is

$$L(x, y, u, v) = xy + u(50 - 5x - 4y) + v(40 - 3x - 6y).$$

The Kuhn-Tucker necessary maximum conditions are:

$$\frac{\partial L}{\partial x} = y - 5u - 3v \leq 0 \tag{5.2}$$

$$\frac{\partial L}{\partial y} = x - 4u - 6v \leq 0 \tag{5.3}$$

$$\frac{\partial L}{\partial u} = 5x + 4y \leq 50 \tag{5.4}$$

$$\frac{\partial L}{\partial v} = 3x + 6y \leq 40 \tag{5.5}$$

$$x\frac{\partial L}{\partial x} = x(y - 5u - 3v) = 0 \tag{5.6}$$

$$y\frac{\partial L}{\partial y} = y(x - 4u - 6v) = 0 \tag{5.7}$$

$$u\frac{\partial L}{\partial u} = u(50 - 5x - 4y) = 0 \tag{5.8}$$

$$v\frac{\partial L}{\partial v} = v(40 - 3x - 6y) = 0 \tag{5.9}$$

$$x \geq 0, y \geq 0 \qquad u \geq 0, v \geq 0 \tag{5.10}$$

In order to solve this system, we divide it into four states which are fully independent and cover all possibilities:

- $u = v = 0$
  From conditions (5.2) and (5.3) we get $x \leq 0$ and $y \leq 0$. However, condition (5.10) leaves us with only one possible candidate for optimum: $x = 0$ and $y = 0$.
- $u > 0$ and $v = 0$
  From (5.6), (5.7) and (5.8) we get the system:
  $$x(y - 5u) = 0 \tag{5.11}$$
  $$y(x - 4u) = 0 \tag{5.12}$$
  $$50 - 5x - 4y = 0 \tag{5.13}$$
  Equation (5.11) implies two possible cases:

    i) $x = 0 \Rightarrow y = 0$ because $u > 0$ and $y(-4u) = 0$ (from (5.12)). However, $x = y = 0$ contradicts (5.13).

    ii) $y - 5u = 0 \Rightarrow y \neq 0 \Rightarrow x = 4u$ (from (5.12)). Now we can plug y and x into (5.13) and compute $u = \frac{5}{4}$. Then we are able to evaluate the next candidate for our optimum: $x = 5$ and $y = \frac{25}{4}$. However, this case contradicts condition (5.5).

- $v > 0$ and $u = 0$

  From (5.6), (5.7) and (5.9) we get the system:

$$x(y - 3v) \quad = \quad 0 \qquad\qquad (5.14)$$
$$y(x - 6v) \quad = \quad 0 \qquad\qquad (5.15)$$
$$40 - 3x - 6y \quad = \quad 0 \qquad\qquad (5.16)$$

  Equation (5.14) implies two possible cases:

    i) $x = 0 \Rightarrow y = 0$ because $v > 0$ and $y(-6v) = 0$ (from (5.15)). However, $x = y = 0$ contradicts (5.16).

    ii) $y - 3v = 0 \Rightarrow y \neq 0 \Rightarrow x = 6v$ (from (5.12)). Now we can plug y and x into (5.16) and find $v = \frac{10}{9}$. Then we are able to count the next candidate for our optimum: $x = \frac{20}{3}$ and $y = \frac{10}{3}$. Fortunately, it satisfies all conditions.

- $v > 0$ and $u > 0$

  Now we have the full system of four equations (5.6)–(5.9). Solving it we get $x = \frac{70}{9}$, $y = \frac{25}{9}$, $u = -\frac{10}{27}$, and $v = \frac{125}{81}$. However, u cannot be negative.

Thus, we have two candidates for an optimum: $[\frac{20}{3}, \frac{10}{3}]$ and $[0,0]$. The first one leads to a higher value of the objective function and therefore it is also the optimal solution.

9. Consider the following maximization problem:

$$\max_{x,y} x$$

$$\text{s.t.} \quad y - (1 - x)^3 \;\leq\; 0$$
$$y \;\geq\; 0$$

Can you refer to Kuhn-Tucker conditions to find an optimal solution to this problem?

Solution: This example illustrates the case when the Kuhn-Tucker conditions provide no solution, but the problem indeed has a solution. In such a case, the constraint is neither a convex nor concave function.

One can directly see that the solution to the optimization program is (1,0). On the other, with the Lagrangian function

$$L(x, y, \lambda) = x + \lambda((1 - x)^3 - y),$$

the Kuhn-Tucker necessary conditions for a maximum become

$$\frac{\partial L}{\partial x} = 1 - 3\lambda(1 - x)^2 \leq 0$$

$$\frac{\partial L}{\partial y} = -\lambda \leq 0$$

$$\frac{\partial L}{\partial \lambda} = (1 - x)^3 - y \geq 0$$

$$x\frac{\partial L}{\partial x} = x(1 - 3\lambda(1 - x)^2) = 0$$

$$y\frac{\partial L}{\partial y} = y(-\lambda) = 0$$

$$\lambda\frac{\partial L}{\partial \lambda} = \lambda((1 - x)^3 - y) = 0$$

$$y \geq 0 \qquad \lambda \geq 0$$

and the system above has no solution.

10. Minimize $x^2 + y^2 + z^2$, subject to $x + y + z \leq -3$.

    Solution:

    Let us write the Lagrangian and the Kuhn-Tucker conditions first.[11]

$$L(x, y, z, \lambda) = x^2 + y^2 + z^2 + \lambda(z + y + z + 3)$$

---

[11] Note, that if there are no nonnegativity constraints on variable x, the conditions w.r.t. x reduce just to one: $\frac{\partial L}{\partial x} = 0$

$$\frac{\partial L}{\partial x} = 2x + \lambda = 0 \qquad \frac{\partial L}{\partial y} = 2y + \lambda = 0$$

$$\frac{\partial L}{\partial z} = 2z + \lambda = 0 \qquad \lambda \geq 0$$

$$\frac{\partial L}{\partial \lambda} = x + y + z + 3 \leq 0 \qquad \lambda(x + y + z + 3) = 0$$

From the first three conditions we have that $x = y = z = -\frac{\lambda}{2}$. If $\lambda = 0$ then $x = y = z = 0$, which does not satisfy the constraint. So, $\lambda > 0$ and thus $x + y + z = -3$. Now we find that the only optimal point is $x = y = z = -1$.

11. Maximize $3x - x^2 + xy - y^2$, subject to $3x + 4y \leq 6, 4y^2 - x \leq 2, x \geq 0, y \geq 0$.

Solution:

First, we write the Lagrangian function and the Kuhn-Tucker conditions for the maximum:

$$L(x, y, \lambda, \mu) = 3x - x^2 + xy - y^2 +$$

$$+\lambda(-3x - 4y + 6) + \mu(-4y^2 + x + 2)$$

and

$$
\begin{array}{rcl}
3 - 2x + y - 3\lambda + \mu & \leq & 0 \quad (1) \\
x(3 - 2x + y - 3\lambda + \mu) & = & 0 \quad (2) \\
x - 2y - 4\lambda - 8\mu y & \leq & 0 \quad (3) \\
y(x - 2y - 4\lambda - 8\mu y) & = & 0 \quad (4) \\
-3x - 4y + 6 & \geq & 0 \quad (5) \\
\lambda(-3x - 4y + 6) & = & 0 \quad (6) \\
-4y^2 + x + 2 & \geq & 0 \quad (7) \\
\mu(-4y^2 + x + 2) & = & 0 \quad (8) \\
x, y, \lambda, \mu & \geq & 0 \quad (9)
\end{array}
$$

If $x = 0$ then our objective function becomes $f(x, y) = -y^2$ which has maximum 0 at $y = 0$, which is not a global maximum in this case, as we can see. If $y = 0$ and $x > 0$ then the objective function takes the form $f(x, y) = 3x - x^2$ and attains its maximum at $x = 3/2$. From (7) we have $2 + x > 0$, so $\mu = 0$, from (2) we have $3 - 2 \cdot 3/2 - 3\lambda = 0$, so that $\lambda = 0$. But the latter contradicts (3), thus we conclude that both $x, y \neq 0$. Now we need to check four various cases (Note that we have to check all cases even if we found a solution because the concavity of the objective function alone is not a sufficient condition for the uniqueness of a solution).

- $\lambda = \mu = 0$: Here we can solve (1) and (3) for x and y and get $x = 2$, $y = 1$, which is in contradiction to (5).

- $\lambda = 0, \mu \neq 0$: Now we can obtain two solutions to (2), (4) and (8), but in both we get $\mu < 0$, which is a contradiction.

- $\mu = 0, \lambda \neq 0$: If we solve (2), (4), (6), we get three various solutions; two of them have $\lambda < 0$, the third solution is $x = 54/37, y = 15/37, \lambda = 6/37$, which satisfies all conditions.

- $\mu \neq 0, \lambda \neq 0$: In this case again no suitable solution can be found; when solving (2), (4), (6) and (8) we find that $4\mu < 0$. Summing up, the maximum of our objective function is $x = 54/37, y = 15/37$.

12. Consider the Kuhn-Tucker conditions for the standard maximization problem

$$\max f(x_1, x_2, \ldots, x_n)$$
$$\text{subject to } g^i(x_1, \ldots, x_n) \leq b_i, \quad i = 1, 2, \ldots, m,$$
$$x_1 \geq 0, \ldots, x_n \geq 0.$$

How will these Kuhn-Tucker conditions change if the non-negativity restriction on $x_1, \ldots, x_n$ is replaced with the constraints $0 \leq x_i \leq 10, i = 1, \ldots, n$?

Solution: Let us write the condition for the original case first:

$$L(x, \lambda) = f(x) + \sum_{i=1}^{m} \lambda_i(b_i - g^i(x))$$

$$\frac{\partial L}{\partial x_i} \leq 0 \quad \frac{\partial L}{\partial x_i} x_i = 0 \qquad x_i \geq 0 \quad i = 1, \ldots, n$$

$$\frac{\partial L}{\partial \lambda_i} \geq 0 \quad \frac{\partial L}{\partial \lambda_i} \lambda_i = 0 \qquad \lambda_i \geq 0 \quad i = 1, \dots, m,$$

where x is a vector of $\{x_1, \dots, x_n\}$ and λ is a vector of $\{\lambda_1, \dots, \lambda_m\}$.

The Kuhn-Tucker conditions for the new problem are:

$$L_N(x, \lambda, \mu) = f(x) + \sum_{i=1}^{m} \lambda_i(b_i - g^i(x)) + \sum_{i=1}^{n} \mu_i(10 - x_i)$$

$$\frac{\partial L_N}{\partial x_i} = \frac{\partial L}{\partial x_i} - \mu_i \leq 0 \qquad \frac{\partial L_N}{\partial x_i} x_i = 0 \qquad x_i \geq 0 \quad i = 1, \dots, n$$

$$\frac{\partial L_N}{\partial \lambda_i} = \frac{\partial L}{\partial \lambda_i} \geq 0 \qquad \frac{\partial L_N}{\partial \lambda_i} \lambda_i = 0 \qquad \lambda_i \geq 0 \quad i = 1, \dots, m$$

$$\frac{\partial L_N}{\partial \mu_i} = (10 - x_i) \geq 0 \qquad \frac{\partial L_N}{\partial \mu_i} \mu_i = 0 \qquad \mu_i \geq 0 \quad i = 1, \dots, n$$

We can see that the second row of the K.-T. conditions stays the same, the first row changes, and there are quite new conditions in the third row.

13. Check whether a concave function is also quasi-concave. Is the inverse true?

    Solution: Function f is concave iff $\forall x, y$ and $\forall \alpha \in [0, 1]$: $f(\alpha x + (1-\alpha)y) \geq \alpha f(x) + (1-\alpha)f(y)$. Without loss of generality, let us assume that $f(x) \geq f(y)$. Then we have:

$$f(\alpha x + (1 - \alpha)y) \geq \alpha f(x) + (1 - \alpha)f(y) \geq$$

$$\geq \alpha f(y) + (1 - \alpha)f(y) = f(y),$$

so that $f(\alpha x + (1 - \alpha)y) \geq f(y)$ for $f(x) \geq f(y)$, which is the definition of quasi-concavity. The inverse is not true. Let us take function $f(x) = x^2$. This function is convex and quasi-concave, as one can check, meaning that concavity does not follow from quasi-concavity.

14. Consider the following non-linear programming problem

$$\max_{x,y} ax^2 + bxy$$

$$\text{subject to} \quad x^2 + cy^2 \le 1, \quad x \ge 0, \quad y \ge 0,$$

where a, b, c are parameters.

(a) For $a = 0$ and $b = c = 1$ find the solution geometrically.

(b) Obtain the Kuhn-Tucker conditions in the general case. For what values of parameters does a solution exist?

Solution:

(a) We can see that in this case the problem is symmetrical w.r.t. x and y, therefore, in the optimal solution $x = y$. The constraint $x^2 + y^2 \le 1$ is part of a unit circle in the positive quadrant and the objective function is the hyperbola $y = k/x$. Thus the solution is such k, where this hyperbola is tangent to the circle and $x = y$. It is clear that the point of tangency is $[1/\sqrt{2}, 1/\sqrt{2}]$.

(b) The Kuhn-Tucker conditions are written as:

$$
\begin{aligned}
L(x, y, \lambda) &= ax^2 + bxy + \lambda(1 - x^2 - cy^2) \\
2ax + by - 2\lambda x &\le 0 \\
(2ax + by - 2\lambda x)x &= 0 \\
bx - 2\lambda cy &\le 0 \\
(bx - 2\lambda cy)y &= 0 \\
1 - x^2 - cy^2 &\ge 0 \\
(1 - x^2 - cy^2)\lambda &= 0 \\
x, y, \lambda &\ge 0
\end{aligned}
$$

Now, we either have to check for what values of parameters the necessity and sufficiency conditions are valid, or we can examine a few cases depending on the values of parameters. Let us try the second approach. Let $c > 0$. Then the set $\{[x, y] : x \ge 0, y \ge 0, x^2 + cy^2 \le 1\}$ is closed and bounded, and according to Weierstrass's theorem the objective function (which is continuous) reaches its maximum in this set, i.e., the optimization program has a solution in this case. Now let $c < 0$. Then the set $\{[x, y] : x \geq 0, y \geq 0, x^2 + cy^2 \leq 1\}$ is unbounded, so that if $a > 0$ or $b > 0$ then the objective function does not reach finite maximum, i.e., a finite solution does not exists. If $a \leq 0$ and $b \leq 0$, then the origin $[0, 0]$ is one of the solutions. Finally, let us proceed to the last option $c = 0$. Now the feasible set is $\{[x, y] : 0 \leq x \leq 1, y \geq 0\}$. As the set is unbounded for y, a finite solution exists for $b \leq 0$ (in particular, $[1, 0]$ is a solution). Summing up,

$c > 0$		solution exists
$c = 0$	$b \leq 0$	solution exists
	otherwise	solution does not exist
$c < 0$	$a \leq 0, b \leq 0$	solution exists
	otherwise	solution does not exist.

15. Find the conditional extrema of the function and classify them:

$$u(x, y, z) = xy^2 z^3,$$

$$\text{s.t. } x + 2y + 3z = a, \quad x > 0, \ y > 0, \ z > 0, \ a > 0.$$

(HINT: You may try monotonic transformation $v = \ln u$.)

Solution: As the transformation $v = \ln u$ is monotonic, the conditional extrema of $u(x, y, z)$ will be the same as the conditional extrema of $v(x, y, z) = \ln(u, x, y, z)$. So we can write the Lagrangian and the F.O.C.:

$$
\begin{aligned}
L &= \ln x + 2\ln y + 3\ln z + \lambda(a - x - 2y - 3z) \\
\frac{\partial L}{\partial x} &= \frac{1}{x} - \lambda = 0 \\
\frac{\partial L}{\partial y} &= \frac{2}{y} - 2\lambda = 0 \\
\frac{\partial L}{\partial z} &= \frac{3}{z} - 3\lambda = 0,
\end{aligned}
$$

which have one solution $x^* = y^* = z^* = \frac{a}{6}$. To classify this stationary point, we need to write a bordered Hessian matrix:

$$H_3 = \begin{pmatrix} 0 & 1 & 2 & 3 \\ 1 & -\frac{1}{x^2} & 0 & 0 \\ 2 & 0 & -\frac{2}{y^2} & 0 \\ 3 & 0 & 0 & -\frac{3}{z^2} \end{pmatrix}$$

Now it is enough to check for the sign of determinants $|H_3|$ and $|H_2|$, evaluated at $[x^*, y^*, z^*]$. We obtain that $|H_3| < 0$ and $|H_2| > 0$, so that $[x^*, y^*, z^*]$ is a local maximum.

5.2.4 Dynamics

1. Consider the following system of differential equations:

$\dot{X} = AX$ where

$$X(t) = (x_1(t), x_2(t), x_3(t))'$$

and

$$A = \begin{pmatrix} 1 & -1 & 1 \\ 1 & 1 & -1 \\ 2 & -1 & 0 \end{pmatrix}.$$

(a) Find the eigenvalues and corresponding eigenvectors of A.

(b) Solve the homogeneous system $\dot{X} = AX$, given the initial values $X(0) = (3, 2, 3)'$.

(c) Find a solution to the non-homogeneous system $\dot{X} = AX + B$ where A is as above and $B = (1, 1, 1)'$.

Solution:

(a) $\det(A - \lambda I) = -\lambda^3 + 2\lambda^2 + \lambda - 2 = -(\lambda - 1)(\lambda + 1)(\lambda - 2) = 0$, thus the eigenvalues are $\lambda_1 = 1$, $\lambda_2 = -1$, $\lambda_3 = 2$, and the corresponding eigenvectors $v_1 = (1, 1, 1)'$, $v_2 = (1, -3, -5)'$, $v_3 = (1, 0, 1)'$.

(b) The general solution to the homogeneous system is $X_h(t) = c_1 v_1 e^t + c_2 v_2 e^{-t} + c_3 v_3 e^{2t}$. Since $X_h(0) = (3,2,3)'$, the coefficients c_1, c_2, c_3 can be found from the linear system $c_1 v_1 + c_2 v_2 + c_3 v_3 = (3,2,3)'$ as $c_1 = 2$, $c_2 = 0$, $c_3 = 1$.

(c) The general solution to the non-homogeneous system is $X_n h(t) = X_h(t) + X_p$, where the particular solution X_p is a constant vector such that $X_p = -A^{-1}B$. Solving this system of linear equation we find $X_p = (1,1,1)'$.

2. (a) Solve the homogeneous differential equation
$y''(x) + 2y'(x) + y(x) = 0$.

(b) Solve the non-homogeneous differential equation
$y''(x) + 2y'(x) + y(x) = 2x$.

(c) Solve the non-homogeneous differential equation
$y''(x) + 2y'(x) + y(x) = 2x + 3e^{-x}\sqrt{x+1}$.

Solution:

(a) The characteristic equation $\lambda^2 + 2\lambda + 1 = 0$ has a repeated root $\lambda_{1,2} = -1$, therefore the general solution to the homogeneous equation is $y_h(x) = c_1 e^{-x} + c_2 x e^{-x}$.

(b) The general solution to the non-homogeneous equation is a linear combination $y_{nh}(x) = y_h(x) + y_{p_1}(x)$. Since the right-hand side is polynomial of degree 1, the particular solution $y_{p_1}(x)$ is searched in the form $y_{p_1}(x) = ax + b$. Direct substitution reveals $a = 2$, $b = -4$.

(c) The general solution is $y_{nh}(x) = y_h(x) + y_{p_1}(x) + y_{p_2}(x)$ where $y_h(x)$ and $y_{p_1}(x)$ are as above, and $y_{p_2}(x)$ can be found using the method of undetermined coefficients: Assume $y_{p_2}(x) = (u_1(x) + x u_2(x))e^{-x}$, and find $u_1(x)$, $u_2(x)$ from

$$u_1'(x)e^{-x} + u_2'(x)xe^{-x} = 0,$$
$$-u_1'(x)e^{-x} + u_2'(x)(e^{-x} - xe^{-x}) = 3e^{-x}\sqrt{x+1}.$$

Clearly, $u_2'(x) = 3\sqrt{x+1}$, thus $u_2(x) = 2(x+1)^{3/2}$. Next, $u_1'(x) = -xu_2'(x) = -3x\sqrt{x+1} = -3(x+1)^{3/2} + 3\sqrt{x+1}$, therefore $u_1(x) = -\frac{6}{5}(x+1)^{5/2} + 2(x+1)^{3/2}$. Finally,

$$y_{p_2}(x) = \left(-\frac{6}{5}(x+1)^{5/2} + 2(x+1)^{3/2} + 2x(x+1)^{3/2}\right)e^{-x}$$

$$= \frac{4}{5}(x+1)^{5/2}e^{-x}.$$

3. Solve the differential equation $(x^2 + y)dx = xdy$.

   <u>Solution:</u> $x \equiv 0$ is a special solution.

   If $x \neq 0$ then the equation can be re-written as $\dfrac{dy(x)}{dx} - \dfrac{1}{x}y(x) = x$ (linear differential equation of the first order). It has a solution $y(x) = Cx + x^2$, where C is an arbitrary constant.

4. Consider a second order linear differential equation $y''(x)+ay'(x)+by(x) = 0$, where a and b are constants.

   Find all a and b such that the equation has *at least one* solution converging to 0 as $x \to +\infty$.

   <u>Solution:</u>   Roots of the characteristic equation $\lambda^2 + a\lambda + b = 0$ should be either complex with negative real parts (i.e. $a > 0$, $a^2 - 4b < 0$), or, if both real, at least one characteristic root must be negative ($b = \lambda_1 \cdot \lambda_2 < 0$, or $b = 0$, $\lambda_1 + \lambda_2 = -a < 0$). Summing up, either $b < 0$, or $b \geq 0$, $a > 0$.

5. Solve the difference equation $y_{t+2} - 3y_{t+1} + 2y_t = 3^t$.

   <u>Solution:</u>   The characteristic equation is $\lambda^2 - 3\lambda + 2 = 0$, the characteristic roots are $\lambda_1 = 1, \lambda_2 = 2$.

   The general solution to the homogeneous equation is $y_t^h = C_1 + C_2 2^t$.

   The particular solution to the non-homogeneous equation is $y_t^p = \frac{1}{2}3^t$. (We use the method of undetermined coefficients and try to find a solution of the form $y_t^p = A3^t$, where the value of A has to be determined by substitution.)

   The general solution to the initial non-homogeneous equation is $y_t = C_1 + C_2 2^t + \frac{1}{2}3^t$.

6. Find equilibria of the system of differential equations and classify them

$$\begin{cases} \frac{dx}{dt} = x(x + y - 2), \\ \frac{dy}{dt} = y(1 - x). \end{cases}$$

<u>Solution:</u>   There are three equilibria: $O_1(0,0)$, $O_2(1,1)$, and $O_3(2,0)$. The characteristic equation at (x_0, y_0) is

$$\det \begin{pmatrix} 2x_0 + y_0 - 2 - \lambda & x_0 \\ -y_0 & 1 - x_0 - \lambda \end{pmatrix} =$$

$$= \lambda^2 - (x_0 + y_0 - 1)\lambda + 4x_0 - 2x_0^2 + y_0 - 2 = 0.$$

$x_0 = y_0 = 0 \Rightarrow \lambda_1 = -2, \lambda_2 = 1 \Rightarrow O_1$ is a saddle.

$x_0 = y_0 = 1 \Rightarrow \text{Re}(\lambda_{1,2}) = 1/2 > 0 \Rightarrow O_2$ is an unstable focus.

$x_0 = 2, y_0 = 0 \Rightarrow \lambda_1, \lambda_2$ are real, of different signs $\Rightarrow O_3$ is a saddle.

7. Find a solution to the following system of linear differential equations:

$$\begin{cases} \dot{x} = 3x - y + z, \\ \dot{y} = x + y + z, \\ \dot{z} = 4x - y + 4z, \end{cases}$$

satisfying the initial conditions $x(0) = 2$, $y(0) = -1$, $z(0) = 0$.

<u>Solution:</u>   Eigenvalues are $\lambda_1 = 1, \lambda_2 = 2, \lambda_3 = 5$.

The corresponding eigenvectors are

$$V_1 = \begin{pmatrix} 1 \\ 1 \\ -1 \end{pmatrix}, \qquad V_2 = \begin{pmatrix} 1 \\ -2 \\ -3 \end{pmatrix}, \qquad V_3 = \begin{pmatrix} 1 \\ 1 \\ 3 \end{pmatrix}.$$

The general solution is

$$\begin{pmatrix} x(t) \\ y(t) \\ z(t) \end{pmatrix} = C_1 \begin{pmatrix} 1 \\ 1 \\ -1 \end{pmatrix} e^t + C_2 \begin{pmatrix} 1 \\ -2 \\ -3 \end{pmatrix} e^{2t} + C_3 \begin{pmatrix} 1 \\ 1 \\ 3 \end{pmatrix} e^{5t}.$$

To solve Cauchy's problem, we need to solve the linear system of three equations in three unknowns C_1, C_2 and C_3:

$$\begin{cases} C_1 + C_2 + C_3 & = & 2 \\ C_1 - 2C_2 + C_3 & = & -1 \\ -C_1 - 3C_2 + 3C_3 & = & 0 \end{cases}$$

This system has a unique solution $C_1 = 0$, $C_2 = C_3 = 1$.

8. Solve the following first-order differential equations with $y = y(x)$:
$x^2 y' + y^2 = 0$

Solution:

$$x^2 \frac{dy}{dx} + y^2 = 0$$

$$x^2 dy + y^2 dx = 0$$

$$\frac{1}{y^2} dy + \frac{1}{x^2} dx = 0$$

$$-\frac{1}{y} + c - \frac{1}{x} = 0$$

Finally,

$$y = \frac{x}{xc - 1}$$

9. Find the solutions to the following first-order differential equations that pass through given points (here $y = y(x)$):

a) $xy' + 2y = xyy';$ $(e, 1)$

Solution:

$$x\frac{dy}{dx} + 2y = xy\frac{dy}{dx}$$

$$\frac{1-y}{y} dy + \frac{2}{x} dx = 0$$

$$\left(\frac{1}{y} - 1\right) dy + \frac{2}{x} dx = 0$$

$$\ln y - y + \ln x^2 + c = 0$$

plug in the initial conditions
for x and y to find c

$$\ln y - y + \ln x^2 - 1 = 0 \qquad\qquad c = -1$$

Implicitly,

$$x^2 = \frac{e^{y+1}}{y}$$

b) $(x+1)y = x^2 y'; \quad (1,1)$

Solution:

$$(x+1)y = x^2 \frac{dy}{dx}$$

$$\frac{x+1}{x^2}dx = \frac{1}{y}dy$$

$$\left(\frac{1}{x} + \frac{1}{x^2}\right)dx = \frac{1}{y}dy$$

$$\ln x - \frac{1}{x} + c = \ln y$$

plug in the initial conditions

for x and y to find c

$$\ln x - \frac{1}{x} + 1 = \ln y \qquad c = 1$$

Finally

$$y = xe^{1-\frac{1}{x}}$$

10. Solve the following first-order differential equations (here $y = y(x)$ is a dependent variable, x is an independent variable):

(a) $y' + 2xy^2 = 2xy$.

Solution: This is a simple variables separable case:

$$y' = 2x(y - y^2)$$

$$\int \frac{dy}{y(1-y)} = \int 2x dx, \quad \text{for } y \notin \{0, 1\}$$

$$\ln\left|\frac{y}{1-y}\right| = x^2 + c$$

$$y = \frac{\exp(x^2 + c)}{1 + \exp(x^2 + c)} = \frac{\exp(x^2)}{\exp(-c) + \exp(x^2)}$$

We can also check that $y \equiv 1$ and $y \equiv 0$ are special solutions.

(b) $y' = \sqrt{2x + y + 1}$

(HINT: Introduce new variable $z = 2x + y + 1$, so that $z' = y' + 2$

Solution: After the substitution we have:

$$z' - 2 = \sqrt{z}$$

$$\int \frac{dz}{\sqrt{z} + 2} = \int dx.$$

By integration (we can use, for example, a substitution $t = \frac{1}{\sqrt{z}+2}$) we get

$$2\sqrt{z} - 4\ln\left(\sqrt{z} + 2\right) = x + c$$

so that the solution to our differential equation is given by the implicit function

$$2\sqrt{2x + y + 1} - 4\ln\left(\sqrt{2x + y + 1} + 2\right) - x = c.$$

(c) $2xyy' = x^2 + y^2$

Solution: Let us use the substitution $z = y^2$. Then we have $z' = 2yy'$, so that $xz' = x^2 + z$. We obtain a first-order linear differential equation $z' - \frac{1}{x}z - x = 0$, with solution:

$$z(x) = e^{-\int -\frac{dx}{x}}\left(c + \int xe^{\int -\frac{dx}{x}}dx\right)$$

$$= x\left(c + \int xx^{-1}dx\right)$$

or

$$z(x) = x(c + x)$$

or

$$y^2(x) = x^2 + cx.$$

If we use a substitution $y = tx$, the same solution can be obtained as follows:

$$2xy\frac{dy}{dx} = x^2 + y^2$$

$$2xydy = (x^2 + y^2)dx$$

$$\text{substitution } \begin{bmatrix} y = tx \\ dy = tdx + xdt \end{bmatrix}$$

$$2x^2t(tdx + xdt) = (x^2 + t^2x^2)dx$$

$$2t^2dx + 2xtdt = dx + t^2dx$$

$$(t^2 - 1)dx + 2xtdt = 0$$

$$-\frac{dx}{x} = \frac{2t}{t^2 - 1}dt$$

$$\text{substitution} \begin{bmatrix} t^2 - 1 = z \\ d2tdt = dz \end{bmatrix}$$

$$-\frac{dx}{x} = \frac{dz}{z}$$

$$-\ln x + c = \ln z$$

$$c = xz$$

plug substitutions back in

$$c = x\left(\frac{y^2}{x^2} - 1\right)$$

$$x(c + x) = y^2$$

(d) $xy' = x^3 + 2y$

Solution: The given differential equation is equivalent to the following first-order linear differential equation:

$$y' - \frac{2}{x}y = x^2.$$

Its solution is given by the formula:

$$y(x) = e^{\int \frac{2}{x}dx}\left(C + \int x^2 e^{-\int \frac{2}{x}dx}dx\right)$$

$$= x^2\left(C + \int x^2 x^{-2}dx\right)$$

$$= x^2(C + x)$$

(e) $(x^2 + e^y)dx + (xe^y + 2y)dy = 0$

Solution: We can check that $\frac{\partial(x^2 + e^y)}{\partial y} = \frac{\partial(xe^y + 2y)}{\partial x}$, thus we deal with an exact differential equation, and we try to find such function $F(x, y)$ that $\frac{\partial F}{\partial x} = x^2 + e^y$ and $\frac{\partial F}{\partial y} = xe^y + 2y$.

$$F'_x = x^2 + e^y \implies F(x, y) = \frac{x^3}{3} + e^y x + \phi(y)$$

and

$$F'_y = xe^y + 2y = e^y x + \phi'(y)$$

$$\implies \phi(y) = \int 2ydy = y^2 + c.$$

Therefore, all solutions are given by the implicit function
$\frac{x^3}{3} + e^y x + y^2 = C.$

11. For what values of (x_0, y_0) does the differential equation $xy' = \sqrt{x - y}$ have a unique solution $y = \phi(x)$ such that $\phi(x_0) = y_0$?

Solution: After rewriting the equation in the form $y' = f(x, y)$ we obtain

$$y' = \frac{\sqrt{x - y}}{x}.$$

By the uniqueness theorem, in order to have a unique solution $y = \phi(x)$, the functions f and $\frac{\partial f}{\partial y}$ must be continuous in an open domain D, such that $y_0 = \phi(x_0)$, for $(x_0, y_0) \in D$. Note that continuity of f satisfies the existence of a solution. In our case, the functions

$$f(x, y) = \frac{\sqrt{x - y}}{x} \quad \text{and} \quad \frac{\partial f}{\partial y}(x, y) = -\frac{1}{2x} \cdot \frac{1}{\sqrt{x - y}}$$

must be continuous. This condition holds in the open domain

$$D = \{(x, y) \in \mathbb{R} : x > y, \ x \neq 0\}.$$

Therefore a unique solution $y = \phi(x)$ such that $y_0 = \phi(x_0)$ exists iff $(x_0, y_0) \in D$.

12. Solve the following linear differential equations with constant coefficients:

(a) $y''' - 2y'' + y' = 0$

Solution: Let us write the characteristic equation:

$$\lambda^3 - 2\lambda^2 + \lambda = \lambda(\lambda - 1)^2 = 0.$$

It has one simple root $\lambda_1 = 0$ and one repeated root of degree two $\lambda_2 = \lambda_3 = 1$. Therefore, the general solution to our equation has the following form:

$$y(x) = c_1 + c_2 e^x + c_3 x e^x.$$

(b) $y^{(4)} - y = 0$

Solution: The characteristic equation now takes the following form:

$$\begin{aligned}
\lambda^4 - 1 &= (\lambda - 1)(\lambda + 1)(\lambda^2 + 1) \\
&= (\lambda - 1)(\lambda + 1)(\lambda - i)(\lambda + i) = 0,
\end{aligned}$$

and the solution is:

$$y(x) = c_1 e^x + c_2 e^{-x} + c_3 \cos x + c_4 \sin x$$

(c) $y^{(4)} + 2y'' + y = 0$

Solution: The characteristic equation $\lambda^4 + 2\lambda^2 + 1 = 0$ can be solved with the substitution $\lambda^2 = \mu$. Then we get:

$$\mu^2 + 2\mu + 1 = (\mu + 1)^2 = 0, \quad \text{implying} \quad \mu_{1,2} = -1.$$

Thus for λ we have the following roots:

$$\lambda_{1,2} = i, \quad \lambda_{3,4} = -i$$

In this case the following solution is generated:

$$y(x) = c_1 \cos x + c_2 x \cos x + c_3 \sin x + c_4 x \sin x$$

(d) $y'' + 2y' + 5y = e^{-x}$

Solution: The characteristic equation of the homogeneous equation has the following root:

$$\lambda^2 + 2\lambda + 5 = 0, \quad \text{thus} \quad \lambda_{1,2} = -1 \pm 2i.$$

This gives us the following general solution:

$$y_G(x) = c_1 e^{-x} \cos 2x + c_2 e^{-x} \sin 2x.$$

We will try to find the particular solution in the following form

$$y_p(x) = A e^{-x}$$

Then, after substituting into the non-homogenous equation we obtain

$$A e^{-x} - 2A e^{-x} + 5A e^{-x} = e^{-x},$$

so that

$$4A = 1, \quad \text{or} \quad a = \frac{1}{4}.$$

Therefore,

$$y(x) = y_G(x) + y_p(x) = c_1 e^{-x} \cos 2x + c_2 e^{-x} \sin 2x + \frac{1}{4}e^{-x}.$$

(e) $y'' - y' - 2y = e^{2x}$

Solution: The characteristic equation has the following form:

$$\lambda^2 - \lambda - 2 = (\lambda - 2)(\lambda + 1) = 0,$$

and the general solution is

$$y_G(x) = c_1 e^{2x} + c_2 e^{-x}.$$

If λ were not equal to 2, the particular solution would be $y_p(x) = Ae^{2x}$; however, this is not our case and the particular solution in the form $y_p(x) = Ae^{2x}$ can not be found (just substitute it into the differential equation and check what happens). Instead, we have to try $y_p(x) = Axe^{2x}$:

$$(2Ae^{2x} + 2Ae^{2x} + 4Axe^{2x}) -$$

$$-(Ae^{2x} + 2Axe^{2x}) - 2Axe^{2x} = e^{2x},$$

which yields $3Ae^{2x} = e^{2x}$, or $A = \frac{1}{3}$. The solution thus has the form:

$$y(x) = c_1 e^{2x} + c_2 e^{-x} + \frac{1}{3}xe^{2x}.$$

(f) $y'' + y' = xe^{-x} - x^2$

Solution: The characteristic equation has two roots: $\lambda_1 = 0$ and $\lambda_2 = -1$, and the general solution is $y_G(x) = c_1 + c_2 e^{-x}$. For the particular solution we will use the method of undetermined coefficients:

$$y_p(x) = u_1(x)y_1(x) + u_2(x)y_2(x).$$

We have

$$u_1'(x) + u_2'(x)e^{-x} = 0,$$
$$-u_2'(x)e^{-x} = xe^{-x} - x^2.$$

Therefore, $u_1'(x) = xe^{-x} - x^2$, so that $u_1(x) = -(x+1)e^{-x} - \frac{x^3}{3}$. Furthermore, $u_2'(x) = -u_1'(x)e^x = x^2e^x - x$ and the latter implies $u_2(x) = e^x(x^2 - 2x + 2) - \frac{x^2}{2}$.

Thus the solution to the non-homogeneous equation is

$$y(x) = c_1 + c_2e^{-x} - \left(\frac{x^2}{2} + x\right)e^{-x} -$$

$$-\frac{x^3}{3} + x^2 - 2x - e^{-x} + 2.$$

Note that the last two terms (i.e. $-e^{-x} + 2$) are redundant since they are already accounted for in the general solution to the homogeneous equation.

Alternatively, $y_p(x)$ can be found as a sum $y_p(x) = y_{p1}(x) + y_{p2}(x)$, where $y_{p1}(x)$ and $y_{p2}(x)$ are particular solutions to non-homogeneous equations $y'' + y' = xe^{-x}$ and $y'' + y' = -x^2$, respectively.

(g) $y'' + 3y' + 2y = \frac{1}{e^x + 1}$

(HINT: Use the method of undetermined coefficients).

Solution: The homogeneous equation has the roots $\lambda_1 = -1$ and $\lambda_2 = -2$, therefore the general solution is $y_G(x) = c_1e^{-x} + c_2e^{-2x}$. Let us use the method of undetermined coefficients:

$$y_p(x) = u_1(x)e^{-x} + u_2(x)e^{-2x},$$

yielding

$$u_1'(x)e^{-x} + u_2'(x)e^{-2x} = 0$$

$$-u_1'(x)e^{-x} - 2u_2'(x)e^{-2x} = \frac{1}{e^x + 1}$$

Adding up,

$$-u_2'(x)e^{-2x} = \frac{1}{e^x + 1}$$

$$u_2'(x) = \frac{-e^{2x}}{e^x + 1}$$

$$u_2(x) = \ln(1 + e^x) - e^x$$

$$u_1'(x) = -u_2'(x)e^{-x} = \frac{e^{2x}}{e^x + 1}e^{-x} = \frac{e^x}{e^x + 1}$$

$$u_1(x) = \ln(1 + e^x)$$

Thus the solution is summarized as

$$
\begin{aligned}
y(x) &= c_1 e^{-x} + c_2 e^{-2x} + \ln(1 + e^x)e^{-x} + \\
&\quad + (\ln(1 + e^x) - e^x)e^{-2x} \\
&= \tilde{c}_1 e^{-x} + c_2 e^{-2x} + \ln(1 + e^x)(e^{-x} + e^{-2x}).
\end{aligned}
$$

13. Consider the following first-order differential equations (unless stated otherwise, $y = y(x)$ is a dependent variable, x an independent variable):

(a) Solve $(xy+1)dx - (x^2+1)dy = 0$. Show that for any pair of real numbers (x_0, y_0) there exists a unique solution $y = \phi(x)$ such that $\phi(x_0) = y_0$.

Solution: After rearranging terms the equation becomes linear:

$$
y'(x) - \frac{x}{x^2 + 1}y(x) = \frac{1}{x^2 + 1}.
$$

Using the well-known formula, the solution can be expressed as

$$
y(x) = e^{-\int -\frac{x}{x^2+1}dx}\left(C + \int \frac{1}{x^2 + 1}e^{\int -\frac{x}{x^2+1}dx}dx\right) \quad (5.17)
$$

First let us compute

$$
\begin{aligned}
\int \frac{x}{x^2 + 1}dx &= \frac{1}{2}\int \frac{2x}{x^2 + 1}dx \\
&= \frac{1}{2}\ln(x^2 + 1) = \ln\sqrt{x^2 + 1}.
\end{aligned}
$$

Thus (5.17) becomes

$$
\begin{aligned}
y(x) &= \sqrt{x^2 + 1}\left(C + \int \frac{1}{(x^2 + 1)^{3/2}}dx\right) \\
&= x + C\sqrt{x^2 + 1}.
\end{aligned}
$$

The integral in brackets can be evaluated by substitution $t^2 = 1/(1 + x^2)$ which yields $-\int t/\sqrt{1 - t^2}dt$. The latter can be integrated either by substitution $u = \sqrt{1 - t^2}$ that reduces $-\int t/\sqrt{1 - t^2}dt$ to $\int du$, or simply by noting that $d\sqrt{1-t^2}/dt = -t/\sqrt{1-t^2}$; thus $\displaystyle\int \frac{1}{(x^2+1)^{3/2}}dx = \dfrac{x}{\sqrt{x^2+1}}$.

To show the existence and uniqueness of the solution, let us refer to the existence theorem. We write the equation in the form $y'(x) = f(x,y)$, where $f(x,y) = (1+xy)/(1+x^2)$. This function is obviously continuous on $\mathbf{R}^2$ and so is $\partial f/\partial y = x/(1+x^2)$. The assumptions of the theorem are satisfied, therefore for any given (x_0, y_0) there exists a unique solution $y = \phi(x)$ such that $\phi(x_0) = y_0$.

(b) Find the solution to $y'(x) = xy^2 - x$, passing through the point $(x = 2, y = 3)$.

Solution: This is the variables separable case. First, note two special solutions $y(x) \equiv 1$ and $y(x) \equiv -1$. All other solutions are obtained by integration of separated variables:

$$dy = x(y^2-1)dx,$$

$$\int \frac{dy}{(y-1)(y+1)} = \int x\,dx, \quad \text{for } y \notin \{-1,1\}$$

$$\frac{1}{2}\int \frac{1}{(y-1)} - \frac{1}{(y+1)}dy = \frac{1}{2}(x^2+c), \quad \text{for } y \notin \{-1,1\}$$

$$\ln\left|\frac{y-1}{y+1}\right| = x^2+c$$

Furthermore, at $x = 2$, $y(2) = 3$. Therefore we can evaluate c from the equation

$$\ln\left|\frac{3-1}{3+1}\right| = 2^2 + c,$$

$$\text{or } c = \ln 1/2 - 4\,.$$

Finally, the solution to the given Cauchy's problem becomes

$$y = \frac{1 + \exp(x^2+c)}{1 - \exp(x^2+c)} = \frac{2 + \exp(x^2-4)}{2 - \exp(x^2-4)}.$$

(c) Solve $x^2 dy = (xy + y^2 e^{-x/y})dx$. (HINT: Try the substitution $y = tx$.)

Solution: Substitution $y = tx$ implies $x/y = 1/t$ and $dy = t\,dx + x\,dt$. After plugging it into the original equation, the variables can be separated:

$$
\begin{aligned}
x^2(tdx + xdt) &= (x(tx) + (tx)^2 e^{-1/t})dx, \\
x^3 dt &= x^2 t^2 e^{-1/t} dx, \\
\int \frac{e^{1/t}}{t^2} dt &= \int \frac{dx}{x}, \\
-e^{1/t} &= \ln|x| + c, \\
-e^{x/y} &= \ln|x| + c.
\end{aligned}
$$

From the latter equation we find

$$
y = \frac{x}{\ln(-\ln|x| - c)}.
$$

(d) Find all solutions to $y'(x) = y^2 - \dfrac{2}{x^2}$.   (HINT: Try the substitution $y = t/x$.)

Solution: With $y = t/x$, $dy = (xdt - tdx)/x^2$. Such a substitution again brings the original equation into variables separable form:

$$
\begin{aligned}
\frac{xdt - tdx}{x^2} &= \left(\frac{t^2}{x^2} - \frac{2}{x^2} \right) dx, \\
\frac{dt}{t^2 + t - 2} &= \frac{dx}{x}, \quad \text{for } t \notin \{1, -2\} \\
\frac{1}{3} \int \frac{1}{t-1} - \frac{1}{t+2} dt &= \int \frac{dx}{x}, \\
\frac{1}{3} \ln \left| \frac{t-1}{t+2} \right| &= \ln|x| + c,
\end{aligned}
$$

so that the solution is given implicitly by the following formula:

$$
\frac{xy - 1}{xy + 2} = cx^3.
$$

Additionally, $t = 1$ and $t = -2$ give rise to two special solutions $y(x) = \dfrac{1}{x}$ and $y = -\dfrac{2}{x}$.

(e) Solve $y' = \sqrt{4x + 2y - 1}$.

(HINT: Introduce the new variable $z = 4x + 2y - 1$, so that $z' = 2y' + 4$ and $y' = \sqrt{z}$.)

<u>Solution:</u>   After substitution the equation becomes:

$$\frac{z' - 4}{2} = \sqrt{z}$$

$$\frac{1}{2} \int \frac{dz}{\sqrt{z} + 2} = \int dx$$

To integrate the latter expression we can use, for instance, the substitution $\sqrt{z} + 2 = t$. Integration yields

$$\sqrt{z} + 2 - 2\ln\left(\sqrt{z} + 2\right) = x + c,$$

so that the solution to the original differential equation is given by the following implicit function

$$\sqrt{4x + 2y - 1} + 2 - 2\ln\left(\sqrt{4x + 2y - 1} + 2\right) - x = c.$$

(f) For what values of (x_0, y_0) does the differential equation $e^{x^2} y' = y + \sqrt{y^2 - x^2}$ have a unique solution $y = \phi(x)$ such that $\phi(x_0) = y_0$?

<u>Solution:</u>   According to the uniqueness theorem, if the functions f and $\frac{\partial f}{\partial y}$ are continuous in an open domain D, then for any pair $(x_0, y_0) \in D$ there exists a unique solution $\phi(x) = y$ of $y'(x) = f(x, y)$, such that $y_0 = \phi(x_0)$. Note that the continuity of f guarantees the existence of a solution. In our case, the functions

$$f(x, y) = \frac{y + \sqrt{y^2 - x^2}}{e^{x^2}}$$

and

$$\frac{\partial f}{\partial y}(x, y) = \frac{1}{e^{x^2}} \cdot \left(1 + \frac{y}{\sqrt{y^2 - x^2}}\right)$$

are continuous in all points where the square root is positive, i.e. in the open domains

$$D^+ = \{(x, y) \in \mathbf{R} \times \mathbf{R}^+ : |x| < y\},$$

$$D^- = \{(x, y) \in \mathbf{R} \times \mathbf{R}^- : |x| < -y\}.$$

Therefore a unique solution $y = \phi(x)$ such that $y_0 = \phi(x_0)$ exists iff $|x_0| < |y_0|$.

(g) Solve $y = (2x + y^3)y'$.

(HINT: Take $x = x(y)$ as a dependent variable, y an independent variable.)

Solution: First note that $y \equiv 0$ is a solution. In what follows, let us assume $y \neq 0$. We rewrite the equation as

$$
\begin{aligned}
ydx &= (2x + y^3)dy, \\
\frac{dx}{dy} &= \frac{2x + y^3}{y}, \\
x'(y) - x\frac{2}{y} &= y^2.
\end{aligned}
$$

As we can see, this equation is linear, and the usual formula applies:

$$
x(y) = e^{-\int -\frac{2}{y}dy}\left(C + \int y^2 e^{\int -\frac{2}{y}dy}dy\right).
$$

We can easily find $\int -\frac{2}{y}dy = -\ln y^2 = \ln 1/y^2$. Therefore,

$$
\begin{aligned}
x(y) &= e^{\ln y^2}\left(C + \int y^2 e^{\ln 1/y^2}dy\right) \\
&= {}^2\left(C + \int dy\right) = Cy^2 + y^3.
\end{aligned}
$$

The above formula defines function $y(x)$ implicitly.

(h) Solve $xy'(x) + y = y^2\ln(x)$.   (HINT: Try the substitution $z = 1/y$.)

Solution: If $z = 1/y$ then $dz = \frac{-dy}{y^2} = -z^2 dy$, so that $\frac{dy}{dx} = -\frac{1}{z^2}\frac{dz}{dx}$. Plugging into the original equation we obtain

$$
\begin{aligned}
\frac{-x}{z^2}\frac{dz}{dx} + \frac{1}{z} &= \frac{1}{z^2}\ln x, \\
\frac{xdz - zdx}{x^2} &= -\frac{\ln x}{x^2}dx, \\
d\left(\frac{z}{x}\right) &= -\frac{\ln x}{x^2}dx, \\
\frac{z}{x} &= -\int \frac{\ln x}{x^2}dx = \frac{\ln x + 1}{x}.
\end{aligned}
$$

From the preceding equation we find

$$y = \frac{1}{z} = \frac{1}{\ln x + 1}.$$

(i) Solve $2x(1 + \sqrt{x^2 - y})dx - \sqrt{x^2 - y}\,dy = 0$.

Solution: Let us make a substitution $t = \sqrt{x^2 - y}$. Thus $dt = \frac{1}{2t}(2xdx - dy)$, and $dy = 2xdx - 2tdt$. In $(t - x)$ variables the original equation becomes

$$
\begin{aligned}
2x(1 + t)dx &= t(2xdx - 2tdt), \\
2xdx &= -2t^2 dt, \\
\int xdx &= -\int t^2 dt, \\
\frac{1}{2}x^2 + c &= \frac{-t^3}{3} = \frac{-\left(\sqrt{x^2 - y}\right)^3}{3}.
\end{aligned}
$$

Expressing y as a function of x,

$$y = x^2 - \left(\frac{3}{2}(x^2 + c)\right)^{\frac{2}{3}}.$$

Another way to find the solution is to note that the differential equation is exact (and solve accordingly).

(j) Solve $(y'(x))^3 - (\sqrt{x} + 2^x)(y'(x))^2 + y'(x) = \sqrt{x} + 2^x$.
(HINT: Try to factor the equation first.)
Solution: Let denote $m = y'(x)$ and $n = \sqrt{x} + 2^x$. Then the original equation

$$m^3 - m^2 n + m - n = 0$$

can be rewritten as

$$(m - n)(m^2 + 1) = 0.$$

Obviously, $m^2 + 1$ is always positive, therefore $m = n$, or

$$y'(x) = \sqrt{x} + 2^x.$$

The solution to this simple variables separable equation is

$$y(x) = \int \sqrt{x} + 2^x = \frac{2}{3}x^{3/2} + \frac{2^x}{\ln 2} + c.$$

14. Consider a first order linear differential equation $y'(x) + y = f(x)$ with bounded right-hand side, i.e. $|f(x)| \leq Const \ \forall x \in \mathbf{R}$. Show that this equation has a unique solution $y = \phi(x)$ that is bounded for all real x.

Solution: Assume that $-M \leq f(x) \leq M$ for all real x. As we can see, this equation is linear, so it solution is

$$y(x) = e^{-x} \left(C + \int f(x)e^x dx \right),$$

where $y_h(x) = Ce^{-x}$ is a general solution to the homogeneous equation and

$$y_p(x) = e^{-x} \int f(x)e^x dx$$

is a particular solution to the non-homogeneous equation.

Obviously there cannot be two solutions which are bounded. Otherwise their difference would be a solution to the homogeneous equation, i.e., of the form $y_h(x) = Ce^{-x}$. This is bounded only for $C = 0$ (because $e^{-x} \to \infty$ as $x \to -\infty$).

Now it remains to be shown that there exists a solution which is bounded. If $f(x)$ is continuous, then there exists a solution to the original equation; we can set, for example, $y(0) = 0$. The obvious candidate for the bounded solution is the above particular solution $y_p(x)$, if we write it in the form $y_p(x) = e^{-x} \int_{-\infty}^{x} f(t)e^t dt$. Then we can use the inequality

$$\int_{-\infty}^{x} f(t)e^t dt \leq \int_{-\infty}^{x} Me^t dt = Me^x.$$

This yields $y_p(x) \leq M$ for all real x. Analogically we can show that $y_p(x) \geq -M$. Hence $y_p(x)$ is bounded.

5.3   Economics Applications

1. In economics, the two most common ways of averaging values $x_1, x_2, \ldots, x_n$ are to compute the arithmetic average $\bar{x} = \frac{1}{n}(x_1 + $

$x_2 + \ldots + x_n$), and the geometric average $\hat{x} = \sqrt[n]{x_1 x_2 \ldots x_n}$. Show that $\hat{x} \leq \bar{x}$.

<u>Solution:</u> First note that for the inequality $\hat{x} \leq \bar{x}$ to hold, we must have $x_1, \ldots, x_n$ positive. To prove the inequality, consider the function

$$F(x_1, \ldots, x_n) = \frac{x_1 + \ldots + x_n}{\sqrt[n]{x_1 x_2 \ldots x_n}}.$$

If we find its minimum and show that the minimal value of F is greater than or equal to n, we are done. First note, that

$$F(tx_1, \ldots, tx_n) = F(x_1, \ldots, x_n), \quad \text{for any } t > 0. \qquad (5.18)$$

So, to find minimum, we can restrict ourselves to finding the minimum for such $(x_1, \ldots, x_n)$ that $x_1 x_2 \ldots x_n = 1$. Expressing x_n from this and substituting it into the formula for F, we obtain

$$
\begin{aligned}
G(x_1, \ldots, x_{n-1}) &= F\left(x_1, \ldots, x_{n-1}, \frac{1}{x_1 x_2 \ldots x_{n-1}}\right) \\
&= x_1 + \ldots + x_{n-1} + \frac{1}{x_1 x_2 \ldots x_{n-1}}.
\end{aligned}
$$

To minimize the function G, we use the methods of calculus. Differentiating w.r.t. x_i we obtain the F.O.C.

$$\frac{\partial G}{\partial x_i} = 1 - \frac{1}{x_i} \cdot \frac{1}{x_1 x_2 \ldots x_{n-1}} = 0.$$

From this we can easily see that $x_1 = \ldots = x_{n-1} = 1$.

To check whether this is really a minimum, we compute the second-order derivatives for $x_1 = \ldots = x_{n-1}$: $\partial^2 G / \partial x_i^2 = 2$ and $\partial^2 G / \partial x_i \partial x_j = 1$ (for $i \neq j$). Then the Hessian matrix is

$$
\begin{pmatrix}
2 & 1 & \ldots & 1 \\
1 & 2 & \ldots & 1 \\
\vdots & \vdots & \ddots & \vdots \\
1 & 1 & \ldots & 2
\end{pmatrix}
$$

Denote D_n determinant of such $[n \times n]$ matrix. Using the Laplace expansion with respect to the first column, we can prove that

$$D_n = D_{n-1} - (n - 1).$$

Moreover we know that $D_1 = 2$. Using mathematical induction, we can easily show that $D_n = n + 1$. Therefore, all principal minors have a positive determinant, which means that $x_1 = \ldots = x_{n-1} = 1$ is a (global) minimum.

So we have proved that $G(x_1, \ldots, x_{n-1}) \geq G(1, \ldots, 1) = n$, for any $x_1, \ldots, x_{n-1}$. Therefore, also

$$F(x_1, \ldots, x_n) \geq F(1, \ldots, 1) = n$$

for all $x_1, \ldots, x_n$ such that $x_1 x_2 \ldots x_n = 1$. Using (5.18), we get $F(x_1, \ldots, x_n) \geq n$ for all $x_1, \ldots, x_n$, which proves the inequality $\hat{x} \leq \bar{x}$.

<u>Another solution:</u> (without methods of calculus)
First we prove so-called *Young's inequality* claiming that

$$px + qy \geq x^p y^q, \tag{5.19}$$

for any $x, y > 0$ and any $p, q \in [0, 1]$ such that $p + q = 1$. An elegant proof can be done as follows. Consider $x = e^a$, $y = e^b$. Then, the inequality can be rewritten as $pe^a + qe^b \geq e^{pa+qb}$, which is equivalent to convexity of exponential function. Clearly, exponential function is convex (use either a second-order derivative or just look at the graph), so the inequality holds, and therefore, the inequality (5.19) also holds.

We prove the inequality

$$\frac{x_1 + \ldots + x_n}{n} \geq \sqrt[n]{x_1 x_2 \ldots x_n}$$

using mathematical induction.

For $n = 1$, the inequality holds with equality. For $n = 2$, the inequality $x_1 + x_2 \geq 2\sqrt{x_1 x_2}$ is equivalent to $(\sqrt{x_1} - \sqrt{x_2})^2 \geq 0$, which clearly holds.

Now, assume that the inequality holds for $n - 1$, i.e., for any $y_1, \ldots, y_{n-1} > 0$, we have

$$\frac{y_1 + \ldots + y_{n-1}}{n - 1} \geq \sqrt[n-1]{y_1 y_2 \cdots y_{n-1}}. \tag{5.20}$$

Consider arbitrary $x_1, \ldots, x_n > 0$. We will substitute $y_i = \frac{n-1}{n} x_i + \frac{1}{n} x_n$ (for $i = 1, \ldots, n-1$) into (5.20). Then its left-hand side will be

$$\frac{\left(\frac{n-1}{n} x_1 + \frac{1}{n} x_n\right) + \ldots + \left(\frac{n-1}{n} x_{n-1} + \frac{1}{n} x_n\right)}{n - 1} = \frac{x_1 + \ldots + x_{n-1} + x_n}{n}.$$

Moreover, using Young's inequality (5.19), we get $y_i = \frac{n-1}{n} x_i + \frac{1}{n} x_n \geq x_i^{(n-1)/n} x_n^{1/n}$, which for the left-hand side means:

$$\sqrt[n-1]{\left(\frac{n-1}{n} x_1 + \frac{1}{n} x_n\right) \cdots \left(\frac{n-1}{n} x_{n-1} + \frac{1}{n} x_n\right)} \geq$$

$$\geq \sqrt[n-1]{x_1^{(n-1)/n} x_n^{1/n} \cdots x_{n-1}^{(n-1)/n} x_n^{1/n}} = \sqrt[n]{x_1 \cdots x_{n-1} x_n}.$$

This gives us the inequality for n which we wanted to prove. The proof is complete.

2. Consider the following Nordhaus-type optimal control problem

$$\max_{E(t)} \int_0^\infty U(C(t)) e^{-rt} dt$$

s.t. $\dot{M} = aE - bM, \quad M(0) = M_0, \quad M(t) \leq M_{max} \; \forall t \geq 0,$

where r, a and b are positive constants, the utility function $U(C)$ is linear in its argument, and $C(t) = f(E(t)) - h(M(t))$ with $f(\cdot)$ being an increasing and concave function and $h(\cdot)$ being an increasing and convex function.

(a) Write down the first-order conditions for optimality.

(b) Is there an equilibrium in the model? If so, can you classify its type?

Solution:

(a) With the utility function $U = \alpha(f(E) - h(M))$, the current-value Hamiltonian takes the form $H = U + \lambda(aE - bM) = \alpha(f(E) - h(M)) + \lambda(aE - bM)$. The F.O.C. for optimality thus become

$$\frac{\partial H}{\partial E} = \alpha f'(E) + a\lambda = 0,$$

$$\frac{\partial H}{\partial M} = -\alpha h'(M) - b\lambda = -\dot{\lambda} + r\lambda,$$

plus the transversality condition.

(b) We can express $\dot{E}$ from the equation $\dot{M} = aE - bM$ as $E = (\dot{M} + bM)/a$ and construct a phase diagram in the $(M - \lambda)$ plane.

Alternatively, in the $(E - M)$ plane the F.O.C. imply

$$\dot{M} = aE - bM, \tag{5.21}$$

$$\dot{E} = -\frac{ah'(M)}{f''(E)} + (b + r)\frac{f'(E)}{f''(E)}. \tag{5.22}$$

In the equilibrium, (5.21) yields $M = \frac{b}{a}E$ that is upward sloping, and (5.22) implies $ah'(M) = (b + r)f'(E)$. In the latter, $dM/dE = (b + r)f''(E)/ah''(M) < 0$, i.e. the isocline $ah'(M) = (b + r)f'(E)$ is downward sloping. Therefore, the equilibrium exists and is unique. Checking the eigenvalues of the linearized system in the equilibrium, we find that the equilibrium is saddle.

3. It has been stated by an analyst that the number of inventions in the IT area in $Year_t$ is comparable to the aggregate number of inventions made in this area since the first appearance of the IBM PC in the early 80s till the end of $Year_{t-1}$. If we believe the statement to be true and such a trend to be persistent, what would the time profile of this invention activity look like?

Mathematically speaking, the problem can be refined as follows: solve the integral equation

$$I(t) = \alpha \int_0^t I(u)du + rt + I_0$$

for $I(t)$, the number of inventions at time t. Here $t = 0$ demarcates the beginning of the PC revolution, I_0 denotes the number of inventions at $t = 0$, α is a positive parameter, and rt captures the linear trend of technological progress.

(HINT: Differentiate with respect to t and solve the differential equation. Do not forget about the initial values.)

<u>Solution:</u>   Differentiating the integral equation we obtain the following differential equation: $\dot{I} = \alpha I + r$. Substituting $t = 0$ into the integral equation, the value of the integral will be zero and we obtain the following initial condition: $I(0) = I_0$. Solving the homogeneous equation we obtain $I(t) = Ce^{\alpha t}$. We can find (after recognizing that we are looking for a constant) a particular solution $I_p(t) = -\frac{r}{\alpha}$. Therefore the general solution is $I(t) = Ce^{\alpha t} - \frac{r}{\alpha}$. Imposing the initial condition we obtain $C = I_0 + \frac{r}{\alpha}$, and so

$$I(t) = \left(I_0 + \frac{r}{\alpha}\right)e^{\alpha t} - \frac{r}{\alpha}.$$

4. Consider the following discrete-time system describing a closed economy:

$$
\begin{aligned}
y_t &= i_t + c_t, \\
i_t &= i^A + \beta\psi(y_{t-1} - \psi y_{t-2}), \\
c_t &= c^A + \gamma\psi y_{t-1}.
\end{aligned}
$$

Here y_t, i_t, c_t denote output, investment, and consumption at time t, respectively; i^A and c^A are constant autonomous investment and consumption, the inverse growth parameter $\psi \in (0,1)$, the accelerator $\beta > 0$, and the marginal propensity of consumption $\gamma \in (0,1)$.

(a) Derive the difference equation for y_t and solve it.

(b) Can you find the conditions under which there exists a unique stable equilibrium in this economy?

Solution:

(a) $y_t = i^A + \beta\psi(y_{t-1} - \psi y_{t-2}) + c^A + \gamma\psi y_{t-1}$, or $y_t - \psi(\beta + \gamma)y_{t-1} + \psi^2\beta y_{t-2} = i^A + c^A$. Therefore $y_t = c_1\lambda_1^t + c_2\lambda_2^t + y^*$, where

$$y^* = \frac{i^A + c^A}{1 - \psi(\beta + \gamma) + \psi^2\beta} \text{ and } \lambda_{1,2} \text{ solve the characteristic}$$

equation $\lambda^2 - \psi(\beta + \gamma)\lambda + \psi^2\beta = 0$.

(b) The equilibrium is always unique; it is stable iff $|\lambda_1| < 1$ and $|\lambda_2| < 1$. Since $\lambda_1\lambda_2 = \beta\psi^2 > 0$ and $\lambda_1 + \lambda_2 = \psi(\beta + \gamma) > 0$, both eigenvalues are positive. Therefore, to secure stability we have to require $\lambda_1 < 1$ and $\lambda_2 < 1$.

5. The following result is important in econometrics theory: Show that $(x^a + y^a)^{1/a} > (x^b + y^b)^{1/b}$ for $x, y > 0$ and $0 < a < b$.

   Solution: Dividing the inequality by x, we obtain an equivalent inequality

   $$\left[1 + \left(\frac{y}{x}\right)^a\right]^{1/a} > \left[1 + \left(\frac{y}{x}\right)^b\right]^{1/b}.$$

   To prove it, we will prove the inequality $(1 + z^a)^{1/a} > (1 + z^b)^{1/b}$. Substituting $z = y/x$, we obtain the above inequality. The last inequality contains only one variable. To prove it, we minimize the function

   $$f(z) = (1 + z^a)^{1/a} - (1 + z^b)^{1/b}$$

   for $z \geq 0$. Taking the derivative, we obtain the F.O.C.

   $$\begin{aligned} f'(z) &= (1 + z)^{1/a-1}z^{a-1} - (1 + z)^{1/b-1}z^{b-1} = \\ &= (1 + z)^{1/b-1}z^{b-1}\left((1 + z)^{1/ab}(1 + 1/z)^{b-a} - 1\right), \end{aligned}$$

   which is positive, because $(1 + z)^{1/ab} > 1$ and $(1 + 1/z)^{b-a} > 1$ for $z > 0$ and $b > a$. Therefore, f is increasing on $[0, +\infty)$, which means that $f(z) > f(0) = 0$ for any $z > 0$. This completes the proof.

6. The *Constant Elasticity of Substitution* production function (CES) is defined to be $Y(L, K) = A(\alpha L^{-\rho} + (1 - \alpha)K^{-\rho})^{-1/\rho}$, where A is a positive constant, $0 < \alpha < 1$ and $\rho \geq -1$. It includes a wide

range of production functions, from linear-additive ($\rho = -1$) to Cobb-Douglas ($\rho = 0$) to Leontief ($\rho = +\infty$).

Check that as $\rho \to 0$, the CES production function indeed takes Cobb-Douglas specification. Next, show that as ρ approaches $+\infty$, the CES production function becomes Leontief, or a fixed-proportion production function: $Y(L, K) = \min\left(\dfrac{L}{a_1}, \dfrac{K}{a_2}\right)$, where a_1 and a_2 are constants (i.e. no effective substitution between factors of production is possible).

<u>Solution:</u>   We can rewrite the CES function as follows:

$$A(\alpha L^{-\rho} + (1-\alpha)K^{-\rho})^{-1/\rho} = AL\left[\alpha + (1-\alpha)\left(\frac{K}{L}\right)^{-\rho}\right]^{-1/\rho}$$

Define $z = K/L$ and consider now the function $g_\rho(z) = [\alpha + (1-\alpha)z^{-\rho}]^{-1/\rho}$. As $\rho \to 0$, then we get a limit of the type $\frac{0}{0}$, so we use L'Hôpital's Rule (note that we differentiate with respect to ρ):

$$
\begin{aligned}
\lim_{\rho \to 0} \ln g_\rho(z) &= -\lim_{\rho \to 0} \frac{\ln\left[\alpha + (1-\alpha)z^{-\rho}\right]}{\rho} \\
&= -\lim_{\rho \to 0} \frac{-(1-\alpha)z^{-\rho}\ln z}{\alpha + (1-\alpha)z^{-\rho}} = (1-\alpha)\ln z.
\end{aligned}
$$

Therefore, $e^{g_\rho(z)} \to z^{1-\alpha}$ as $\rho \to 0$, and thus

$$Y(L, K) \to AL(K/L)^{1-\alpha} = AK^{1-\alpha}L^{\alpha}$$

as $\rho \to 0$.

When taking $\rho \to +\infty$, we again use the function $g_\rho(z)$. We discriminate three cases:

- If $z > 1$, i.e, $L < K$, then $z^{-\rho} \to 0$. Therefore, $g_\rho(z) \to 1$ and so $Y(L, K) \to AL$.

- If $0 < z < 1$, i.e., $K < L$, then $g_\rho(z) = z[\alpha z^{\rho} + (1-\alpha)]^{-1/\rho} \to z$. Therefore, $Y(L, K) \to AL(K/L) = AK$.

- If $z = 1$, then $g_\rho(z) = 1$, so $Y(L, K) = AK = AL$.

In summary, we can write $Y(L, K) \to A\min(L, K)$ as $\rho \to \infty$.

7. Consider a *homothetic* production function $F(u,v)$ (that is $F(u,v) = G(h(u,v))$, where $G'(\cdot) > 0$ and $h(u,v)$ is linear homogeneous). Find the elasticity of substitution σ between u and v. Here, *Elasticity of Substitution* is defined as the percentage change in the input ratio per percentage change in the marginal rate of substitution MRS$= F_u/F_v$ along the production isoquant, i.e.

$$\sigma = -\frac{F_u/F_v}{u/v} \cdot \frac{d(u/v)}{d(F_u/F_v)} = -\frac{F_u v}{F_v u} \cdot \frac{d(u/v)/du}{d(F_u/F_v)/du},$$

where $v = v(u)$ and $dv/du = -F_u/F_v$.

Solution: First note that linear homogeneous means homogeneous of degree 1. If h is such, then $h(tu, tv) = th(u,v)$ for any $t > 0$. After differentiating and setting $t = 1$, we obtain $uh_u + vh_v = h$, which is Euler's theorem. Taking partial derivatives w.r.t. u and v, we obtain $uh_{uu} + vh_{uv} = 0$ and $uh_{uv} + vh_{vv} = 0$.

Moreover, we have $F_u = G'h_u$ and $F_v = G'h_v$. Then $v' = -F_u/F_v = -h_u/h_v$, so $h_u + v'h_v = 0$. Writing v as a function of u in the last equality, we have $h_u(u, v(u)) + v'(u)h_v(u, v(u)) = 0$. Differentiating w.r.t. u leads to $v''h_v - \frac{1}{uv}(v - uv')^2 h_{uu} = 0$.

Moreover, we can compute that $d(u/v)/du = \frac{1}{v^2}(v - uv')$ and $d(F_u/F_v)/du = -v''$. Substituting the last three equations into the given formula gives

$$\sigma = \frac{h_u}{h_v} \cdot \frac{v}{u} \cdot \frac{v - uv'}{v^2} \cdot \frac{h_v}{h_{uv}} \cdot \frac{uv}{(v - uv')^2} = \frac{h_u h_v}{h_{uv}h}.$$

In the last equality we used $h = h_v(v - uv')$, which can be obtained from $uh_u + vh_v = h$ and $u + vh_v = 0$.

Moreover, we can claim that this depends only on u/v, not on (u, v), i.e., σ for (u, v) is the same as for (tu, tv), where $t > 0$. We can prove this statement in the following way: If we write $h(u, v) = uh(1, \frac{v}{u}) = ug(\frac{v}{u})$, where $g(z) = h(1, z)$, then we can compute $h_u = g(\frac{v}{u}) - \frac{v}{u}g'(\frac{v}{u})$, $h_v = g'(\frac{v}{u})$, and $h_{uv} = g'(\frac{v}{u}) - \frac{v}{u}g''(\frac{v}{u})$. We see that $\sigma = h_u h_v/(h_{uv}h)$ can be written as the function of one variable $\frac{v}{u}$.

8. Consider the Cobb-Douglas production optimization problem

$$\max_{L,K} Y(K, L) = K^\alpha L^{1-\alpha}$$

$$\text{s.t. } rK + wL \leq M, \quad L \geq 0, \quad K \geq 0.$$

(a) Show that the inequality constraint can be replaced with the equality constraint $rK + wL = M$ (here constant M is income, r rental price of capital, w wage).

(b) Find the optimal capital/labor ratio. How would you interpret your results economically? What is the meaning of the Lagrangian multiplier?

(c) If you are told that for $r = w = 1$ and $M = 600$ the optimal allocation of labor and capital is $K = 200, L = 400$, what is the value of α?

Solution:

(a) Let us assume that in the optimal solution is $rK^* + wL^* < M$. This means that there exists such $K' > 0$ that $r(K^* + K') + wL^* \leq M$, so that $[K^* + K', L^*]$ is a feasible point. And because $Y(K, L)$ is increasing in K ($\frac{\partial Y}{\partial K} = \alpha K^{\alpha-1} L^{1-\alpha} > 0$, it has to be $Y(K^* + K', L^*) > Y(K^*, L^*)$, which contradicts the assumption that $[K^*, L^*]$ is the optimal point. The latter means that the constraint has to be satisfied with the equality.

(b) We can write the Lagrangian and the F.O.C. (the constraint is equality):

$$\begin{aligned}
\mathcal{L}(K, L, \lambda) &= K^\alpha L^{1-\alpha} - \lambda(M - wL - rK) \\
\frac{\partial \mathcal{L}}{\partial K} &= \alpha K^{\alpha-1} L^{1-\alpha} + \lambda r = 0 \\
\frac{\partial \mathcal{L}}{\partial L} &= (1 - \alpha)K^\alpha L^{-\alpha} + \lambda w = 0 \\
M &= rK + wL
\end{aligned}$$

If we divide the first equality by the second, we get

$$\frac{\alpha L}{(1 - \alpha)K} = \frac{r}{w} \Rightarrow \frac{K}{L} = \frac{\alpha}{1 - \alpha}\frac{w}{r}$$

The economic interpretation of this result is that the higher the interest rate, the lower the ratio of capital to labor. If the cost of capital increases, the producers shift resources from capital to labor. As with the wage rate, the higher the w, the higher the capital/labor ratio is. The usual interpretation of the multiplier λ is that it is the shadow price of the constraint, i.e. how much the producer is willing to pay for the additional unit of income.

(c) If we put $r = w = 1, K = 200, L = 400$ into the equality $\frac{K}{L} = \frac{\alpha}{1-\alpha}\frac{w}{r}$ and solve it for α, we get that $\alpha = \frac{1}{3}$.

5.4   Written Assignments

Written Assignment 1

1. *Drill Square*

   (a) Calculate, where possible, AB, $A'B$, BA, $B'A$, $C'-B$, $C'-AC'B'$, given

   $$A = \begin{pmatrix} 3 & 5 \\ 1 & 1 \end{pmatrix}, \quad B = \begin{pmatrix} 6 & 7 & 2 \\ 0 & 3 & 0 \end{pmatrix}, \quad A = \begin{pmatrix} 9 & 2 \\ 1 & 7 \\ 0 & 0 \end{pmatrix}.$$

   (b) Evaluate the determinant of the following matrix:

   $$\begin{pmatrix} 4 & 3 & 1 & 9 & 2 \\ 0 & 3 & 2 & 4 & 2 \\ 0 & 3 & 4 & 6 & 4 \\ 1 & -1 & 2 & 2 & 2 \\ 0 & 0 & 3 & 3 & 3 \end{pmatrix}.$$

   (c) Find all real numbers x such that:

   $$\det \begin{pmatrix} 1 & 2 & x \\ 3 & (4-x) & 5 \\ 1 & 7 & 8 \end{pmatrix} \geq 0.$$

2. *Art of Proof: The Very First Steps*

    (a) If a matrix A is invertible and $AB = I$, prove $BA = I$. (Here I denotes the unit matrix.)

    (b) If A and $A + B$ are invertible prove $I + A^{-1}B$ is invertible. And *visa versa*, if A and $I + A^{-1}B$ are invertible prove $A + B$ is invertible.

        (HINT: Let $C = (I + A^{-1}B)^{-1}$. Show that $(A + B)^{-1} = CA^{-1}$.)

    (c) If $A^2 = A$, prove that either $A = I$ or $\det A = 0$.

3. *Open-Ended Questions*

    (a) If A and B are both invertible, is $A+B$ necessarily invertible?

    (b) Suppose $AB = \mathbf{0}$, and neither A nor B is the null matrix. What would be your conclusion about the value of the determinants of A and B? Justify your answer.

    (c) Suppose $A^2 = \mathbf{0}$. Does the matrix $A + I$ have an inverse?

        (HINT: Recall the good old rule: for any given real numbers a and b, $(a + b)(a - b) = a^2 - b^2$.)

Solutions:

1.   (a) $AB = \begin{pmatrix} 18 & 36 & 6 \\ 6 & 10 & 2 \end{pmatrix}$

                $A'B = \begin{pmatrix} 18 & 25 & 6 \\ 30 & 38 & 10 \end{pmatrix}$

                $B'A = (A'B)'$

            $C' - B' = \begin{pmatrix} 3 & -6 & -2 \\ 2 & 4 & 0 \end{pmatrix}$

    It is not possible to calculate the others.

    (b) $\det A = 4 \cdot \begin{vmatrix} 3 & 2 & 4 & 2 \\ 3 & 4 & 6 & 4 \\ -1 & 2 & 2 & 2 \\ 0 & 3 & 3 & 3 \end{vmatrix} - \begin{vmatrix} 3 & 1 & 9 & 2 \\ 3 & 4 & 6 & 4 \\ 3 & 4 & 6 & 4 \\ 0 & 3 & 3 & 3 \end{vmatrix}$

The first determinant is equal to zero, since the second and the fourth column are identical. Next, if in the second determinant we subtract the second row from the third one we get

$$
\det A = - \begin{vmatrix} 3 & 1 & 9 & 2 \\ 3 & 2 & 4 & 2 \\ 0 & 2 & 2 & 2 \\ 0 & 3 & 3 & 3 \end{vmatrix} = 0
$$

(the third row is equal to the fourth row multiplied by $3/2$).

(c) Determinant of the given matrix is equal to $x^2 + 9x - 41$. The roots of this quadratic polynomial are:

$$
x_1 = \frac{-9 - 7\sqrt{5}}{2} \qquad x_2 = \frac{-9 + 7\sqrt{5}}{2}
$$

Therefore, the determinant is non-negative for $x \in (-\infty; x_1] \cup [x_2; \infty)$.

2. (a) We should multiply the identity $AB = I$ by A^{-1} from the left (since $\exists A^{-1}$) and by A from the right.

   (b) Let $C = (I + A^{-1}B)^{-1} \Rightarrow C = (A^{-1}(A+B))^{-1} = (A+B)^{-1} \cdot A$

   (c) $A^2 = A \Rightarrow \det(A^2) = (\det A)^2 = \det A$. Therefore, either $\det A = 0$ or $\det A = 1$. In the latter case we multiply the identity $A^2 = A$ by A^{-1} to get $A = I$.

3. (a) Not necessarily. Consider the following counterexample

$$
A = \begin{pmatrix} 1 & 0 \\ 0 & 1 \end{pmatrix} = I
$$

$$
B = \begin{pmatrix} -1 & 0 \\ 0 & -1 \end{pmatrix} = -I
$$

and, therefore, $A + B = 0$.

   (b) Since $\det AB = \det A \cdot \det B = 0$, then either $\det A = 0$ or $\det B = 0$ or both determinants are equal to zero. Let us assume that one of the two determinants (for instance, $\det A$) is non-zero. This assumption implies that $A^{-1}AB = A^{-1}0 \Rightarrow B = 0$. The latter contradicts the initial condition $B \neq 0$. Thus, both A and B must be singular ($\det A = \det B = 0$).

(c) Yes, the matrix $A + I$ has an inverse. Using the identity $(I - A)(I + A) = I^2 - A^2 = I$ we find $(A + I)^{-1} = I - A$.

Written Assignment 2

1. *Warm-up*

   (a) Consider an $[m \times n]$ matrix A, $m > n$.

      i. Show that $\text{rank}A \le n$.
      ii. Prove: If $\text{rank}A = n$ then $AX \ne 0 \ \forall X \ne 0$, but $\exists Y \ne 0$ such that $A'Y = 0$.

   (b) Prove or disprove:

      i. If A is positive definite then A^{-1} is negative definite.
      ii. If A is positive semi-definite then $-A$ is negative semi-definite.

   (c) Diagonalize the matrix $A = \begin{pmatrix} 0 & 1 & 0 \\ 1 & 0 & 1 \\ 0 & 1 & 0 \end{pmatrix}$, i.e. find an orthogonal matrix U such that $U'AU$ is diagonal.

2. *Determinants, Inverses and Linear Systems*

   (a) When possible, find the inverse of the matrix

   $$A = \begin{pmatrix} 0 & 0 & 2 & 4 \\ 1 & -2 & 0 & 0 \\ 0 & 0 & 1 & c \\ 4 & 7 & 0 & 0 \end{pmatrix}$$

   at $c = 2$, $c = -2$, and $c = 0$.

   (b) *(Whisper of the Past)*

   Solve the system of linear equations

   $$\begin{cases} x + y + cz & = 1 \\ x + cy + z & = 1 \\ cx + y + z & = 1 \end{cases}$$

   where c is a real parameter. Consider all possible cases!

(c) *Once Again!*

Solve the system of three linear equations in four unknowns

$$\begin{cases} 2x - y + z + u & = 1 \\ x + 2y - z + 4u & = 2 \\ x + 7y - 4z + 11u & = c \end{cases}$$

where c is a real parameter. Consider all possible cases!

3. *Rank Story*

   (a) Show that $\text{rank}(AB) = \text{rank}(A)$ for any two square matrices A and B, such that B has full rank.

   (b) Use the result in (a) to prove that the rank of a symmetric matrix equals the number of its non-zero eigenvalues.

4. *How Definite Are You?*

   (a) Given a quadratic form $Q = 5x_1^2 + x_2^2 + 7x_3^2 + 4x_1x_2 + 6x_1x_3 + 8x_2x_3$, find its matrix representation and check for sign definiteness.

   (b) *(Econometric Application: Regression Analysis)*
   Let an $m \times n$ matrix A (with $m > n$) have full rank (i.e. $\text{rank}A = n$). Show that $A'A$ is positive definite, and AA' is positive semi-definite (but not necessarily positive definite). (HINT: Use the result in 1a(ii).)

5. *Econometric Application: Theory of Stability of Simultaneous-Equations Models*

   A matrix A is called *nilpotent* if $\lim_{K \to \infty} A^K = \mathbf{0}$. (here $A^K = \underbrace{A \cdot A \cdot \ldots \cdot A}_{K\text{ times}}$.) Prove that a necessary and sufficient condition for a symmetric matrix to be nilpotent is that all its characteristic roots are less than one in absolute value.

Solutions:

1. (a) i. By definition, $\text{rank}(A)$ is the maximum number of linearly independent rows (columns), therefore $\text{rank}(A) \leq \min(m, n) = n$.

ii. Let $\exists X \neq \mathbf{0} : AX = \mathbf{0}$. Clearly, $AX = \mathbf{0}$ is a system of m equations in n unknowns. Since rank$(A) = n$, A contains n independent rows, and the rest $m - n$ rows are linearly dependent on them and thus can be excluded. Therefore, $AX = \mathbf{0}$ is equivalent to a system of n equations in n unknowns, and the matrix of this system is non-singular$\Rightarrow$ it should have only a trivial solution $(X = \mathbf{0})$. We've gotten a contradiction $\Rightarrow \forall X \neq \mathbf{0} : \quad AX \neq \mathbf{0}$.

The system $A'Y = 0$ has n equations and m unknowns $(m > n)$. Let usassume that the first n columns of A' are linearly independent. In such a case we can take $y_{n+1}, \dots, y_{m-n}$ as parameters and leave them on the right-hand side as a column-vector b. To this end, the system $A'Y$ is equivalent to a system $\bar{A}'\bar{Y} = b$, where matrix $\bar{A}'$ is composed of the first n columns of A' and $\bar{Y} = (y_1, \dots, y_n)$. Since $\det(\bar{A}') \neq 0$, there exists vector $\bar{Y} \neq \mathbf{0} : \bar{A}'\bar{Y} = b$.

(b) i. Since A is PD, its eigenvalues $\lambda_i > 0, \forall i$. We also know that $1/\lambda_i > 0$ are eigenvalues of A^{-1} (that was proven at the exercise session). Therefore, A^{-1} should also be PD.

ii. From the identity $0 = \det(A - \lambda I) = \det[(-1)(-A - (-\lambda)I)] = (-1)^n \det(-A - (-\lambda)I)$ it follows that if λ is an eigenvalue of A then $-\lambda$ is an eigenvalue of $-A$. Since A is PSD, $\lambda_i \geq 0 \quad \forall i$ and, therefore $-\lambda_i \leq 0 \quad \forall i$. It means that $-A$ is NSD.

(c) Solving the characteristic equation $\det(A - \lambda I)$ we find that

$$\lambda_1 = 0, \lambda_2 = \sqrt{2}, \lambda_3 = -\sqrt{2}.$$

The corresponding eigenvectors are

$$v_1 = (-1, 0, 1)', v_2 = (1, \sqrt{2}, 1)', v_3 = (1, -\sqrt{2}, 1)'.$$

Therefore, the orthogonal transformation matrix (formed by normalized eigenvectors) is

$$U = \begin{pmatrix} -1/\sqrt{2} & 1/2 & 1/2 \\ 0 & 1/\sqrt{2} & -1/\sqrt{2} \\ 1/\sqrt{2} & 1/2 & 1/2 \end{pmatrix}.$$

2. (a) Gauss elimination method implies that

$$\begin{pmatrix} 0 & 0 & 2 & 4 & | & 1 & 0 & 0 & 0 \\ 1 & -2 & 0 & 0 & | & 0 & 1 & 0 & 0 \\ 0 & 0 & 1 & c & | & 0 & 0 & 1 & 0 \\ 4 & 7 & 0 & 0 & | & 0 & 0 & 0 & 1 \end{pmatrix} \sim$$

$$\sim \begin{pmatrix} 1 & -2 & 0 & 0 & | & 0 & 1 & 0 & 0 \\ 4 & 7 & 0 & 0 & | & 0 & 0 & 0 & 1 \\ 0 & 0 & 1 & c & | & 0 & 0 & 1 & 0 \\ 0 & 0 & 2 & 4 & | & 1 & 0 & 0 & 0 \end{pmatrix} \sim$$

$$\sim \begin{pmatrix} 1 & 0 & 0 & 0 & | & 0 & \frac{7}{15} & 0 & \frac{2}{15} \\ 0 & 1 & 0 & 0 & | & 0 & -\frac{4}{15} & 0 & \frac{1}{15} \\ 0 & 0 & 1 & c & | & 0 & 0 & 1 & 0 \\ 0 & 0 & 2 & 4 & | & 1 & 0 & 0 & 0 \end{pmatrix} \sim$$

$$\sim \begin{pmatrix} 1 & 0 & 0 & 0 & | & 0 & \frac{7}{15} & 0 & \frac{2}{15} \\ 0 & 1 & 0 & 0 & | & 0 & -\frac{4}{15} & 0 & \frac{1}{15} \\ 0 & 0 & 1 & c & | & 0 & 0 & 1 & 0 \\ 0 & 0 & 0 & 4-2c & | & 1 & 0 & -2 & 0 \end{pmatrix}.$$

If $c = 2$ then A^{-1} does not exist, otherwise we can divide the last row by $4 - 2c$. Thus, for $c = 0$

$$A^{-1} = \begin{pmatrix} 0 & 7/15 & 0 & 2/15 \\ 0 & -4/15 & 0 & 1/15 \\ 0 & 0 & 1 & 0 \\ 1/4 & 0 & -1/2 & 0 \end{pmatrix};$$

for $c = -2$ we subtract the 4th row from the 3rd row to get

$$A^{-1} = \begin{pmatrix} 0 & 7/15 & 0 & 2/15 \\ 0 & -4/15 & 0 & 1/15 \\ 1/4 & 0 & 1/2 & 0 \\ 1/8 & 0 & -1/4 & 0 \end{pmatrix}.$$

(b)
$$\begin{pmatrix} 1 & 1 & c & | & 1 \\ 1 & c & 1 & | & 1 \\ c & 1 & 1 & | & 1 \end{pmatrix} \sim \begin{pmatrix} 2+c & 2+c & 2+c & | & 3 \\ 1 & c & 1 & | & 1 \\ c & 1 & 1 & | & 1 \end{pmatrix}.$$

If $c = -2$ then $\operatorname{rank}(A) \neq \operatorname{rank}(A|b) \Rightarrow$ no solutions. Otherwise, dividing the 1st row by $2 + c$ and subtracting it from the 2nd and 3rd rows we get

$$\begin{pmatrix} 1 & 1 & 1 & | & 3/(2+c) \\ 1 & c & 1 & | & 1 \\ c & 1 & 1 & | & 1 \end{pmatrix} \sim$$

$$\sim \begin{pmatrix} 1 & 1 & 1 & | & 3/(2+c) \\ 0 & c-1 & 0 & | & 1-3/(2+c) \\ c-1 & 0 & 0 & | & 1-3/(2+c) \end{pmatrix}.$$

If $c \neq 1$ then A has full rank $\Rightarrow$ unique solution.

If $c = 1 \Rightarrow \operatorname{rank}(A) = \operatorname{rank}(A|b) = 1 \Rightarrow$ infinite number of solutions.

Summing up,

i. if $c = -2$ then there is no solution;

ii. if $c = 1$ then there is an infinite number of solutions: $x = 1 - y - z$;

iii. if $c \neq 1$ and $c \neq -2$ then there is a unique solution: $x = y = z = 1/(c+2)$.

(c) Using the Gauss elimination method we get

$$\begin{pmatrix} 2 & -1 & 1 & 1 & | & 1 \\ 1 & 2 & -1 & 4 & | & 2 \\ 1 & 7 & -4 & 11 & | & c \end{pmatrix} \sim$$

$$\sim \begin{pmatrix} 2 & -1 & 1 & 1 & | & 1 \\ 1 & 2 & -1 & 4 & | & 2 \\ 0 & 5 & -3 & 7 & | & c-2 \end{pmatrix} \sim$$

$$\sim \begin{pmatrix} 2 & -1 & 1 & 1 & | & 1 \\ 0 & 5 & -3 & 7 & | & 3 \\ 0 & 5 & -3 & 7 & | & c-2 \end{pmatrix} \sim$$

$$\sim \begin{pmatrix} 2 & -1 & 1 & 1 & | & 1 \\ 0 & 5 & -3 & 7 & | & 3 \\ 0 & 0 & 0 & 0 & | & c-5 \end{pmatrix}.$$

Thus, if $c = 5 \Rightarrow \operatorname{rank}(A) = \operatorname{rank}(A|b) = 2 \Rightarrow$ infinite number of solutions $x = 4 - z - 6u$ and $y = \dfrac{3 + 3z - 7u}{5}$; otherwise there is no solution.

3. (a) We know that $\operatorname{rank}(AB) \leq \min(\operatorname{rank}(A), \operatorname{rank}(B))$. Since $\operatorname{rank}(B) = n$, $\operatorname{rank}(AB) \leq \operatorname{rank}(A)$. We should show that $\operatorname{rank}(AB)$ cannot be strictly less than $\operatorname{rank}(A)$. Let $k = \operatorname{rank}(A) \leq n$. It means that A contains k linearly independent rows. Without loss of generality, let these rows be $a_1, \ldots, a_k$. Let $C = AB$, with its rows denoted by $c_1, \ldots, c_n$. Independence of $a_1, \ldots, a_k$, combined with the full-rank property of B, implies that for any real numbers $\alpha_1, \ldots, \alpha_k$ not all zero, a linear combination of $c_1, \ldots, c_k$ should also be non-zero:

$$\forall \alpha = (\alpha_i, \ldots \alpha_k) \neq \mathbf{0}:$$

$$\sum_{i=1}^{k} \alpha_i c_i = \sum_{i=1}^{k} \alpha_i a_i B = \left(\sum_{i=1}^{k} \alpha_i a_i \right) B \neq \mathbf{0}.$$

Thus, rows $c_1, \ldots, c_k$ are also independent $\Rightarrow \operatorname{rank}(AB)$ cannot be smaller than $\operatorname{rank}(A)$.

(b) To prove the statement, just notice that $A = U \Lambda U'$, where U is an orthogonal matrix (U has full rank) and Λ is a diagonal matrix of eigenvalues.

4. (a) Matrix representation is $Q = x'Ax$, where

$$A = \begin{pmatrix} 5 & 2 & 3 \\ 2 & 1 & 4 \\ 3 & 4 & 7 \end{pmatrix} \text{ and } x = \begin{pmatrix} x_1 \\ x_2 \\ x_3 \end{pmatrix}.$$

The leading principal minors are $D_1 = 5 > 0, D_2 = 1 > 0, D_3 = -34 < 0$, indicating that given the quadratic form is indefinite.

(b) Let $X \neq 0$. Then $X'(A'A)X = (AX)'(AX) = Z'Z = \sum z_i^2 > 0$, where $Z = AX \neq 0$ from 1a(ii). Therefore, by definition $A'A$ is PD.

By the same token, for $\forall Y \neq 0$, $Y'(AA')Y = (A'Y)'(A'Y) = Z'Z = \sum z_i^2 \geq 0$, where $Z = A'Y$ and Z can be a zero vector (from 1a(ii)). Therefore, AA' is PSD.

5. A is symmetric $\Rightarrow A = U\Lambda U'$, where U is orthogonal $(UU' = I)$ and Λ is a diagonal matrix of the eigenvalues of A, i.e.

$$\Lambda = \begin{pmatrix} \lambda_1 & \cdots & 0 \\ 0 & \ddots & 0 \\ 0 & \cdots & \lambda_n \end{pmatrix}.$$

We compute $A^k = U\Lambda U' \cdot U\Lambda U' \cdot \ldots \cdot U\Lambda U' = U\Lambda^k U'$, where

$$\Lambda^k = \begin{pmatrix} \lambda_1^k & \cdots & 0 \\ 0 & \ddots & 0 \\ 0 & \cdots & \lambda_n^k \end{pmatrix}.$$

Thus $\lim_{k \to \infty} A^k = 0 \Leftrightarrow |\lambda_i| < 1, \forall i = 1, \ldots, n.$

Written Assignment 3

1. *Basic Calculus: Painful but Necessary Drills...*

   (a) Prove that for any real x, e^x can be evaluated as $e^x = \lim_{n \to \infty} \left(1 + \dfrac{x}{n}\right)^n$.

   (b) Find the limit $\lim_{x \to \infty} \dfrac{x^\alpha}{a^x}$, where $\alpha > 0$, $a > 1$.

   (c) Find $\dfrac{dy}{dx}$, $\dfrac{d^2y}{dx^2}$ if x and y are defined parametrically:

   $$x = x(t) = 2t - 1, \ y = y(t) = t^4 - 1$$

   (d) Find u_x, u_y, u_z if $u(x, y, z) = x^y y^z z^x$.
   Find u_{xx}, u_{xy}, u_{yy} if $u(x, y) = xy \ln(xy)$.

   (e) If y is given by $y(x) = \frac{1}{2}(e^x + e^{-x})$, check that $\dfrac{dx}{dy} = \dfrac{1}{\sqrt{y^2 - 1}}$.

   (f) Find the local minima and maxima of y by the second-derivative test, where

   $$y = x \ln x.$$

(g) Draw the graph of the function $y = \dfrac{x^2 - 2x + 2}{x - 1}$. Please be specific!

(h) Without using a calculator, find approximately the value of $\ln(e + 0.272)$.

(i) Evaluate the indefinite integrals:

$$a) \int \sqrt[3]{(x^3 - 8)}\, x^2 dx, \qquad b) \int \sqrt{x}\ln x\, dx$$

(j) Evaluate the area defined by the following inequalities:

$$x^2 + y^2 \le 4, \quad y \ge \frac{1}{2}(x + 2)^2$$

2. *Your Name is Leibniz...*

   Try to prove a simplified version of Leibniz's formula:

   Given an integrable function over $[a, b,]$ $f(x)$, let define $A(x) = \int_a^x f(u)du$, $x \in (a, b)$. Show that $A(x)$ is differentiable and $A'(x) = f(x)$ for every $x \in (a, b)$.

3. *Economic Application: Social Inequality, Lorenz Curve, Gini Co-efficient and Pareto Distribution of Income*

   A distribution of incomes across individuals in a society determines the degree of social inequality. One function that is frequently used to model income distribution in a population is the so-called *Pareto distribution*.

   (a) The *Pareto distribution* of income across a population is defined on the interval $[b, \infty)$ by the density function $f(x) = ab^a x^{-a-1}$, where a and b are positive parameters. Compute the following key characteristics of Pareto income distribution:

   i. The *distribution function* for income $F(x)$,
   $$F(x) = \int_b^x f(u)du;$$

   ii. *Mean income* $\mu = \int_b^\infty uf(u)du$; (by the way, what condition do you need to impose on a to secure that μ does indeed exist?)

iii. The *Lorenz curve* for the income distribution, $l(p)$, that is defined for each p as the proportion of total income accruing to the bottom $100p\%$ of income recipients. (Mathematically, the Lorenz curve is constructed as follows: for each $0 \le p \le 1$ let $p = F(y)$. Then an individual with income y is ranked $100p\%$ of the way up the income distribution. And the Lorenz curve is defined for each p as $l(p) = \dfrac{1}{\mu} \displaystyle\int_b^y uf(u)du$, where p and y are related through the identity $p = F(y)$);

iv. The *Gini coefficient* that measures the extent of inequality. Mathematically, the Gini coefficient is defined to be $G = 1 - 2\int_0^1 l(p)dp$. Its geometric interpretation is the area between the Lorenz curve and the 45^0-line or *line of complete equality*;

v. Check whether the Lorenz curve is concave, or convex.

(b) Consider now a perfectly egalitarian society where all incomes of income recipients are equal. Show that in a perfectly egalitarian society the Lorenz curve coincides with the line of complete equality.

Solutions:

1. (a) By definition $e = \displaystyle\lim_{n \to \infty} \left(1 + \frac{1}{n}\right)^n$

If $x \ne 0 \Rightarrow$ we can introduce another variable $k = \dfrac{n}{x}$.
Therefore,

$$\lim_{n \to \infty} \left(1 + \frac{x}{n}\right)^n = \left(\lim_{k \to \infty} \left(1 + \frac{1}{k}\right)^k\right)^x = e^x.$$

If $x = 0 \Rightarrow e^0 = \lim_{n \to \infty} 1^n = 1$.

(b) There exists a natural number k such that $k < \alpha \le k + 1$ $(k + 1 - \alpha \ge 0) \Rightarrow$ applying L'Hôpital's rule $k + 1$ times,

$$\lim_{x \to \infty} \frac{x^\alpha}{a^x} = \lim_{x \to \infty} \frac{\alpha x^{\alpha-1}}{a^x \ln x} = \dots$$

$$= \lim_{x \to \infty} \frac{\alpha \cdot (\alpha - 1) \cdot \ldots \cdot (\alpha - k)}{x^{k+1-\alpha} \cdot a^x (\ln a)^{k+1}} = 0.$$

(c) $\dot{x} = 2$; $\ddot{x} = 0$; $\dot{y} = 4t^3$; $\ddot{y} = 12t^2$.

$$\frac{dy}{dx} = \frac{\dot{y}}{\dot{x}} = 2t^3;$$

$$\frac{d^2y}{dx^2} = \frac{\ddot{y}\dot{x} - \ddot{x}\dot{y}}{\dot{x}^3} = -3t^2.$$

(d) $\quad u(x, y, z) = x^y y^z z^x = e^{y\ln(x) + z\ln(y) + x\ln(z)}$

$$u_x = x^y y^z z^x \left(\frac{y}{x} + \ln z\right);$$

$$u_y = x^y y^z z^x \left(\frac{z}{y} + \ln x\right);$$

$$u_x = x^y y^z z^x \left(\frac{x}{z} + \ln y\right).$$

$u(x, y) = xy\ln(xy)$

$$u_x = y\ln(xy) + y; \quad u_y = x\ln(xy) + x;$$

$$u_{xx} = \frac{y}{x}; \quad u_{xy} = 2 + \ln(xy); \quad u_{yy} = \frac{x}{y}.$$

(e) $\dfrac{dy}{dx} = \dfrac{1}{2}\left(e^x - e^{-x}\right)$.

Using the formula for inverse function derivative,

$$\frac{dx}{dy} = \frac{1}{dy/dx} = \frac{2}{e^x - e^{-x}} = \sqrt{\frac{4}{(e^x - e^{-x})^2}}$$

$$= \sqrt{\frac{4}{e^{2x} - 2 + e^{-2x}}}$$

$$= \sqrt{\frac{4}{(e^x + e^{-x})^2 - 4}} = \frac{1}{\sqrt{y^2 - 1}}.$$

(f) F.O.C.: $y' = \ln x + 1 = 0 \Rightarrow x^* = 1/e$;
S.O.C. $y'' = 1/x \Rightarrow y''(x^*) = e > 0 \Rightarrow x^*$ is a local min.

(g) $y(x) = \dfrac{x^2 - 2x + 2}{x - 1} = x - 1 + \dfrac{1}{x - 1}$
$x = 0 \Rightarrow y = -2$; $y \neq 0$.
F.O.C. $y' = 1 - \dfrac{1}{(x-1)^2} = 0 \Rightarrow x = 2$ or $x = 0$; $y(2) = 2$.

$y'' = \dfrac{2}{(x-1)^3} \Rightarrow y'' > 0$ if $x > 1$ and $y'' < 0$ if $x < 1$;

$x = 2$ is min and $x = 0$ is max.

$\lim\limits_{x\to-\infty} y(x) = -\infty; \quad \lim\limits_{x\to\infty} y(x) = \infty;$

$\lim\limits_{x\to 1-} y(x) = -\infty; \quad \lim\limits_{x\to 1+} y(x) = -\infty.$

$\lim\limits_{x\to\infty} \dfrac{y(x)}{x} = 1; \quad \lim\limits_{x\to\infty} y(x) - x = -1 \Rightarrow$ the line $y = x - 1$ is an asymptote.

(h) The linear approximation of $\ln(1+x)$ around $x_0 = 0$ implies that $\ln(1+x) \approx x$ for small x. Therefore, $\ln(e+0.272) \approx \ln(e(1+0.1)) = 1 + \ln(1+0.1) \approx 1.1$. For comparison, the exact answer is $\ln(e+0.272) = 1.09536764....$

(i) a) $\displaystyle\int \sqrt[3]{(x^3-8)}x^2\,dx = \dfrac{1}{3}(x^3-8)^{1/3}d(x^3-8) = \dfrac{1}{4}(x^3-8)^{4/3} + C$

b) $\displaystyle\int \sqrt{x}\ln x\,dx = \dfrac{2}{3}\int \ln x\,d(x^{3/2}) =$

$= \dfrac{2}{3}\left(x^{3/2}\ln x - \int x^{3/2}\dfrac{1}{x}\,dx\right) = \dfrac{2}{3}x^{3/2}\ln x - \dfrac{4x^{3/2}}{9} + C$

(j) If you draw the graphs of $x^2 + y^2 = 4$ (a circle of radius 2 that is centered in the origin) and $y = 0.5(x+2)^2$ (an upward-directed parabola with minimum at $x = -2, y = 0$), you will easily see that

$$S = \int_{-2}^{0} \sqrt{4-x^2}\,dx - \int_{-2}^{0} \dfrac{1}{2}(x+2)^2\,dx = I_1 - I_2$$

I_1 equals $1/4$ of the area of a circle with $r = 2$. Thus, $I_1 = \pi$. (If you would like to be more 'scientific', evaluate I_1 by substitution: $x = 2\cos t \Rightarrow dx = -2\sin t\,dt$ and $4 - x^2 = 4\sin^2 t$.)

$$I_2 = \dfrac{1}{2}(x+2)^2 d(x+2) = \dfrac{(x+2)^3}{2\cdot 3}\bigg|_{-2}^{0} = \dfrac{4}{3}$$

Therefore, $S = \pi - \dfrac{4}{3}$.

2. By the definition of indefinite integral, $\displaystyle\int f(x)\,dx = F(x) + C,$

where $F(x)$ is such that $F'(x) = f(x)$ and C is an arbitrary constant.

The Newton-Leibniz formula says that $\forall x \in [a,b] \Rightarrow A(x) = \int_a^x f(u)du = F(x) - F(a)$. Therefore, $\dfrac{d}{dx}[A(x)] = F'(x) - \dfrac{F(a)}{dx} = f(x)$.

Alternatively, the statement can be proven by using the definition of the first derivative and by applying the mean-value theorem:

$$
\begin{aligned}
A'(x) &= \lim_{\Delta x \to 0} \frac{A(x + \Delta x) - A(x)}{\Delta x} \\
&= \lim_{\Delta x \to 0} \frac{1}{\Delta x} \int_x^{x+\Delta x} f(u)du \\
&= \lim_{\Delta x \to 0} \frac{1}{\Delta x} f(x + \theta \Delta x) \cdot \Delta x = f(x) \quad \text{(here } \theta \in [0,1]\text{).}
\end{aligned}
$$

3. (a) i. $F(x) = \displaystyle\int_b^x ab^a u^{-a-1} du = ab^a \left. \frac{u^{-a}}{(-a)} \right|_b^x = 1 - \left(\frac{b}{x}\right)^a$

     ii. $\mu = \displaystyle\int_b^\infty u ab^a u^{-a-1} du = ab^a \int_b^\infty u^{-a} du =$

$$
= \begin{cases} ab^a \left. \dfrac{u^{-a+1}}{1-a} \right|_b^\infty = \dfrac{ab}{a-1}, & \text{if } a > 1; \\[2mm] \text{does not exist, otherwise.} \end{cases}
$$

     iii.

$$
\begin{aligned}
l(p) &= \frac{1}{\mu} \int_b^y u f(u) du = \frac{a-1}{ab} \int_b^y ab^a u^{-a} du \\
&= \frac{a-1}{ab} ab^a \left. \frac{u^{-a+1}}{1-a} \right|_b^y = 1 - \left(\frac{b}{y}\right)^{a-1}
\end{aligned}
$$

$$
p = F(y) = 1 - \left(\frac{b}{y}\right)^a \Rightarrow \left(\frac{b}{y}\right)^{a-1} = (1-p)^{\frac{a-1}{a}}.
$$

Therefore,

$$
l(p) = 1 - (1-p)^{\frac{a-1}{a}}.
$$

     iv.

$$
\begin{aligned}
G &= 1 - 2 \int_0^1 \left(1 - (1-p)^{\frac{a-1}{a}}\right) dp \\
&= 1 - 2 - 2 \left. \frac{(1-p)^{1-1/a+1}}{2 - 1/a} \right|_0^1 = \frac{1}{2a - 1}
\end{aligned}
$$

Note that $a > 1$ from (ii). Therefore, $G < 1$, i.e. Pareto income distribution can never be perfectly egalitarian.

v.

$$l'(p) = \frac{a-1}{a}(1-p)^{-1/a}$$

$$l''(p) = \frac{a-1}{a^2}(1-p)^{-1/a-1} \geq 0 \text{ for } a \geq 1$$

$$\Rightarrow \text{ convex.}$$

(b) Since we consider a perfect egalitarian society (all incomes are equal), the Gini coefficient should be equal to zero. We also know that Gini coefficient is the area between the Lorenz curve and the line of complete equality. This means that in a perfectly egalitarian society the Lorenz curve coincides with the line of complete equality.

Written Assignment 4

1. *Classified*

   Find and classify the stationary points of

   (a) $f(x,y) = 3x^2y - x^3 - y^4$
   (b) $f(x,y,z) = 2x^2 - xy + 2xz - y + y^3 + z^2$

2. *Dear Cobb and Douglas...*

   Consider a Cobb-Douglas production function $Y(L, K) = AL^\alpha K^\beta$, where Y is output, L labor input, K capital input, positive constant A measures the level of technology, and α and β are positive parameters.

   For what values of α and β is the production function concave? Can you find α and β such that the production function is convex?

3. *Game of Functions*

   (a) Let $z = f(x,y)$ be a homogeneous function of degree n, i.e. $f(tx, ty) = t^n f(x,y)$ for any t. Prove that $x\frac{\partial z}{\partial x} + y\frac{\partial z}{\partial y} = nz$.

(b) Let $f_1(x_1, \ldots, x_n), \ldots, f_m(x_1, \ldots, x_n)$ be convex functions. Prove that a linear combination $\sum_{i=1}^{m} \alpha_i f_i(x_1, \ldots, x_n)$ is a convex function if and only if all constants α_i are non-negative.

(c) Show that the implicit function $z = z(x, y)$, defined as $x - mz = \phi(y - nz)$ (where m, n are parameters, ϕ is an arbitrary differentiable function), solves the differential equation

$$m\frac{\partial z}{\partial x} + n\frac{\partial z}{\partial y} = 1.$$

(d) Suppose the equation $F(x, y) = C$ defines y as a function of x, $y = y(x)$. You already know that $\dfrac{dy}{dx} = -\dfrac{F_x}{F_y}$, provided $F_y \neq 0$. Now, find the expression for $\dfrac{d^2y}{dx^2}$.

4. *Sensitivity Analysis*

The objective function $f(x, y) = \alpha x + y$ is to be maximized under the constraints

$$x^2 + \alpha y^2 \leq 1, \ x \geq 0, \ y \geq 0,$$

where α is a positive real parameter.

Check how the optimal solution (x^*, y^*) changes with small variations in the value of α, provided that the optimal solution does exist.

5. *Economics Application: The Famous Ordinary Least Square Regression Method (OLS, for short)*

In the method of least squares in regression theory the objective is to determine the best fit of the straight line $y = a + bx$ to the data (x_i, y_i), $i = 1, 2, \ldots, n$ by minimizing the sum of squared errors:

$$S(a, b) = \sum_{i=1}^{n} (y_i - (a + bx_i))^2$$

by choice of two parameters a (the intercept) and b (the slope).

Minimize $S(a, b)$ by choosing optimal a and b, and show that at this point of optimality the sufficient conditions for minimum are met.

6. *In Case You Miss It at the Exercise Session*

  (a) Using the Lagrange multipliers method find the stationary points of

   (a1) $U(x, y) = xy$ subject to $x + y = 6$;

   (a2) $V(x, y) = x^2 + y^2$ subject to $x + 4y = 2$.

   (HINT: Just form the Lagrangian function and solve the first-order conditions.)

  (b) Illustrate your findings in (a1) and (a2) graphically, i.e. sketch the graphs of indifference curves of the objective function and the graphs of the constraints.

  (c) Looking at the graphs in (b), can you argue that in (a1) the objective function $(U(x, y)$ at the stationary point reaches its constrained maximum, while in (a2) the objective function $V(x, y)$ reaches its constrained minimum?

  (d) Prove your arguments in (c) analytically, i.e. check the second-order conditions.[12]

---

[12] To classify the stationary points of the constrained maximization problem
  max $f(x, y)$ subject to $g(x, y) = C$,
you should perform the following steps:
First, construct the Lagrangian function $L(x, y, \lambda) = f(x, y) - \lambda(g(x, y) - C)$. Let (x^*, y^*, λ^*) denote its stationary point.
Next, construct the bordered Hessian matrix:

$$H = \begin{pmatrix} 0 & \frac{\partial g(x^*, y^*)}{\partial x} & \frac{\partial g(x^*, y^*)}{\partial y} \\ \frac{\partial g(x^*, y^*)}{\partial x} & \frac{\partial^2 L(x^*, y^*, \lambda^*)}{\partial x^2} & \frac{\partial^2 L(x^*, y^*, \lambda^*)}{\partial x \partial y} \\ \frac{\partial g(x^*, y^*)}{\partial y} & \frac{\partial^2 L(x^*, y^*, \lambda^*)}{\partial y \partial x} & \frac{\partial^2 L(x^*, y^*, \lambda^*)}{\partial y^2} \end{pmatrix}$$

(Note that H is a matrix with constant coefficients.)
Finally, the second-order conditions read as follows:

• If $\det(H) > 0$ then $f(x, y)$ at (x^*, y^*) reaches its maximum.
• If $\det(H) < 0$ then $f(x, y)$ at (x^*, y^*) reaches its minimum.

7. *Dear Kuhn and Tucker...*

   Derive the Kuhn-Tucker conditions for the constrained minimization problem

   $$\min f(x_1, \ldots, x_n)$$

   $$\text{s.t. } x_i \geq 0, \; i = 1, 2, \ldots, n,$$

   $$g^j(x_1, \ldots, x_n) \geq b_j, \; j = 1, \ldots, m.$$

8. *What Happens When Sufficient Conditions Do Not Hold*

   Consider the maximization problem $\max x + y$ subject to the constraints $4x - x^2 - y \leq 0$ and $2x + 3y \leq 8$.

   Write down the Kuhn-Tucker conditions. Show that there is a unique solution to the K.-T. conditions, given by $(x = 2/3, y = 20/9, \lambda_1 = 0.1, \lambda_2 = 11/30)$; but no finite solution to the optimization problem. (If necessary, illustrate your argument graphically.)

Solutions:

1. (a) There are two stationary points: $(0, 0)$ and $(6, 3)$. The point $(6, 3)$ is a saddle. In order to check point $(0, 0)$ we should check the behavior of the first derivatives $f_x(x, 0)$ and $f_y(0, y)$.

   (b) Stationary points: $\left(-\dfrac{1}{4}, -\dfrac{1}{2}, \dfrac{1}{4}\right)$ — saddle;

   $\left(\dfrac{1}{3}, \dfrac{2}{3}, -\dfrac{1}{3}\right)$ — minimum.

2. Y is concave $\Leftrightarrow d^2Y$ is NSD (or $-d^2Y$ is PSD).

   We can use eigenvalues test (all eigenvalues for d^2Y should be non-positive). Therefore, the production function is concave for $\alpha + \beta \leq 1$.

   There are no real α and β, for which the production function is convex.

3. (a) Since $f(tx, ty) = t^n f(x, y)$ $\forall t$, we can differentiate given equality withe respect to t and then take $t = 1$.

   (b) Use the definition of a convex function.

   (c) Use the implicit function theorem:

   $$\frac{\partial z}{\partial x} = \frac{1}{m - \phi'n};$$

   and

   $$\frac{\partial z}{\partial y} = -\frac{1}{m - \phi'n}.$$

   (d) $$\frac{d^2y}{dx^2} = \frac{2F_x F_y F_{xy} - F_{xx}(F_y)^2 - F_{yy}(F_x)^2}{(F_y)^3}.$$

4. We can see from the graph that the first constraint is binding. From the F.O.C. we can find

$$\frac{dx^*}{d\alpha} = \frac{3\alpha^2}{(\alpha + \alpha^4)^{3/2}}; \quad \frac{dy^*}{d\alpha} = -\frac{1 + 4\alpha^3}{2(\alpha + \alpha^4)^{3/2}}$$

(applying the implicit function theorem or making a substitution).

5. F.O.C.:

$$\frac{\partial S}{\partial a} = \sum_{i=1}^{n} [2(y_i - (a + bx_i))(-1)] = 0$$

$$\frac{\partial S}{\partial b} = \sum_{i=1}^{n} [2(y_i - (a + bx_i))(-b_i)] = 0;$$

therefore

$$a^* = \bar{y} - b\bar{x}; \quad b^* = \frac{\sum_{i=1}^{n}(x_i - \bar{x})(y_i - \bar{y})}{\sum_{i=1}^{n}(x_i - \bar{x})^2},$$

where

$$\bar{x} = \frac{1}{n}\sum_{i=1}^{n} x_i, \quad \bar{y} = \frac{1}{n}\sum_{i=1}^{n} y_i.$$

The Hessian matrix $H = \begin{pmatrix} 2n & 2\sum x_i \\ 2\sum x_i & 2\sum x_i^2 \end{pmatrix}$ is positive definite.

6. (a1) Stationary point $(3,3)$ is maximum, since $|H| > 0$, where

$$H = \begin{pmatrix} 0 & 1 & 1 \\ 1 & 0 & 1 \\ 1 & 1 & 0 \end{pmatrix}.$$

    (a2) Stationary point $(2/17; 8/17)$ is minimum, since $|H| < 0$, where

$$H = \begin{pmatrix} 0 & 1 & 4 \\ 1 & 2 & 0 \\ 4 & 0 & 2 \end{pmatrix}.$$

7. The minimization problem can be reformulated as

$$\max f(x_1, ..., x_n)$$

s.t. $-g^j(x_1, ..., x_n) \leq -b_j, \qquad j = 1, ..., m;$

           $x_i \geq 0, \qquad\qquad\qquad i = 1, ..., n.$

The Lagrangian function is $L = f(x_1, ..., x_n) - \sum \lambda_j(b_j - g^j(x_1, ..., x_n))$.

The K.-T. conditions are:

$$\frac{\partial L}{\partial x_i} \geq 0, \quad x_i \geq 0 \quad \text{and} \quad x_i\frac{\partial L}{\partial x_i} = 0 \quad i = 1, ..., n;$$

$$\frac{\partial L}{\partial \lambda_j} \leq 0, \quad \lambda_j \geq 0 \quad \text{and} \quad \lambda_j\frac{\partial L}{\partial \lambda_j} = 0, \quad j = 1, ..., m.$$

8. Our maximization problem is

$$\max(x + y)$$

s.t. $4x - x^2 - y \leq 0,$

          $2x + 3y \leq 8.$

(note that there are no non-negativity constraints).

The Lagrangian function is $L = (x + y) - \lambda(4x - x^2 - y) - \mu(2x + 3y - 8)$.

The K.-T. conditions are:

$$\frac{\partial L}{\partial x} = 1 - 4\lambda + 2\lambda x - 2\mu = 0, \tag{5.23}$$

[because of no non-negativity constraints for x]

$$\frac{\partial L}{\partial y} = 1 + \lambda - 3\mu = 0, \tag{5.24}$$

[because of no non-negativity constraints for y]

$$\frac{\partial L}{\partial \lambda} = 4x - x^2 - y \geq 0, \tag{5.25}$$

$$\lambda \geq 0, \tag{5.26}$$

$$\lambda \frac{\partial L}{\partial \lambda} = \lambda(4x - x^2 - y) = 0, \tag{5.27}$$

$$\frac{\partial L}{\partial \mu} = 2x + 3y - 8 \geq 0, \tag{5.28}$$

$$\mu \geq 0, \tag{5.29}$$

$$\mu \frac{\partial L}{\partial \lambda} = \mu(2x + 3y - 8) = 0. \tag{5.30}$$

Condition (5.24) implies $\mu = \frac{1+\lambda}{3} > 0$; if $\lambda = 0$ then (5.23) $\Rightarrow$ $\mu = \frac{1}{2}$, while (5.24) $\Rightarrow \mu = \frac{1}{3}$, which is a contradiction.

Thus, $\lambda > 0$ and $\mu > 0$.

Let's consider the case $\lambda > 0, \mu > 0$ in more detail: (5.27) and (5.30) imply $(x_1^*, y_1^*) = (4, 0)$; $(x_2^*; y_2^*) = (2/3, 20/9)$.

However, in case $(x_1^*, y_1^*) = (4, 0)$ we can notice that (5.23) and (5.24) yield $\mu = \frac{3}{10}$, $\lambda = -0.1 < 0$, which is a contradiction.

Thus, the point of maximum — derived from the K.-T. conditions — is $(x_2^*; y_2^*) = (2/3, 20/9)$. However, even at (4,0) the value of the objective function is greater. Furthermore, if $x > 4$ then the only binding constraint is $2x + 3y = 8$, that is, $y = (8 - 2x)/3$, and the objective function becomes $x + (8 - 2x)/3 = (x+8)/3$. Obviously, the objective function monotonically increases as $x \to \infty$.

Written Assignment 5

1. *Continuous Time Dynamics at a Glance*

    (a) Solve $\dfrac{dy}{dx} = xy^2 + 2xy$

    (HINT: Variables separable case. Do not forget about special solutions of the form $y \equiv constant$.)

    (b) Solve $\dfrac{dy}{dx} - 2xy = 3x^2 - 2x^4$.

    (HINT: Linear differential equation of the first order.)

    (c) Solve $\dfrac{dy}{dx} + 2y = e^x y^2$.

    (HINT: Introduce new variable $z = 1/y$. Then $\dfrac{dz}{dx} = \dfrac{dz}{dy}$.

    $\dfrac{dy}{dx} = -\dfrac{1}{y^2}(e^x y^2 - 2y) = -e^x + 2z$.)

    (d) Find all solutions of the differential equation $3y'''(x)+7y''(x)+2y'(x) = f(x)$ in the following cases:

        i. $f(x) = e^x$

        ii. $f(x) = \sin x$

        iii. $f(x) = x^2$

        iv. $f(x) = e^{-2x}$

    (e) Find all a and b such that all solutions of the differential equation $y''(x) + ay'(x) + by(x) = 0$ converge to 0 as x goes to infinity.

    (f) Solve the following system of linear differential equations:

    $$\begin{cases} \dot{x} = x - 2y - z, \\ \dot{y} = y - x + z, \\ \dot{z} = x - z. \end{cases}$$

    (g) Find the equilibria of the system and classify them:

    $$\begin{cases} \frac{dx}{dt} = \ln(2 - y^2), \\ \frac{dy}{dt} = e^x - e^y. \end{cases}$$

2. *Economics Application: Company's Profit*

    The growth of a company's profit $dP(t)/dt$ at time t is proportional to a current level of profit $P(t)$ (which, if negative, represents a loss) and also depends on exogenously given function $g(t)$

that captures intertemporal development of the market environment. To this end, the profit is growing according to the following differential equation $\dfrac{dP(t)}{dt} = g(t) + KP(t)$, where $K > 0$ is a constant. Suppose that $g(t) = e^t$ and $P(0) = 0$. Derive the time profile of $P(t)$.

3. *Economics Application: Solow's Growth Model in Macroeconomics*

   In his famous paper, Robert Solow explained the dynamics of labor-adjusted capital $k(t) \equiv \frac{K(t)}{L(t)}$ (i.e. capital per effective worker) by means of the following differential equation (so-called *Solow's equation*)

   $$\dot{k} = \sigma f(k) - \theta k,$$

   where σ is the savings rate and constant θ captures various exogenous factors (such as population growth, technological progress, capital depreciation, etc.), and at the beginning of the period $(t = t_0)$, $k(t_0) = k_0$.

   Let $f(k)$ be a standard decreasing returns to scale production function $f(k) = A(k(t))^\alpha$, constants $\alpha \in (0,1)$ and $A > 0$.

   Using the analytical solution of Solow's equation in this case which was derived in Economics Application 20, describe the dynamics of $k(t)$ (i.e. what happens to $k(t)$ over time). Consider two cases: (i) the initial capital stock k_0 is fairly small, and (ii) k_0 is fairly large. Could you argue that in the long run the economy reaches a steady-state (equilibrium) rate of capital accumulation (the so-called 'Golden Rule' of capital accumulation, defined in Economics Application 17)?

4. *Learning By Doing...*

   (a)   i. Solve the difference equation $y_{t+1} - y_t = 0$.
      ii. Solve the difference equation $y_{t+1} + ay_t = 0$ where a is an arbitrary constant.
      iii. Find the solution to $y_{t+1} + ay_t = 0$ if you know that $y_0 = 10$.
      iv. Now solve the difference equation $y_{t+1} + ay_t = b$, when a, b are constants.

    v. Finally, assume that $b = b_t$ is a given function. Find the solution to the difference equation $y_{t+1} - y_t = b_t$, $t = 0, 1, 2 \ldots$.

(b) *Economic Application: National Income Growth*

Following Samuelson (c.f. P.A.Samuelson, "Interaction Between the Multiplier Analysis and the Principle of Acceleration", *Review of Economic Statistics,* 21, 75-58), the difference equation for the national income function Y can be derived as

$$Y_{t+2} - \alpha(1 + \beta)Y_{t+1} + \alpha\beta Y_t = 1, \qquad t = 0, 1, 2, \ldots$$

where $\alpha > 0$ is the *marginal propensity to consume*, and $\beta > 0$ is the so-called *relation.*

Solve this difference equation for $\alpha = 1$, $\beta = 2$.

(c) Consider the following optimal control problem:

$$\min \int_0^1 u^2(t)dt$$

$$\text{subject to } \frac{dy(t)}{dt} = y(t) + u(t), \quad y(0) = 1, \quad y(1) = 0$$

    i. Set up the Hamiltonian function and write down the F.O.C.

    ii. Find the solution to this optimal control problem.

Solutions:

1. (a) Rearranging and integrating we get

$$\int \frac{dy}{y(y + 2)} = \int x dx$$

$$\Rightarrow \ln \frac{y}{y + 2} = x^2 + c \Rightarrow \frac{y}{y + 2} = e^{x^2 + c}$$

Substituting $c_1 = e^c$ and solving for y we obtain

$$y = \frac{2c_1 e^{x^2}}{1 - c_1 e^{x^2}}$$

(b) Using the formula for the solution to the linear differential equation of the first order we get

$$y = e^{x^2}\left[c + \int (3x^2 - 2x^4)e^{-x^2}\,dx\right]$$

(c) Let $z = \frac{1}{y}$. Thus

$$\frac{dz}{dx} = \frac{dz}{dy}\frac{dy}{dx} = -\frac{1}{y^2}(e^x y^2 - 2y) = -e^x + 2z$$

Rewriting the expression above we get the following linear non-homogeneous differential equation with constant coefficients

$$z' - 2z = -e^x$$

The general solution to the homogeneous equation is $z_h = ce^{2x}$, and the particular solution to the non-homogeneous equation is of the form $z_p = Ae^x$. Substituting the latter into the equation for z we find that $A = 1$ and the general solution to the non-homogeneous equation in z-variable is

$$z = ce^{2x} + 1$$

Substituting and rearranging this expression for y we find that

$$y = \frac{1}{ce^{2x} + 1}$$

(d) The characteristic equation for the homogeneous equation is $3k^3 + 7k^2 + 2k = 0$, with the roots $k_{1,2,3} = 0, -\frac{1}{3}, -2$.

Thus the solution to the homogeneous equation is

$$y_0 = c_1 + c_2 e^{-\frac{1}{3}x} + c_3 e^{-2x}$$

i. $f(x) = e^x$

We are looking for the particular solution in the form $y_p = Ae^x$. Substituting this solution into the original (nonhomogeneous) equation and solving for A we compute $A = \frac{1}{12}$. Thus the solution to the non-homogeneous equation is

$$y = y_0 + y_p = c_1 + c_2 e^{-\frac{1}{3}x} + c_3 e^{-2x} + \frac{1}{12}e^x$$

ii. $f(x) = \sin x$

The particular solution has the form:

$$y_p = a\sin x + b\cos x$$

where a and b are constants that have to be determined. Substitution of y_p into the original equation yields

$$-(10a + 2b)\sin x + (2a - 4b)\cos x = \sin x$$

Now let us equate the multipliers of sin and cos, respectively:

$$\sin: \qquad -10a - 2b = 1,$$
$$\cos: \qquad -4b + 2a = 0.$$

Solving this system of equations we compute $a = -\frac{1}{11}$, $b = -\frac{1}{22}$. Therefore, the solution to the initial non-homogeneous differential equation appears as

$$
\begin{aligned}
y &= y_0 + y_p \\
&= c_1 + c_2 e^{-\frac{1}{3}x} + c_3 e^{-2x} - \frac{1}{11}\sin x - \frac{1}{22}\cos x
\end{aligned}
$$

iii. With $f(x) = x^2$, we can express $f(x)$ as $f(x) = x^2 e^{0x}$. Since 0 is a root of the characteristic equation we will have to additionally multiply the form of the particular solution by x. Thus the particular solution becomes

$$y_p = (ax^2 + bx + c)x.$$

Taking the derivatives and substituting this form into the original equation we get

$$18a + 42ax + 14b + 6ax^2 + 4bx + 2c = x^2.$$

Equating the multipliers of x^2, x, and 1 respectively we obtain the following system of equations

$$
\begin{cases}
6a = 1 \\
42a + 4b = 0 \\
18a + 14b + 2c = 0
\end{cases}
\Rightarrow
\begin{cases}
a = \frac{1}{6} \\
b = -\frac{7}{4} \\
c = \frac{43}{4}
\end{cases}
$$

and the solution to the original non-homogeneous equation is

$$
\begin{aligned}
y &= y_0 + y_p \\
&= c_1 + c_2 e^{-\frac{1}{3}x} + c_3 e^{-2x} + \frac{1}{6}x^3 - \frac{7}{4}x^2 + \frac{43}{4}x
\end{aligned}
$$

iv. $f(x) = e^{-2x}$

Since -2 is a root of the characteristic equation, we look for a particular solution in the following form

$$y_p = axe^{-2x}$$

Taking the derivatives, substituting them into the original equation and factoring out e^{-2x} we find that

$$36a - 24ax - 28a + 28ax + 2a - 4ax = 1.$$

The multipliers of x cancel out, and a is evaluated as $a = \frac{1}{10}$. To this end, the solution to the original non-homogeneous equation is

$$y = y_0 + y_p = c_1 + c_2 e^{-\frac{1}{3}x} + c_3 e^{-2x} + \frac{1}{10}xe^{-2x}$$

(e) The characteristic equation is $k^2 + ak + b = 0$, and its roots are $k_{1,2} = \dfrac{-a \pm \sqrt{a^2 - 4b}}{2}$

There are three cases possible:

i. $k_{1,2}$ are real and distinct, which holds if $a^2 - 4b > 0$ or $|a| > 2b$. In this case the solutions to the equation are

$$y = c_1 e^{k_1 x} + c_2 e^{k_2 x}.$$

The condition so that the all the solutions converge to 0 is

$$k_1 < 0 \text{ and } k_2 < 0,$$

that is, $k_{1,2} < 0$ iff $\frac{-a+\sqrt{a^2-4b}}{2} < 0 \Longrightarrow a > \sqrt{a^2 - 4b} \Longrightarrow$
$\begin{cases} a > 0 \\ b > 0 \end{cases}$

Summing up, in this case the following conditions should be satisfied:

$$\begin{cases} a > 0 \\ b > 0 \\ |a| > 2b \end{cases}$$

ii. $k_{1,2}$ are real and equal, i.e., $k_{1,2} = k$. The latter holds if $a^2 - 4b = 0$ or $|a| = 2b \Longrightarrow b > 0$. In this case the solutions to the equation are

$$y = c_1 e^{kx} + c_2 x e^{kx}.$$

The condition so that all the solutions converge to 0 is $k < 0$ (you can check that $\lim_{x \to \infty} xe^{kx} = 0$ iff $k < 0$ using L'Hopital's rule's rule)

Furthermore, $k < 0$ iff $\dfrac{-a + \sqrt{a^2 - 4b}}{2} < 0$, but since $a^2 - 4b = 0$, this condition simplifies to $a > 0$.

So for this case we get the following conditions:

$$\begin{cases} a > 0 \\ b > 0 \\ |a| = 2b \end{cases}$$

iii. $k_{1,2}$ are complex, which holds if $a^2 - 4b < 0$ or $|a| < 2b$ and $b > 0$. Thus all solutions are given by

$$y = c_1 e^{-\frac{a}{2}x} \cos(x\sqrt{-a^2 + 4b}) + \\ + c_1 e^{-\frac{a}{2}x} \sin(x\sqrt{-a^2 + 4b}).$$

Since $\cos(\cdot)$ and $\sin(\cdot)$ are bounded functions, all the solutions converge to 0 if the exponents converge to 0 (recall that if we multiply the function which converges to 0 by a bounded function, the product will converge to 0 as well). Therefore, to provide convergence we need $-\frac{a}{2} < 0 \implies a > 0$.

Summing up, in this case convergence requires

$$\begin{cases} a > 0 \\ b > 0 \\ |a| < 2b \end{cases}$$

Putting three cases together, we obtain the criterion for convergence:

$$\begin{cases} a > 0 \\ b > 0 \end{cases}$$

Alternative Solution: Depending on the roots λ_1, λ_2 of characteristic equation $\lambda^2 + a\lambda + b = 0$ we have 3 cases:

- $y(x) = c_1 e^{\lambda_1 x} + c_2 e^{\lambda_2 x}$ for real $\lambda_1 \neq \lambda_2$
- $y(x) = c_1 e^{\lambda_1 x} + c_2 x e^{\lambda_1 x}$ for real $\lambda_1 = \lambda_2$
- $y(x) = c_1 e^{\alpha x} \cos \beta x + c_2 e^{\alpha x} \sin \beta x$ for complex $\lambda_1 = \alpha + i\beta$, $\lambda_2 = \alpha - i\beta$.

In the first and second case the solution converges to zero for x going to infinity only if both $\lambda_1, \lambda_2 < 0$. In the third case only if the real part of λ_1 is negative. Since

$$\lambda_1 + \lambda_2 = -a \quad \text{and} \quad \lambda_1 \lambda_2 = b,$$

the above conditions are equivalent to $a > 0$ and $b > 0$.

(f) The system can be rewritten in the form $\dot{X} = AX$, where

$$A = \begin{pmatrix} 1 & -2 & -1 \\ -1 & 1 & 1 \\ 1 & 0 & -1 \end{pmatrix}, \quad X = \begin{pmatrix} x \\ y \\ z \end{pmatrix}$$

Since $\det A = -1$, the system is regular. First we find eigenvalues from the characteristic equation $\det(A - \lambda I) = 0$ as $\lambda_{1,2,3} = 0, 2, -1$. Then for each eigenvalue $\lambda_1, \lambda_2, \lambda_3$ we find corresponding eigenvectors V_1, V_2, V_3 from the identity $(A - \lambda_i I)V_i = 0$:

$$V_1 = \begin{pmatrix} 1 \\ 0 \\ 1 \end{pmatrix}, \quad V_2 = \begin{pmatrix} 3 \\ -2 \\ 1 \end{pmatrix}, \quad V_3 = \begin{pmatrix} 0 \\ 1 \\ -2 \end{pmatrix}.$$

Therefore, the solution X can be expressed in vector form as

$$X(t) = c_1 V_1 e^{\lambda_1 t} + c_2 V_2 e^{\lambda_2 t} + c_3 V_3 e^{\lambda_3 t},$$

or, substituting numerical values of eigenvalues and eigenvectors, as

$$\begin{pmatrix} x \\ y \\ z \end{pmatrix} = \begin{pmatrix} c_1 + 3c_2 e^{2t} \\ -2c_2 e^{2t} + c_3 e^{-t} \\ c_1 + c_2 e^{2t} - 2c_3 e^{-t} \end{pmatrix}.$$

(g) By the definition of equilibrium,

$$\begin{cases} \frac{dx}{dy} = 0 \\ \frac{dy}{dt} = 0 \end{cases},$$

or

$$\begin{cases} \ln(2 - y^2) = 0 \\ e^x - e^y = 0 \end{cases} \implies \begin{cases} y = \pm 1 \\ x = y \end{cases}.$$

There are two equilibria: $\begin{cases} y = 1 \\ x = 1 \end{cases}$ and $\begin{cases} y = -1 \\ x = -1 \end{cases}$.

The Jacobian matrix is evaluated at (x, y) as

$$J(x, y) = \begin{pmatrix} 0 & \frac{-2y}{2 - y^2} \\ e^x & -e^y \end{pmatrix}$$

Let us investigate the eigenvalues of the Jacobian matrix at the points of equilibrium.

i. At $\begin{cases} y = 1 \\ x = 1 \end{cases}$, the eigenvalues for $J(1,1)$ are

$$\lambda_{11}, \lambda_{12} = \frac{-e \pm \sqrt{e^2 - 8e}}{2} \approx -1.36 \pm 1.89i$$

The eigenvalues are complex, $Re(\lambda_{11}), Re(\lambda_{12}) < 0$, thus the equilibrium $(1,1)$ is a stable focus.

ii. At $\begin{cases} y = -1 \\ x = -1 \end{cases}$, the eigenvalues for $J(-1,-1)$ are

$$\lambda_{21}, \lambda_{22} = \frac{-e^{-1} \pm \sqrt{e^{-2} + 8e^{-1}}}{2};$$

approximately, $\lambda_{21} \approx -1.05$, $\lambda_{22} \approx 0.69$.
The eigenvalues are real and of different signs, thus the equilibrium $(-1,-1)$ is a saddle.

2. In the language of mathematics we just need to solve the differential equation $\dot{P} - KP = e^t$ with the initial condition $P(0) = 0$. Solving the homogeneous equation (it is a first-order linear differential equation), we obtain the solution $P(t) = Ce^{Kt}$.

We try to find a particular solution in the form $P_p(t) = ae^t$. Substituting into the equation we obtain $a(1 - K)e^t = e^t$, which implies $a(1 - K) = 1$, so $a = \frac{1}{1-K}$, for $K \neq 1$. Then the general solution is

$$P(t) = Ce^{Kt} + \frac{1}{1 - K}e^t.$$

Imposing the initial condition, i.e., setting $t = 0$ we obtain $0 = P(0) = C + \frac{1}{1-K}$. Therefore $C = -\frac{1}{1-K}$ and so

$$P(t) = \frac{1}{1 - K}(e^t - e^{Kt}), \quad \text{for } K \neq 1.$$

For $K = 1$, we try to find a particular solution in the form $P_p(t) = (a + bt)e^t$. Then $be^t = e^t$, which means that $b = 1$ and a can be arbitrary. So, we can take the particular solution in the form $P_p(t) = te^t$ and the general solution in the form $P(t) = ce^{Kt} + te^t$. Imposing the initial condition we obtain $C = 0$ and so the solution to the original problem is

$$P(t) = te^t, \quad \text{for } K = 1.$$

3. We have the equation $\dot{k} = \sigma A k^\alpha - \theta k$ with the initial condition $k(t_0) = k_0$. The solution obtained in the Economics Application 20 can be written as

$$k(t) = \left[\frac{A\sigma}{\theta} + \left(k_0^{1-\alpha} - \frac{A\sigma}{\theta} \right) e^{-(1-\alpha)\theta(t-t_0)} \right]^{1/(1-\alpha)}.$$

Assume that $\theta > 0$. Because $-(1-\alpha)\theta < 0$, then $e^{-(1-\alpha)\theta(t-t_0)}$ is a decreasing function. For fairly large k_0, i.e., such that $k_0^{1-\alpha} > \frac{A\sigma}{\theta}$, also $k(t)$ is a decreasing function. Otherwise, $k(t)$ is increasing. Taking $t \to \infty$, we obtain $e^{-(1-\alpha)\theta(t-t_0)} \to 0$, so $k(t) \to \left(\frac{A\sigma}{\theta} \right)^{1/(1-\alpha)}$, the 'Golden Rule' where $f'(k) = \theta$. This is also the equilibrium, because after plugging it into the right-hand side of the original equation we obtain zero, i.e., $\dot{k} = 0$.

4. (a)   i. The characteristic equation is $k - 1 = 0$, implying $k = 1$. Therefore, the solution becomes $y_t = c(1)^t = c$.

   ii. The characteristic equation is: $k + a = 0 \Longrightarrow k = -a$
   The solution is $y_t = c(-a)^t$.

   iii. The characteristic equation is $k + a = 0 \Longrightarrow k = -a$, and the solution is derived as $y_t = c(-a)^t$. From the initial condition, $y_0 = c = 10$. Therefore, $y_t = 10(-a)^t$.

   iv. Similar to the preceding case, the characteristic equation is $k + a = 0 \Longrightarrow k = -a$. The solution to the homogeneous equation is $y_{g_t} = c(-a)^t$, and a particular solution to the non-homogeneous equation should take the form $y_{p_t} = A$. Substituting y_p into the original equation we find $A = \frac{b}{a+1}$. Finally, the general solution to the non-homogeneous equation is evaluated as

   $$y_t = y_{g_t} + y_{p_t} = c(-a)^t + \frac{b}{a+1}.$$

   v. Assume that at time 0, $y_0 = y_0$. Then, recursively

   $$\begin{aligned} y_1 &= y_0 + b_0, \\ y_2 &= y_1 + b_1 = y_0 + b_0 + b_1, \\ y_3 &= y_2 + b_2 = y_0 + b_0 + b_1 + b_2, \\ &\text{etc.} \end{aligned}$$

By the method of induction we conclude that

$$Y_{t+1} = y_0 + b_0 + b_1 + ... + b_t.$$

(b) Consider the difference equation $Y_{t+2} - 3Y_{t+1} + 2Y_t = 1$.

The corresponding characteristic equation is $k^2 - 3k + 2 = 0$ with the roots $k_{1,2} = 2; 1$.

The solution to the homogeneous equation is $y_{g_t} = c_1(2)^t + c_2$.

We are looking for the particular solution in the form $y_{p_t} = At$. Substituting y_p into the original non-homogeneous equation and solving for A we get $A = -1$, and the general solution to the non-homogeneous equation is expressed as

$$y_t = y_{g_t} + y_{p_t} = c_1(2)^t + c_2 - t.$$

(c) To maximize $\int\limits_0^1 -u^2(t)dt$, subject to the equation of motion $\frac{dy(t)}{dt} = y(t) + u(t)$ and initial and terminal conditions $y(0) = 1$, $y(1) = 0$, let us construct the Hamiltonian function $H = -u^2 + \lambda(y + u)$ and derive the F.O.C.:

$$\frac{dH}{du} = -2u + \lambda = 0,$$

$$\frac{dH}{dy} = \lambda = -\dot{\lambda},$$

plus the transversality condition $\lambda(1)y(1) = 0$.

Putting together the derivatives of Hamiltonian and the constraint we obtain the following system of ordinary differential equations:

$$\begin{cases} \lambda = 2u \\ \lambda = -\dot{\lambda} \\ \dot{y} = y + u \end{cases}$$

From the first two equations we get $\dot{u} + u = 0$, which yields $u(t) = c_1 e^{-t}$.

Substituting $u(t)$ into the third equation of the system we find

$$\dot{y} - y = c_1 e^{-t}.$$

Solving the latter for y we get

$$y(t) = c_2 e^t - \frac{c_1}{2} e^{-t}.$$

Substituting the initial and terminal conditions we compute $c_1 = \frac{-2e^2}{e^2-1}$ and $c_2 = \frac{-1}{e^2-1}$.

Finally,

$$y^{opt}(t) = \frac{-1}{e^2-1} e^t - \frac{-e^2}{e^2-1} e^{-t},$$

$$u^{opt}(t) = \frac{-2e^2}{e^2-1} e^{-t},$$

$$\lambda^{opt}(t) = 2u = \frac{-4e^2}{e^2-1} e^{-t}.$$

Note that for the optimal solution the transversality condition

$$\lambda^{opt}(1) y^{opt}(1) = \frac{-4e}{e^2-1} 0 = 0$$

holds.

5.5   Sample Problem Sets

Sample Problem Set–I

1. *Easy Shots:*

   (a) For which values of x, y is the function $f(x,y) = 5x^3 + 4xy + 3y^2$ convex?

   (b) Prove or disprove: If a production function $F(L, K)$ is homogeneous of degree 1, i.e.
   $$F(\lambda L, \lambda K) = \lambda F(L, K) \quad \forall \lambda \geq 0$$
   then $F(L, K) = L \cdot F_L(L, K) + K \cdot F_K(L, K)$.

   (c) Solve the following differential equation: $\sqrt{y^2 + 1}dx = xydy$.

2. Find all a and b such that the following system of linear equations

$$\begin{cases} 3x - 2y + z & = & b \\ 5x - 8y + 9z & = & 3 \\ 2x + y + az & = & -1 \end{cases}$$

has an infinite number of solutions.

3. Sketch the graph of the function $y(x) = x^x$, $x \geq 0$. Please be specific! (In particular, what is the limit of $f(x)$ as x approaches 0 from the right?)

4. Solve the following differential equation:

$$y'' - y' = e^x + 2$$

5. Solve the following difference equation:

$$x_{t+1} + x_t = b_t.$$

6. Find the conditional extrema of the following function and classify them:

$$f(x, y, z) = (x + y)z, \quad \text{s.t.} \quad \frac{1}{x^2} + \frac{1}{y^2} + \frac{2}{z^2} = 4$$

7. Using the Kuhn-Tucker conditions solve the following optimization problem:

$$\max_{x,y} xy$$

subject to

$$5x + 4y \leq 50, \quad 3x + 6y \leq 40 \qquad x \geq 0, \quad y \geq 0.$$

8. Consider the following optimal control problem:

$$\max I = \int_0^1 (x + u)dt$$

subject to

$$\dot{x} = 1 - u^2, \qquad x(0) = 1.$$

(a) Set up the Hamiltonian function and write down the first-order conditions.

(b) Find an optimal solution, if any.

(HINT: In this problem the transversality condition takes the form $\lambda(1) = 0$.)

Bonus problem

B1 Find the equilibria of the system of non-linear differential equations and classify them:

$$\dot{x} = x^2 - y,$$
$$\dot{y} = \ln(1 - x + x^2) - \ln 3.$$

Solutions to Sample Problem Set–I

1. (a) $f_x = 15x^2 + 4y,$ $\qquad f_y = 4x + 6y,$
    $f_{xx} = 30x,$ $\qquad f_{xy} = 4,$ $\qquad f_{yy} = 6.$
    $f(x, y)$ is convex $\Leftrightarrow d^2 f$ is positive semi-definite $\Leftrightarrow$ all principal minors of the matrix

    $$\begin{pmatrix} 30x & 4 \\ 4 & 6 \end{pmatrix}$$

    are non-negative $\Leftrightarrow x \geq 0$ and $180x - 16 \geq 0$. Therefore, the function is convex for $x \geq 4/45$ and $\forall y$.

    (b) $F(\lambda L, \lambda K) = \lambda F(L, K)$ $\quad \forall \lambda \geq 0$. Differentiating this equality with respect to λ and taking $\lambda = 1$ we get
    $L \cdot F_L(L, K) + K \cdot F_K(L, K) = F(K, L)$

    (c) $\qquad \dfrac{dx}{x} = \dfrac{y\,dy}{\sqrt{y^2 + 1}}$

    $$\int \frac{dx}{x} = \frac{1}{2} \int \frac{d(y^2 + 1)}{\sqrt{y^2 + 1}}$$

    $$\ln |x| = \sqrt{y^2 + 1} + c$$

2. The system has an infinite number of solutions $\Leftrightarrow \det A = 0$ and rank$(A) = $ rank(B), where

$$A = \begin{pmatrix} 3 & -2 & 1 \\ 5 & -8 & 9 \\ 2 & 1 & a \end{pmatrix} \quad \text{and} \quad B = \begin{pmatrix} 3 & -2 & 1 & b \\ 5 & -8 & 9 & 3 \\ 2 & 1 & a & -1 \end{pmatrix}.$$

$\det A = -24a + 5 - 36 + 16 - 27 + 10a = 0 \Leftrightarrow a = -3$

We can easily show that rank$(A) = 2$. The second row is a linear combination of the first and the third ones $(a^2 = 3a^1 - 2a^3$, where a^i is the i-th row of A).

To make rank$(B) = 2$ we have to impose on the fourth column of B the same restriction $3 = 3b + 2$.

Therefore, $a = -3, b = 1/3$.

3. $x^x = e^{x \ln x}$

$$\lim_{x \to 0+} x \ln x = \lim_{x \to 0+} \frac{\ln x}{1/x} = \{\text{by L'Hôpital's Rule}\}$$

$$= \lim_{x \to 0+} \frac{1/x}{-1/x^2} = \lim_{x \to 0+} (-x) = 0$$

$$\lim_{x \to 0+} y = e^0 = 1$$

$$y' = x^x (\ln x + 1)$$

$$y'' = x^x (\ln x + 1)^2 + x^x \cdot \frac{1}{x} > 0$$

$\forall x > 0$ (the function is convex)

$y' = 0 \Leftrightarrow x^0 = 1/e$, $y'' > 0 \Rightarrow x^0$ is minimum;
$x < x^0 \Rightarrow y' < 0 \Rightarrow y$ is decreasing;
$x > x^0 \Rightarrow y' > 0 \Rightarrow y$ is increasing.

4. The characteristic equation $\lambda^2 - \lambda = 0$ has the roots $\lambda_1 = 0$ and $\lambda_2 = 1$. Thus the general solution is $y_g = c_1 + c_2 e^x$.

   The particular solution is $y_p = ax + bxe^x$; in order to find coefficients a and b we should plug y_p into the initial equation to obtain $a = -2$, $b = 1$.

   Therefore, the general solution to the non-homogeneous equation becomes $y = y_g + y_p = -2x + xe^x + c_1 + c_2 e^x$

5. Characteristic equation: $\lambda + 1 = 0 \Rightarrow \lambda = -1$

   General solution: $x_t = x_0 \cdot (-1)^t + \displaystyle\sum_{i=0}^{t-1} (-1)^{t-1-i} b_i$

6. Lagrangian function: $L = (x + y)z + \lambda \left(4 - \dfrac{1}{x^2} - \dfrac{1}{y^2} - \dfrac{2}{z^2} \right)$

   F.O.C.

$$\frac{\partial L}{\partial x} = z + \frac{2\lambda}{x^3} = 0 \tag{5.31}$$

$$\frac{\partial L}{\partial y} = z + \frac{2\lambda}{y^3} = 0 \tag{5.32}$$

$$\frac{\partial L}{\partial z} = x + y + \frac{4\lambda}{z^3} = 0 \tag{5.33}$$

$$\frac{1}{x^2} + \frac{1}{y^2} + \frac{2}{z^2} = 4 \tag{5.34}$$

From (5.31), (5.32) it follows that $x = y$.

(5.31) $\Rightarrow$ $2\lambda = -zx^3$, or $\lambda = -zx^3/2$; we can substitute the latter expression into (5.33) to get $z^2 = x^2$.

(5.34) $\Rightarrow$ $4/x^2 = 4 \Rightarrow x^2 = 1$. Thus, we have four stationary points: (1,1,1); (1,1,-1); (-1,-1,-1); (-1,-1,1). In order to classify them we should check the sign of bordered Hessian matrices:

$$|H_2| = \det \begin{pmatrix} 0 & -\frac{2}{x} & -\frac{2}{y} \\ -\frac{2}{x} & -\frac{6\lambda}{x^4} & 0 \\ -\frac{2}{y} & 0 & -\frac{6\lambda}{y^4} \end{pmatrix} = \frac{24\lambda}{x^2 y^2} \left(\frac{1}{x^2} + \frac{1}{y^2} \right)$$

and

$$|H_3| = \det \begin{pmatrix} 0 & -\frac{2}{x} & -\frac{2}{y} & -\frac{4}{z} \\ -\frac{2}{x} & -\frac{6\lambda}{x^4} & 0 & 1 \\ -\frac{2}{y} & 0 & -\frac{6\lambda}{y^4} & 1 \\ -\frac{4}{z} & 1 & 1 & -\frac{12\lambda}{z^4} \end{pmatrix}_{(\pm1,\pm1,\pm1)}$$

$$= -192(6\lambda^2 \pm \lambda).$$

Therefore, $\text{sign}(|H_2|) = \text{sign}(\lambda)$ and given that $\lambda = \pm1/2$, $|H_3|$ is always negative.

For $(1,1,1)$ and $(-1,-1,-1)$ $|H_2| < 0, |H_3| < 0 \Rightarrow$ local minimum; for $(1,1,-1)$ and $(-1,-1,1)$ $|H_2| > 0, |H_3| < 0 \Rightarrow$ local maximum.

7. Lagrangian function:

$$L = xy + \lambda(50 - 5x - 4y) + \eta(40 - 3x - 6y)$$

The Kuhn-Tucker conditions are

$$\frac{\partial L}{\partial x} = y - 5\lambda - 3\eta \le 0 \quad (1) \qquad x(y - 5\lambda - 3\eta) = 0 \quad (5)$$

$$\frac{\partial L}{\partial y} = x - 4\lambda - 6\eta \le 0 \quad (2) \qquad y(x - 4\lambda - 6\eta) = 0 \quad (6)$$

$$\frac{\partial L}{\partial \lambda} = 50 - 5x - 4y \ge 0 \quad (3) \qquad \lambda(50 - 5x - 4y) = 0 \quad (7)$$

$$\frac{\partial L}{\partial \eta} = 40 - 3x - 6y \ge 0 \quad (4) \qquad \eta(40 - 3x - 6y) = 0 \quad (8)$$

$$\lambda \ge 0, \eta \ge 0, x \ge 0, y \ge 0.$$

1) $x = y = 0$
   (1)–(8) are satisfied $(\lambda = \eta = 0)$; $f(x,y) = 0$
2) $x = 0, y > 0$
   $(2) \Rightarrow -4\lambda - 6\eta = 0$ and $\lambda, \eta \ge 0 \Rightarrow \lambda = \eta = 0$
   from $(1) \Rightarrow y \le 0 \Rightarrow y = 0$

3) $x > 0, y = 0$; again, we get $x = y = 0$

4) $x, y > 0$

(5), (6) $\Rightarrow$

$y - 5\lambda - 3\eta = 0$

$x - 4\lambda - 6\eta = 0$

(A) $\lambda, \eta > 0$

$5x + 4y = 50$

$3x + 6y = 40$

Therefore, $x = 70/9, y = 25/9$

From (1),(2) we can compute the values of λ and η and find that $\lambda < 0$ (contradiction).

(B) $\lambda = \eta = 0 \Rightarrow x = y = 0$

(C) $\lambda = 0, \eta > 0$

(5),(6)$\Rightarrow y = 3\eta$ and $x = 6\eta = 2y$. Non-negativity of η implies that (8)$\Rightarrow 40 - 3x - 6y = 0$. Since $x = 2y$, (8)$\Rightarrow y = 10/3, x = 20/3$ and $f(x, y) = 200/9$

(D) $\lambda > 0, \eta = 0$

As in case (C), we compute $x = 5, y = 25/4$ and $f(x, y) = 125/4$. However, the choice of $x = 5, y = 25/4$ contradicts (4).

Therefore, the solution is $(20/3, 10/3)$.

8. Hamiltonian function:

$H = x + u + \lambda(1 - u^2)$; ($x$-state variable, u-control variable).

F.O.C.

$$\frac{\partial H}{\partial u} = 1 - 2\lambda u \;=\; 0 \qquad (5.35)$$

$$\frac{\partial H}{\partial x} = 1 \;=\; -\dot{\lambda} \qquad (5.36)$$

Transversality condition: $\lambda(1) \;=\; 0 \qquad (5.37)$

(5.36) $\Rightarrow \dot{\lambda} = 1 \Rightarrow \lambda = -t + c$;

from (5.37) $c = 1 \Rightarrow \lambda = 1 - t$;

(5.35) $\Rightarrow u = \dfrac{1}{2\lambda} = \dfrac{1}{2(1 - t)}$;

equation of motion $\Rightarrow \dot{x} = 1 - \dfrac{1}{4(1-t)^2} \Rightarrow x = t - \dfrac{1}{4(1-t)} + C$;

since $x(0) = 1$, $C = 5/4$

B1. Equilibria are determined by the system of algebraic equations

$$\begin{cases} x^2 - y = 0, \\ \ln(1 - x + x^2) - \ln 3 = 0 \end{cases}$$

The first equation implies $y = x^2$; the second equation yields $x_1 = 2$, $x_2 = -1$, thus $y_1 = 4$, $y_2 = 1$.

The Jacobian matrix is given by

$$J = \begin{pmatrix} 2x & -1 \\ \frac{2x-1}{1-x+x^2} & 0 \end{pmatrix}$$

At point (-1,1), $\det(J - \lambda I) = 0 \Leftrightarrow \lambda^2 + 2\lambda - 1 \Leftrightarrow$ eigenvalues are real and of different signs $\Leftrightarrow$ saddle.

At point (2,4), $\det(J - \lambda I) = 0 \Leftrightarrow \lambda^2 - 4\lambda + 1 \Leftrightarrow$ eigenvalues are real and positive $\Leftrightarrow$ unstable node.

Sample Problem Set–II

1. *Easy Shots:*

   (a) What are the maximum and minimum values of the function $y = x^3 - 9x^2 + 24x + 1$ on the range $0 \le x \le 5$?

   (b) Derive an approximate value of $\sqrt{103}$ (Use Taylor series, not calculators!).

   (c) Solve the differential equation $y' = ln(x)$, $y(1) = -1$. Is the solution unique?

   (d) Consider a matrix equation $AX = \mathbf{0}$, where A is a $[m \times n]$ matrix and $m < n$. Show that there are an infinite number of solutions.

   (e) A firm producing two goods x and y has profit function $\pi = 32x - x^2 + 2xy - 2y^2 + 16y - 8$. Find the profit-maximizing output levels of x and y.

(f) A producer is known to have a constant return to scale Cobb-Douglas production function $F(K, L) = K^\alpha L^{1-\alpha}$, where α is an unknown parameter. While solving the constrained maximization problem $\max Q(K, L)$ subject to $K + L \leq 500$, the producer chooses $K = 100$, $L = 400$. What is the value of α?

(g) Let $f(x_1, \ldots, x_n)$ be a twice continuously differentiable function. Prove that if $y = f(x_1, \ldots, x_n)$ is strictly concave then $z = -f(x_1, \ldots, x_n)$ is strictly convex, and vice versa.

(h) Find all solutions to the differential equation $y'' - 2y' - 3y = 0$.

2. Given the matrix $A = \begin{pmatrix} 0 & 0 & 1 \\ 0 & 0 & 0 \\ 1 & 0 & 0 \end{pmatrix}$, find a matrix P, $PP' = P'P = I$, such that $P'AP$ is a diagonal matrix. Is matrix A sign definite?

3. Find the stationary values of $z = 3x^2 + 2y^2 - 4y + 1$ subject to $x^2 + y^2 = 16$ and classify them.

4. Let the production function $Y = F(K, L)$ be homogeneous of degree 1.

   (a) Show that the Hessian matrix
   $$\mathcal{H} = \begin{pmatrix} F_{KK}(K, L) & F_{KL}(K, L) \\ F_{LK}(K, L) & F_{LL}(K, L) \end{pmatrix}$$
   is singular.

   (b) Show that the functions $F_K(K, L)$ and $F_L(K, L)$ are functionally dependent. (HINT: Two functions $u(x, y)$ and $v(x, y)$ are *functionally dependent* is there exists a function ϕ such that $\phi(u, v) \equiv 0$.)

   (c) Let at any time t, the output $Y(t) = F(K(t), L(t))$. Prove or disprove: the rate of growth of output is a weighted sum of the relative rates of growth of factor inputs, i.e. at any point in time $\gamma_Y = \alpha\gamma_K + (1-\alpha)\gamma_L$, where $\gamma_Y = \frac{\dot{Y}}{Y}$, $\gamma_K = \frac{\dot{K}}{K}$, $\gamma_L = \frac{\dot{L}}{L}$, a dot over variables denotes time derivative and $\alpha = \alpha(t) \in (0, 1)$.

5. A firm has total revenue $TR = 10Q - Q^2 + \frac{A}{2}$ where Q is its output and A is its advertising expenditures. Its total costs are $TC = \frac{Q^2}{2} + 5Q + 1 + A$. The managers of the firm have to maximize total revenue subject to the minimum profit constraint $\pi \geq 3$; Q and A are required to be non-negative. What levels of Q and A will the managers choose?

6. Minimize the function $(x-1)^2 + (y-1)^2$ subject to the constraint $x^2 \leq y \leq 2$ and $x \geq 0$. Check that the Kuhn-Tucker conditions will be necessary and sufficient, find the optimal solution, and sketch the solution graphically.

7. Suppose you are asked to extremize the function $f(x,y) = e^x + e^y + 2x + 2y$, subject to $10 = x + cy$, where c is a positive real parameter, and let (x^*, y^*) be the optimal values. Can you tell what happens to (x^*, y^*) if c increases by a very small number?

8. For which x is the matrix $A = \begin{pmatrix} 2 & 6 & 0 & 0 \\ 1 & x & 0 & 0 \\ 0 & 0 & 1 & 1 \\ 0 & 0 & 1 & 2 \end{pmatrix}$ non-singular?

   If it is non-singular, find its inverse.

9. Sketch a graph of the function $y = y(x)$ defined parametrically: $x = -1 + 2t - t^2$ and $y = 2 - 3t + t^3$, t is a real parameter. Be specific!

10. Solve the differential equation: $xy' + x^2 + xy - y = 0$.

11. Find all the solutions to the following linear differential equation: $y'' - 2y' - 3y = e^{4x} + 1$. (HINT: Use your result in 1(h).)

12. In the two-dimensional model of optimal growth, the linearized 'Golden Rule' model results in the following system of linear differential equations:

$$\begin{aligned} \dot{c} &= -\beta(k - k^*), \\ \dot{k} &= \theta(k - k^*) - (c - c^*), \end{aligned}$$

where c denotes consumption, k is capital stock, β and θ are positive parameters, and c^* and k^* stand for the steady state levels of consumption and capital stock, respectively.

Find the solution for k such that, starting from some k_0, it converges to k^*.

(HINT: Take the derivative of the second equation with respect to time and substitute the right-hand side of the first equation for $\dot{c}$. Solve the resulting second-order linear differential equation for $k(t)$. What condition have you imposed to provide convergence to k^*?).

Solutions to Sample Problem Set–II

1. 1(a) $y' = 3(x^2 - 6x + 8) = 3(x - 2)(x - 4), \ x \in [0, 5]$

Evaluate the function at the boundary points: $\begin{cases} y(0) = 1 \\ y(5) = 21 \end{cases}$

Evaluate the function at the stationary points: $\begin{cases} y(2) = 21 \\ y(4) = 17 \end{cases}$

The minimum: $y = 1$ at $x = 0$; and maximum: $y = 21$ at $x = 2$ and $x = 5$.

1(b) $\sqrt{x} = \sqrt{x_0} + \frac{1}{2\sqrt{x_0}}(x - x_0) - \frac{1}{8(\sqrt{x_0})^3}(x - x_0)^2 + \cdots$

Take $x_0 = 100$, $\Delta x = x - x_0 = 3$

$\Rightarrow \sqrt{103} \approx \sqrt{100} + \frac{3}{2\sqrt{100}} - \frac{9}{8(\sqrt{100})^3} = 10.148875$

[Please compare with the exact value $\sqrt{103} = 10.14889157$]

1(c) $y' = \ln x \Leftrightarrow \frac{dy}{dx} = \ln(x) \ dy = \ln(x)dx \Leftrightarrow \int dy = \int \ln(x) \ dx$
$\Leftrightarrow y = \int \ln(x) \ dx = x \ln(x) - x + C$, by integration by parts.
Thus $y(x) = x \ln(x) - x + C$.

Using the initial condition $y(1) = -1$, we find $C = 0$.

$f(x, y) \equiv \ln(x)$ satisfies the uniqueness theorem. Therefore, the solution is unique.

1(d) $r = \text{rank}(A) \leq \min(m, n) \leq m < n$

$\Rightarrow$ The dimension of the solution space equals $d = n - r > 0$. In other words, the solution vector $X = (x_1, \ldots, x_n)'$ is comprised of two parts: r $x_i's$ solve the linear system and $n - r$ $x_i's$ are arbitrarily chosen.

1(e) F.O.C.: $\begin{cases} \pi_x = 32 - 2x + 2y = 0 \\ \pi_y = 16 + 2x - 4y = 0 \end{cases} \Rightarrow \begin{cases} x^* = 40 \\ y^* = 24 \end{cases}$. The corresponding Hessian matrix $\begin{pmatrix} -2 & 2 \\ 2 & -4 \end{pmatrix}$ is negative definite, therefore $(x^* = 40, y^* = 24)$ is indeed the point of maximum.

1(f) F.O.C.: MRS_{LK}= - slope of the budget line, i.e. $\dfrac{F_K}{F_L} = \dfrac{r}{w} =$

    1. In the optimum, $\dfrac{\alpha}{1-\alpha} \cdot \dfrac{L^*}{K^*} = \dfrac{\alpha}{1-\alpha} \cdot \dfrac{400}{100} = 1 \Rightarrow \alpha = \frac{1}{5}.$

1(g) By definition, $f(\cdot)$ is concave if and only if $\theta f(u) + (1 - \theta)f(v) \le f(\theta u + (1-\theta)v)$. After multiplication by (-1) this inequality becomes $\theta(-f(u)) + (1 - \theta)(-f(v)) \ge -f(\theta u + (1 - \theta)v)$, i.e. $-f(\cdot)$ satisfies the definition of concavity.

1(h) The characteristic equation $\lambda^2 - 2\lambda - 3 = 0$ has the roots $\lambda_1 = -1$ and $\lambda_2 = 3$. The general solution to a linear homogeneous equation is $y(x) = C_1 e^{-x} + C_2 e^{3x}$, where C_1 and C_2 are arbitrary constants.

2. First, let us find eigenvalues of the matrix A:

$$\det(A - \lambda E) = \det \begin{pmatrix} -\lambda & 0 & 1 \\ 0 & -\lambda & 0 \\ 1 & 0 & -\lambda \end{pmatrix}$$

$$= -\lambda^3 + \lambda = -\lambda(\lambda^2 - 1) = 0,$$

i.e. $\lambda_1 = 0$, $\lambda_2 = -1$, and $\lambda_3 = 1$. Therefore, matrix A is sign indefinite. The corresponding eigenvectors are $v_1 = \begin{pmatrix} 0 \\ 1 \\ 0 \end{pmatrix}$,

$v_2 = \begin{pmatrix} -1 \\ 0 \\ 1 \end{pmatrix}$, and $v_3 = \begin{pmatrix} 1 \\ 0 \\ 1 \end{pmatrix}$. Thus, $P = \begin{pmatrix} 0 & -\frac{1}{\sqrt{2}} & \frac{1}{\sqrt{2}} \\ 1 & 0 & 0 \\ 0 & \frac{1}{\sqrt{2}} & \frac{1}{\sqrt{2}} \end{pmatrix}$.

3. Let us construct the Lagrangian function: $L = 3x^2 + 2y^2 - 4y + 1 + \lambda(16 - x^2 - y^2)$

F.O.C.: $\begin{cases} 6x - 2\lambda x = 0 \\ 4y - 4 - 2\lambda y = 0 \\ 16 - x^2 - y^2 = 0 \end{cases} \Rightarrow \begin{cases} 3x = \lambda x \\ 2y - 2 = \lambda y \\ x^2 + y^2 = 16 \end{cases}$

$x = 0 \Rightarrow y = 4,\ \lambda = \frac{3}{2}$ or $y = -4,\ \lambda = \frac{5}{2}$;

$x \ne 0 \Rightarrow \lambda = 3,\ y = -2,\ x = \pm 2\sqrt{3}.$

Bordered Hessian: $H = \begin{pmatrix} 0 & -2x & -2y \\ -2x & 6 - 2\lambda & 0 \\ -2y & 0 & 4 - 2\lambda \end{pmatrix}$

$\det(H)(x, y, \lambda) = -4x^2(4 - 2\lambda) - 4y^2(6 - 2\lambda);$

$\det(H)(0, 4, \frac{3}{2}) < 0$ — local minimum;

$\det(H)(0, -4, \frac{5}{2}) < 0$ — local minimum;

$\det(H)(\pm 2\sqrt{3}, -2, 3) > 0$ — local maximum.

4. F is homogeneous of degree $1 \Leftrightarrow F(tK, tL) = tF(K, L)$ for $\forall t$.

Euler equation: $F(K, L) = K \cdot F_K + L \cdot F_L$.

4(a) We will first differentiate both sides of Euler's equation
$F(K, L) = K \cdot F_K + L \cdot F_L$:

$\frac{\partial}{\partial L}(*) : F_L = KF_{KL} + F_L + LF_{LL} \Rightarrow F_{LL} = -\frac{K}{L}F_{KL};$

$\frac{\partial}{\partial K}(*) : F_K = F_K + KF_{KK} + LF_{LK} \Rightarrow F_{KK} = -\frac{L}{K}F_{KL}.$

$$
\begin{aligned}
\det(H) &= \det \begin{pmatrix} F_{KK} & F_{KL} \\ F_{LK} & F_{LL} \end{pmatrix} \\
&= F_{KK} \cdot F_{LL} - (F_{KL})^2 \\
&= \left(-\frac{L}{K}F_{KL}\right) \cdot \left(-\frac{K}{L}F_{KL}\right) - (F_{KL})^2 \\
&= (F_{KL})^2 - (F_{KL})^2 = 0.
\end{aligned}
$$

Therefore, the rows of H are linearly dependent.

4(b) $u(K, L) = F_K$, $v(K, L) = F_L$ are functionally dependent $\Leftrightarrow$ the determinant of the Jacobian of u, v is zero, i.e.

$$\det J(u, v) = \det \begin{pmatrix} u_K & u_L \\ v_K & v_L \end{pmatrix} = \det \begin{pmatrix} F_{KK} & F_{KL} \\ F_{LK} & F_{LL} \end{pmatrix} = 0.$$

From Euler's equation, we can take $\phi = KF_K + LF_L - F \equiv 0$.

4(c) With $Y = F(K, L)$, $\dot{Y} = \dot{F}(K, L) = F_K \cdot \dot{K} + F_L \cdot \dot{L}$. Thus

$$
\begin{aligned}
\gamma_Y = \frac{\dot{Y}}{Y} &= \frac{F_K K}{Y} \cdot \frac{\dot{K}}{K} + \frac{F_L L}{Y} \cdot \frac{\dot{L}}{L} \\
&= \alpha \cdot \gamma_K + (1 - \alpha) \cdot \gamma_L,
\end{aligned}
$$

where $\alpha = \dfrac{F_K K}{Y}$ and $(1 - \alpha) = \dfrac{F_L L}{Y}$.

Note, that Euler's equation implies $1 = \dfrac{F_K K}{Y} + \dfrac{F_L L}{Y}$.

5. The setup of the problem:

$$\max_{Q,A} 10Q - Q^2 + \frac{A}{2}$$

$$\text{s.t.} \begin{cases} \pi \geq 3 \\ Q \geq 0 \\ A \geq 0 \end{cases}$$

$\pi = TR - TC \geq 3 \Leftrightarrow \frac{3}{2}Q^2 - 5Q + \frac{1}{2}A + 4 \leq 0.$

The Kuhn-Tucker conditions are satisfied (concave and differentiable objective function; convex and differentiable constraint). Therefore, the F.O.C. produce a unique solution.

For the following maximization problem:

$$\max_{Q,A} 10Q - Q^2 + \frac{A}{2} \qquad \text{s.t.} \qquad \frac{3Q^2}{2} - 5Q + \frac{A}{2} \leq -4$$
$$A \geq 0$$
$$Q \geq 0$$

the Lagrangian function is

$$L(Q, A, \lambda) = 10Q - Q^2 + \frac{A}{2} + \lambda\left(-4 - \frac{3Q^2}{2} + 5Q - \frac{A}{2}\right)$$

The Kuhn-Tucker necessary maximum conditions are:

$$\frac{\partial L}{\partial Q} = 10 - 2Q - 3Q\lambda + 5\lambda \leq 0 \tag{5.38}$$

$$\frac{\partial L}{\partial A} = \frac{1}{2} - \frac{\lambda}{2} \leq 0 \tag{5.39}$$

$$\frac{\partial L}{\partial \lambda} = -4 - \frac{3Q^2}{2} + 5Q - \frac{A}{2} \geq 0 \tag{5.40}$$

$$Q\frac{\partial L}{\partial Q} = Q(10 - 2Q - 3Q\lambda + 5\lambda) = 0 \tag{5.41}$$

$$A\frac{\partial L}{\partial A} = A\left(\frac{1}{2} - \frac{\lambda}{2}\right) = 0 \tag{5.42}$$

$$\lambda\frac{\partial L}{\partial \lambda} = \lambda\left(-4 - \frac{3Q^2}{2} + 5Q - \frac{A}{2}\right) = 0 \tag{5.43}$$

$$Q \geq 0, \ A \geq 0, \ \lambda \geq 0 \tag{5.44}$$

From condition (5.39) we see that $\lambda \neq 0$ and therefore condition (5.42) implies two possible cases:

- $\lambda = 1$: (5.40) implies $Q \neq 0$ which leaves us with $Q = 3$ (from (5.41)). However, in this case $A = -5$ which contradicts the initial condition $A \geq 0$. Therefore, $\lambda > 1$, and from (5.42) it follows that $A = 0$.

- $A = 0$: now from condition (5.43) we get two possible solutions: $Q = \frac{4}{3}$ and $Q = 2$. The first one, however, does not satisfy condition (5.41) because λ would be negative $(-\frac{22}{3})$. Fortunately, the second solution satisfies all conditions.

Thus, managers should choose $Q = 2$ and $A = 0$.

6. The problem is defined as

$$\min \left[(x-1)^2 + (y-1)^2 \right]$$

$$\text{s.t.} \quad \begin{cases} y \geq x^2 \\ y \leq 2 \\ x \geq 0 \\ y \geq 0 \end{cases}$$

As the objective function is concave and differentiable, and the constraints are convex and differentiable,, the solution is unique and the F.O.C. (Kuhn-Tucker) are necessary and sufficient.

Intuitively, minimum is attained at $(x^* = 1, y^* = 1)$. Let us derive the solution analytically.

The Lagrangian function is

$$L(x, y, u, v) = (x-1)^2 + (y-1)^2 + u(-2+y) + v(-y+x^2)$$

(here u and v are Lagrange multipliers).

The Kuhn-Tucker necessary minimum conditions are:

$$\frac{\partial L}{\partial x} = 2x - 2 + 2xv \geq 0 \tag{5.45}$$

$$\frac{\partial L}{\partial y} = 2y - 2 + u - v \geq 0 \tag{5.46}$$

$$\frac{\partial L}{\partial u} = -2 + y \le 0 \tag{5.47}$$

$$\frac{\partial L}{\partial v} = -y + x^2 \le 0 \tag{5.48}$$

$$x\frac{\partial L}{\partial x} = x(2x - 2 + 2xv) = 0 \tag{5.49}$$

$$y\frac{\partial L}{\partial y} = y(2y - 2 + u - v) = 0 \tag{5.50}$$

$$u\frac{\partial L}{\partial u} = u(-2 + y) = 0 \tag{5.51}$$

$$v\frac{\partial L}{\partial v} = v(-y + x^2) = 0 \tag{5.52}$$

$$x \ge 0, \ u \ge 0, \ v \ge 0 \tag{5.53}$$

In order to solve this system, we divide it into four states which are fully independent and cover all possibilities:

- $u = v = 0$
  From (5.49) and (5.50) we get the system:

$$x(2x - 2) = 0$$
$$y(2y - 2) = 0$$

There are four possible solutions: [0,0], [0,1], [1,0] and [1,1]. However, the first three contradict conditions (5.45) and (5.46). The solution [1,1] satisfies all conditions.

- $u > 0$ and $v = 0$
  From (5.49), (5.50) and (5.51) we get the system:

$$x(2x - 2) = 0$$
$$y(2y - 2 + u) = 0$$
$$y = 2$$

However, this system implies $u = -2$ which contradicts the initial assumption.

- $v > 0$ and $u = 0$

  From (5.49), (5.50) and (5.52) we get the system:

  $$
  \begin{aligned}
  x(2x - 2 + 2xv) &= 0 \\
  y(2y - 2 - v) &= 0 \\
  y &= x^2
  \end{aligned}
  $$

  Checking conditions (5.45) and (5.46), we find that with $v > 0$ and $u = 0$, $x \neq 0$ and $y \neq 0$. Thus, provided non-zero x and y, the above system leads to the following solution: $x = y = 1$ and $v = 0$. However, v is assumed to be positive, thus we have a contradiction.

- $v > 0$ and $u > 0$

  From (5.51) and (5.52) we compute $y = 2$ and $x = \sqrt{2}$. Using these two values we can also compute u and v from (5.49) and (5.50). However, the computed values of $u = \frac{1}{2}(-6 + \sqrt{2}) < 0$ and $v = \frac{1}{2}(-2 + \sqrt{2}) < 0$ contradict our initial assumption.

Thus we are left with only one solution $x = y = 1$.

7. Let us construct the Lagrangian function:

   $$L = e^x + e^y + 2x + 2y + \lambda(10 - x - cy)$$

   When writing the conditions for the constrained extreme of our function, we get the following system of equations (λ is the Lagrangian multiplier):

   $$
   \begin{aligned}
   e^x + 2 - \lambda = 0 &= F^1(x, y, \lambda, c) \\
   e^y + 2 - \lambda c = 0 &= F^2(x, y, \lambda, c) \\
   10 - x - cy = 0 &= F^3(x, y, \lambda, c)
   \end{aligned}
   $$

   The Jacobian of this system is

   $$
   J = \begin{pmatrix}
   \frac{\partial F^1}{\partial x} & \frac{\partial F^1}{\partial y} & \frac{\partial F^1}{\partial \lambda} \\
   \frac{\partial F^2}{\partial x} & \frac{\partial F^2}{\partial y} & \frac{\partial F^2}{\partial \lambda} \\
   \frac{\partial F^3}{\partial x} & \frac{\partial F^3}{\partial y} & \frac{\partial F^3}{\partial \lambda}
   \end{pmatrix}
   = \begin{pmatrix}
   e^x & 0 & -1 \\
   0 & e^y & -c \\
   -1 & -c & 0
   \end{pmatrix}
   $$

$|J| = -(c^2 e^x + e^y) < 0$, so that we can use the implicit function theorem for calculating the first-order partial derivatives of x and y withe respect to c.

$$\frac{\partial x}{\partial c} = -\frac{\begin{vmatrix} \frac{\partial F^1}{\partial c} & \frac{\partial F^1}{\partial y} & \frac{\partial F^1}{\partial \lambda} \\ \frac{\partial F^2}{\partial c} & \frac{\partial F^2}{\partial y} & \frac{\partial F^2}{\partial \lambda} \\ \frac{\partial F^3}{\partial c} & \frac{\partial F^3}{\partial y} & \frac{\partial F^3}{\partial \lambda} \end{vmatrix}}{|J|} = -\frac{\begin{vmatrix} 0 & 0 & -1 \\ -\lambda & e^y & -c \\ -y & -c & 0 \end{vmatrix}}{|J|}$$

$$= -\frac{\lambda c + y e^y}{c^2 e^x + e^y} = -\frac{e^y(1+y) + 2}{c^2 e^x + e^y}$$

Similarly, we can calculate

$$\frac{\partial y}{\partial c} = \frac{\lambda - c y e^x}{c^2 e^x + e^y} = \frac{e^x(1 - cy) + 2}{c^2 e^x + e^y}$$

If we want, we can check that the function $e^y(1+y) + 2$ is always positive, and because $c^2 e^x + e^y > 0$, we get that $\frac{\partial x}{\partial c} < 0$. The effect of c on y is ambiguous, as the numerator $e^x(1 - cy) + 2$ can be either negative or positive.

8. If $x = 3$, the matrix is singular. Let $x \neq 3$. Consider $A_{(4 \times 4)} = \begin{pmatrix} P_{(2 \times 2)} & 0 \\ 0 & Q_{(2 \times 2)} \end{pmatrix}$, where $P_{(2 \times 2)} = \begin{pmatrix} 2 & 6 \\ 1 & x \end{pmatrix}$ and $Q_{(2 \times 2)} = \begin{pmatrix} 1 & 1 \\ 1 & 2 \end{pmatrix}$. Then $A^{-1} = \begin{pmatrix} P^{-1} & 0 \\ 0 & Q^{-1} \end{pmatrix}$. We also know that $\begin{pmatrix} a & b \\ c & d \end{pmatrix}^{-1} = \frac{1}{ad-bc} \begin{pmatrix} d & -b \\ -c & a \end{pmatrix}$. Therefore,

$$A^{-1} = \begin{pmatrix} \frac{x}{2(x-3)} & -\frac{3}{x-3} & 0 & 0 \\ -\frac{1}{2(x-3)} & \frac{1}{x-3} & 0 & 0 \\ 0 & 0 & 2 & -1 \\ 0 & 0 & -1 & 1 \end{pmatrix}.$$

9. $x = -1 + 2t - t^2 = -(t-1)^2$, $y = 2 - 3t + t^3 = (t-1)^2(t+2)$

$\frac{dy}{dx} = \frac{\dot{y}}{\dot{x}} = -\frac{3}{2}(t+1) \rightarrow$ increasing for $t < -1$, decreasing for $t > -1$;

$$\frac{d^2y}{dx^2} = \frac{d}{dt}\left(-\frac{3}{2}(t+1)\right)\cdot\frac{1}{\dot{x}} = \frac{3}{4(t+1)} \rightarrow \text{convex for } t > 1, \text{ con-}$$
cave for $t < 1$;

At $t = 1$, $x(t) = y(t) = 0$;

At $t = -1$, $x(t) = -4$, $y(t) = 4$;

$\dot{x} = -2(t-1) \rightarrow$ increasing for $t < 1$, decreasing for $t > 1$;

$\dot{y} = 3(t^2 - 1) \rightarrow$ increasing for $|t| > 1$, decreasing for $|t| < 1$;

$x(t) = 0$ at $t = 1$, $y(t) = 0$ at $t = 1$ and $t = -2$.

10. The original equation can be re-written as $y' + \frac{x-1}{x}y = -x$.

   Homogeneous equation: $y' = -\frac{x-1}{x}y \Rightarrow \int \frac{1}{y}\,dy = -\int \frac{x-1}{x}\,dx$

   Thus $y_H = Ce^{-\int \frac{x-1}{x}\,dx} = Ce^{-x+\ln(x)} = Cxe^{-x}$.

   Considering C as a function of x, let us find a particular solution to the non-homogeneous equation in the form: $y_P = C(x)xe^{-x}$.

   Plugging this expression into the original differential equation, we find $C'(x)(xe^{-x}) + \underbrace{C(x)(xe^{-x})' + \frac{x-1}{x}C(x)xe^{-x}}_{=\,0} = -x$

   $\Rightarrow C'(x) = -e^x \Rightarrow y_P = -x$

   Therefore, $y(x) = Cxe^{-x} - x$.

11. A solution to the equation $y'' - 2y' - 3y = e^{4x} + 1$ is a linear combination of a general solution to the homogeneous equation and a particular solution to the non-homogeneous equation.

   Solving the characteristic equation, we find $y_H = C_1e^{-x} + C_2e^{3x}$. Next, we search for a particular solution in the form $y_P = Ae^{4x} + B$

   $\begin{cases} y_P' = 4Ae^{4x} \\ y_P'' = 16Ae^{4x} \end{cases} \Rightarrow 16Ae^{4x} - 8Ae^{4x} - 3Ae^{4x} - 3B = e^{4x} + 1.$

   Using the method of undetermined coefficients, we find $A = \frac{1}{5}$, $B = -\frac{1}{3}$.

   Therefore, $y_{NH} = C_1e^{-x} + C_2e^{3x} + \frac{1}{5}e^{4x} - \frac{1}{3}$.

12. $\begin{cases} \dot{c} = -\beta(k - k^*) \\ \dot{k} = \theta(k - k^*) - (c - c^*) \end{cases}$

   $\Rightarrow \ddot{k} = \theta\dot{k} - \dot{c}$

$$\Rightarrow \ddot{k} = \theta\dot{k} + \beta(k - k^*)$$
$$\Rightarrow \ddot{k} - \theta\dot{k} - \beta k = -\beta k^*$$

That is, after taking the derivative of the second equation with respect to time and substituting the right-hand side of the first equation for $\dot{c}$, we obtain $\ddot{k} = \theta\dot{k} - \dot{c} = \theta\dot{k} + \beta(k - k^*)$, which can be rewritten as

$$\ddot{k} - \theta\dot{k} - \beta k = -\beta k^*.$$

We also impose the so-called boundary conditions (note that these conditions are somehow different from Cauchy's initial value problem):

$$k(0) = k_0 \quad \text{and} \quad \lim_{t \to \infty} k(t) = k^*.$$

To solve the homogeneous equation, we solve the characteristic equation $\lambda^2 - \theta\lambda - \beta = 0$, which has (recall that $\beta, \theta > 0$) two distinct real solutions $\lambda_{1,2} = \frac{1}{2}(\theta \pm \sqrt{\theta^2 + 4\beta})$. We can find a particular solution (as a constant) $k_p(t) = k^*$. Then the general solution to the above equation is

$$k(t) = k^* + C_1 e^{\frac{1}{2}(\theta + \sqrt{\theta^2 + 4\beta})t} + C_2 e^{\frac{1}{2}(\theta - \sqrt{\theta^2 + 4\beta})t}.$$

Setting $t = 0$ we obtain $C_1 + C_2 = k_0 - k^*$. Taking $t \to \infty$, we get $e^{\frac{1}{2}(\theta - \sqrt{\theta^2 + 4\beta})t} \to 0$, because $\frac{1}{2}(\theta - \sqrt{\theta^2 + 4\beta}) < 0$, and $e^{\frac{1}{2}(\theta + \sqrt{\theta^2 + 4\beta})t} \to \infty$, because $\frac{1}{2}(\theta + \sqrt{\theta^2 + 4\beta}) > 0$. To achieve convergence of the solution to k^*, we must have $C_1 = 0$ (otherwise $k(t)$ converges to $+\infty$ or $-\infty$ as $t \to \infty$). Then $C_2 = k_0 - k^*$, which gives the solution

$$k(t) = k^* + (k_0 - k^*)e^{\frac{1}{2}(\theta - \sqrt{\theta^2 + 4\beta})t}.$$

Sample Problem Set–III

1. Prove or disprove the statements below:

   (a) If A is an $[n \times n]$ matrix and $\text{rank}(A) < n$ then for any matrix B $\text{rank}(AB) < n$.

(b) For any real a and b the definite integral $\int_0^1 x^a(1-x)^b dx$ is convergent.

(c) As x approaches 0 from the right, $\lim_{x \to 0_+} x^x = e^{-1}$.

(d) A concave function is always quasi-concave.

(e) Given a point (x_0, y_0) in the $(X - Y)$ plane, you can always find a solution $y = y(x)$ to the differential equation $\frac{dy}{dx} = 2xy + y^2$ such that $y(x_0) = y_0$.

2. Solve the difference equation $3y_{k+2} + 2y_{k+1} - y_k = 3^k - k - 1$.

3. Solve the differential equation $2xy' + y^2 = 1$.

4. Find the critical value(s) of the function $f(x, y) = x + y + 1/x + 4/y$ and classify them.

5. Let $f \in C^2(R^n)$ be homogeneous of degree 1. Show that the matrix of mixed partials of f is not invertible (HINT: Use Euler's theorem).

6. Consider the two-commodity utility maximization problem

$$\max_{x,y} U(x, y) = xy^2$$

$$\text{s.t.} \quad x + 2y = M$$

where income level M is assumed to be strictly positive.

(a) Write down the F.O.C. for this problem.

(b) Use the Implicit Function Theorem to derive dx/dM, dy/dM, $d\lambda/dM$ at the point of optimality (here λ is the Lagrange multiplier (shadow price)).

7. For the following system of linear differential equations

$$\frac{dx}{dt} = 4x - y - z,$$

$$\frac{dy}{dt} = x + 2y - z,$$

$$\frac{dz}{dt} = x - y + 2z$$

(a) find all eigenvalues and eigenvectors of the matrix of the system;

(b) find the general form of solution;

(c) find the solution that at $t = 0$ passes through the point $(3, 2, 1)$.

8. For the non-linear optimization problem

$$\max_{x,y}(x + \ln y)$$

s.t. $x + y \leq 4, \ x + 2y \leq 6, \ x \geq 0, \ y \geq 0$

(a) check whether this problem has a unique optimal solution;

(b) write down the Kuhn-Tucker conditions and find the optimal solution numerically.

9. Consider the following neoclassical model of optimal growth:

$$\max \int_0^\infty \left(\bar{U} - \frac{1}{b}(c(t))^{-b} \right) e^{-rt} dt$$

s.t. $\dfrac{dk(t)}{dt} = A(k(t))^\alpha - c(t) - (n + \delta)k(t),$

$$k(0) = k_0, \ 0 \leq c(t) \leq A(k(t))^\alpha.$$

Here $u(c(t)) = \bar{U} - \frac{1}{b}(c(t))^{-b}$ is the social utility index function; $\bar{U}$, b, the discount rate r, the rate of population growth n, the depreciation rate δ, and the level of technology A are positive parameters; $\alpha \in (0, 1)$; $c(t)$ is the level of per capita consumption (control variable), and $k(t)$ is the capital-labor ratio (state variable).

(a) Set up the Hamiltonian and derive F.O.C.

(b) Construct the phase diagram and check the type of equilibrium.

(c) Plot the optimal solution.

Solutions to Sample Problem Set–III

1. (a) TRUE. Consider matrix $A = \begin{pmatrix} a_1^T \\ \cdots \\ \cdots \\ a_n^T \end{pmatrix}$, $a_i^T = (a_{i1}, ..., a_{in})$.

Thus $AB = \begin{pmatrix} a_1^T B \\ \cdots \\ \cdots \\ a_n^T B \end{pmatrix}$. Since rank$(A) < n$, there exist

such constants $\beta_1, ..., \beta_n$ (not all equal zero) that $\beta_1 a_1^T + ... + \beta_n a_n^T = 0$ (Recall the definition of linear dependent vectors and its link to the rank of a matri!) Therefore, $\beta_1 a_1^T B + ... + \beta_n a_n^T B = 0$, meaning that rank$(AB) < n$.

   (b) FALSE. $B(\alpha, \beta) = \int\limits_0^1 x^{\alpha-1}(1-x)^{\beta-1}dx$ is the *beta function* that is defined for $\alpha > 0, \beta > 0$. Thus the definite integral

$$\int\limits_0^1 x^a(1-x)^b dx = B(a+1, b+1) < \infty$$

converges for $a > -1, b > -1$.

   (c) FALSE. It is known that $x = e^{\ln(x)}$, therefore, $x^x = e^{x\ln(x)}$. Furthermore, $\lim_{x\to 0} x^x = e^{\lim_{x\to 0}\left(\frac{\ln(x)}{1/x}\right)} = /\text{L'Hôpital's rule}/ = e^{\lim_{x\to 0}\left(\frac{1/x}{-1/x^2}\right)} = e^{\lim_{x\to 0}(-x)} = e^0 = 1$

   (d) TRUE. By definition a function, f, is *concave* iff for any pair of distinct points, u and v in the domain of the function and for any $0 < \alpha < 1$

$$f(\alpha u + (1-\alpha)v) \geq \alpha f(u) + (1-\alpha)f(v)$$

A function, f, is called *quasi-concave* iff for any pair of distinct points, u and v in the domain of the function and for any $0 < \alpha < 1$

$$f(\alpha u + (1-\alpha)v) \geq \min\{f(u), f(v)\}.$$

To this end, a concave function is quasiconcave.

(e) TRUE. Recall the existence theorem: if a function $f(x, y)$ is continuous in open domain D, then for any given point $(x_0, y_0) \in D$ there exists a solution to Cauchy's problem $y'(x) = f(x, y)$ such that $y(x_0) = y_0$. In our case, the domain D is the $(X - Y)$ plain. D is an open set, and the function $2xy + y^2$ is continuous in D. Thus the solution always exists.

2. Roots of the characteristic equation $3m^2 + 2m - 1 = 0$ are $m_1 = 1/3$ and $m_2 = -1$. The solution to the corresponding homogeneous equation is:

$$y_k = C_1(1/3)^k + C_2(-1)^k.$$

In order to construct a *particular* solution, we consider the *trial equation*:

$$y_k = A \cdot 3^k + B \cdot k + C,$$

where coefficients A, B, C are unknown and have to be found by substituting the trial equation into the original equation and equating corresponding coefficients on the left-hand side and the right-hand side. After substitution we find

$$32A \cdot 3^k + 4B \cdot k + 8B + 4C = 3^k - k - 1.$$

Thus,

$$
\begin{aligned}
32A &= 1, \\
4B &= -1, \\
8B + 4C &= -1,
\end{aligned}
$$

so that $A = 1/32$, $B = -1/4$, and $C = 1/4$. Finally, the general solution to the non-homogeneous equation becomes

$$y_k = C_1(1/3)^k + C_2(-1)^k + \frac{3^k}{32} - \frac{k}{4} + \frac{1}{4}.$$

3. This variables separable differential equation can be re-written as $\dfrac{dy}{1-y^2} = \dfrac{dx}{2x}$. Integrating both parts we find a solution:

$$1/2C \ln(x) = \int 1/2 \left(\frac{1}{1-y} + \frac{1}{1+y} \right) dy = 1/2 \ln \left(\frac{1+y}{1-y} \right),$$

or, after simple algebra, $y(x) = \dfrac{Cx-1}{1+Cx}$.

4. Critical values of a function f should solve the first-order conditions:

$$1 - 1/x^2 = 0$$
$$1 - 4/y^2 = 0$$

From the F.O.C., we find four "suspected" points: $(1,2)$, $(-1,2)$, $(1,-2)$, $(-1,-2)$. To classify them, we need to construct a Hessian matrix and check its sign definiteness using, say, the leading principal minors test. The Hessian in its general form is

$$H(x,y) = \begin{pmatrix} 2/x^3 & 0 \\ 0 & 8/y^3 \end{pmatrix}.$$

Since $\det(H(1,2)) = 2 > 0$ and $2 > 0$, the point $(1,2)$ is a (local) minimum.

$\det(H(-1,2)) = -2 < 0$ and $-2 < 0$, therefore the point $(-1,2)$ is a saddle point.

$\det(H(1,-2)) = -2 < 0$ and $2 > 0$, thus the point $(1,2-)$ is again a saddle point.

Finally, $\det(H(-1,-2)) = 2 > 0$ and $-2 < 0$. Therefore, the point $(1,2)$ delivers a (local) maximum.

5. According to Euler's theorem, a homogeneous of degree 1 function $f(x) = \sum_{i=1}^{n} x_i \dfrac{\partial f}{\partial x_i}(x)$. Thus, $\dfrac{\partial f}{\partial x_i} = \dfrac{\partial f}{\partial x_i} + \sum_{j=1}^{n} x_j \dfrac{\partial^2 f}{\partial x_j \partial x_i}(x)$ $\forall i = 1, ..., n$. In matrix notation, the latter reads as $Hx = 0$, where $H = \left(\dfrac{\partial^2 f}{\partial x_j \partial x_i} \right)^n_{i,j=1}$ stands for the matrix of mixed partials of the function. Since vector x is arbitrary, the matrix H is singular.

6. The Lagrangian function is $L = xy^2 + \lambda(M - x - 2y)$, and the F.O.C.

$$
\begin{aligned}
y^2 &= \lambda \\
2xy &= 2\lambda \\
x + 2y &= M
\end{aligned}
$$

yield $x = y = M/3$ and $\lambda = y^2 = M^2/9$.

To derive dx/dM, dy/dM, $d\lambda/dM$ at the point of stationarity, let us use the Implicit Function Theorem. The Jacobian of the system is

$$
J(x, y, \lambda) = \begin{pmatrix} 0 & 2y & -1 \\ y & x & -1 \\ 1 & 2 & 0 \end{pmatrix}.
$$

$\det(J(x, y, \lambda)) = x - 4y$, and the inverse Jacobian

$$
J^{-1}(x, y, \lambda) = \frac{1}{x - 4y} \begin{pmatrix} 2 & -2 & x - 2y \\ -1 & 1 & -y \\ 2y - x & 2y & -2y^2 \end{pmatrix}.
$$

Solving

$$
\begin{pmatrix} dx/dM \\ dy/dM \\ d\lambda/dM \end{pmatrix} = -J^{-1}(M/3, M/3, M^2/9) \cdot \begin{pmatrix} 0 \\ 0 \\ -1 \end{pmatrix}
$$

we find that $dx/dM = 1/3$, $dy/dM = 1/3$, and $d\lambda/dM = 2M/9$.

7. To find eigenvalues, solve $\det(A - \lambda I) = 0$, that is

$$
\begin{vmatrix} 4 - \lambda & -1 & -1 \\ 1 & 2 - \lambda & -1 \\ 1 & -1 & 2 - \lambda \end{vmatrix} =
$$

$$
= (4 - \lambda)((2 - \lambda)^2 - 1) + 2 + 2(2 - \lambda) = 0.
$$

The roots of this cubic equation are $\lambda_1 = \lambda_2 = 3$ and $\lambda_3 = 2$. To find eigenvalues, we need to solve $(A - \lambda I)v = 0$.

For $\lambda_1 = \lambda_2 = 3$, $A - 3I = \begin{pmatrix} 1 & -1 & -1 \\ 1 & -1 & -1 \\ 1 & -1 & -1 \end{pmatrix}$, and we are able

to find two linear independent eigenvalues: $v_1 = (1, 1, 0)'$ and $v_2 = (1, 0, 1)'$.

For $\lambda_3 = 2$, $A - 2I = \begin{pmatrix} 2 & -1 & -1 \\ 1 & 0 & -1 \\ 1 & -1 & 0 \end{pmatrix}$, thus the corresponding

eigenvector is $v_3 = (1, 1, 1)'$.

Finally, the general solution to the system becomes

$$\begin{pmatrix} x(t) \\ y(t) \\ z(t) \end{pmatrix} = C_1 v_1 e^{3t} + C_2 v_2 e^{3t} + C_3 v_3 e^{2t}$$

$$= \begin{pmatrix} e^{3t}(C_1 + C_2) + e^{2t} C_3 \\ e^{3t} C_1 + e^{2t} C_3 \\ e^{3t} C_2 + e^{2t} C_3 \end{pmatrix}.$$

Cauchy's problem in part (c) can be solved from

$$\begin{pmatrix} x(0) \\ y(0) \\ z(0) \end{pmatrix} = \begin{pmatrix} C_1 + C_2 + C_3 \\ C_1 + C_3 \\ C_2 + C_3 \end{pmatrix},$$

thus $C_1 = 2$, $C_2 = 1$, and $C_3 = 0$.

8. Our problem has convex a constraint set: $C = \{(x, y) | \ x + y \leq 4, x + 2y \leq 6, x \geq 0, y \geq 0\}$ and the objective function $f(x, y) = x + \ln(y)$ is concave, since

$$\alpha x_1 + \ln(\alpha y_1) + (1 - \alpha)x_2 + \ln((1 - \alpha)y_1) \geq$$

$$\geq \alpha(x_1 + \ln(y_1)) + (1 - \alpha)(x_2 + \ln(y_2)).$$

Therefore, the conditions of the Kuhn-Tucker sufficiency theorem are satisfied and the problem has a unique global point of optimality (maximum).

To find this point, let us form the Lagrangian function

$$L = x + \ln(y) + \lambda(4 - x - y) + \mu(6 - x - 2y)$$

and derive the first-order conditions:

$$1 - \lambda - \mu \leq 0 \tag{5.54}$$
$$x(1 - \lambda - \mu) = 0 \tag{5.55}$$
$$1/y - \lambda - 2\mu \leq 0 \tag{5.56}$$
$$y(1/y - \lambda - 2\mu) = 0 \tag{5.57}$$
$$4 - x - y \geq 0 \tag{5.58}$$
$$\lambda(4 - x - y) = 0 \tag{5.59}$$
$$6 - x - 2y \geq 0 \tag{5.60}$$
$$\mu(6 - x - 2y) = 0 \tag{5.61}$$
$$\mu \geq 0 \tag{5.62}$$
$$\lambda \geq 0 \tag{5.63}$$

If $y = 0$ then (5.56) is violated. Thus $y \neq 0$ and from (5.57) $y = 1/(\lambda + 2\mu)$.

If both λ and μ are equal to zero, then (5.54) is violated.

If both $\lambda, \mu \neq 0$ then $-x - y = 0$ and $6 - x - 2y = 0$, so that $x_1^* = y_1^* = 2$.

If $\lambda = 0, \mu \neq 0$ them from (5.54) $\mu = 1$ and $y = 1/2$. Next, from (5.61) we find $x = 5$; the latter, however, contradicts (5.58).

If $\lambda \neq 0, \mu = 0$ then $\lambda = 1$, $y_2^* = 1$ and (from (5.58)) $x_2^* = 3$.

Since $x_1^* + \ln y_1^* < x_2^* + \ln y_2^*$, then $(x_2^* = 3, y_2^* = 1)$ is the global maximum.

9. First, write down the present-value Hamiltonian:

$$H = u(c(t))e^{-\rho t} + \lambda(A(k(t))^\alpha - c(t) - (n + \delta)k(t)).$$

The next step is to derive the F.O.C:

$$\partial H/\partial c = 0,$$

$$\partial H/\partial k = -\dot{\lambda}.$$

Here c is *control variable*, k is *state variable* and λ is *co-state variable*.

More specifically, the F.O.C. read as follows:

$$c^{-b-1}e^{-\rho t} = \lambda(t) \qquad (5.64)$$
$$\lambda(t)(\alpha Ak^{\alpha-1} - n - \delta) = -\dot{\lambda}(t) \qquad (5.65)$$

If we differentiate (5.64) with respect to time, and combine with (5.64), (5.65), we arrive at so-called *Euler's Equation*:

$$\frac{\dot{c}(t)}{c(t)} = \frac{1}{\theta}\left(A\alpha k^{\alpha-1}(t) - n - \delta - \rho\right)$$

Euler's Equation and the equation of motion for the state variable $k(t)$, together with the transversality condition $\lim_{t\to\infty}(\lambda(t)k(t)) = 0$, completely describe the behavior of the model and can be used to derive the optimal solution to the problem. There is a unique saddle-type equilibrium in this system. Therefore, there exists a unique optimal solution originating from $(k_0, 0)$ in the $(k - c)$ plane that converges to the equilibrium.[13]

Sample Problem Set–IV

1. Problem group I.

   (a) For what value of the parameter a is the function $y_1 = ax^2$ tangent to the curve $y_2 = \ln x$?

   (b) If x is very small compared to a, show that $\sqrt{a^2 + x} \approx a + x/2a$ $(a > 0)$.

---

[13] For further details, for instance, see R. Barro and X. Sala-i-Martin's textbook on economic growth theory.

(c) Let function $f(x)$ be such that $f(x) = \begin{cases} x^2, & \text{if } x \le x_0 \\ ax + b, & \text{if } x > x_0 \end{cases}$.

Could you choose a and b such that $f(x)$ becomes continuous and differentiable?

(d) What is the rank of the matrix $\begin{pmatrix} 2 & 5 & -1 \\ 1 & a & 2 \\ 5 & 2 & 5 \end{pmatrix}$ where a is a parameter?

(e) If the inverse supply function is given as $P(x) = (x \ln(x))^2$, evaluate the surplus $\int_1^e P(x)dx$.

(f) Find maximum and minimum values of the function $f(x) = |x|e^{-|x-1|}$.

(g) Plot the graph of the function $f(x) = \frac{(x+1)^4}{(x-1)^4}$.

(h) Let continuously differentiable functions f and Φ be such that $f(x, y, z) = \Phi(\alpha, \beta, \gamma)$, where $x^2 = \beta\gamma$, $y^2 = \alpha\gamma$, $z^2 = \alpha\beta$. Show that $xf_x + yf_y + zf_z = \alpha\Phi_\alpha + \beta\Phi_\beta + \gamma\Phi_\gamma$ (Here $f_x = \partial f/\partial x$, $\Phi_\alpha = \partial\Phi/\partial\alpha$, etc.)

(i) Find the limit (here $x \to 3_+$ means 'as x approaches 3 from the right'): $\lim\limits_{x \to 3_+} (x^2 - 5x + 6)^{1/\ln(x-3)}$.

2. Problem group II.

(a) Show that for any sufficiently differentiable functions u and v the function $\phi(x, y) = u[x + v(y)]$ solves the following equation in partial derivatives: $\dfrac{\partial\phi}{\partial x}\dfrac{\partial^2\phi}{\partial x \partial y} = \dfrac{\partial\phi}{\partial y}\dfrac{\partial^2\phi}{\partial x^2}$.

(b) If the function $f(x)$ is differentiable and $\lim\limits_{x\to\infty} f'(x) = 0$, show that $\lim\limits_{x\to\infty} \dfrac{f(x)}{x} = 0$ (i.e., if for large x the marginal value goes to zero then then average value vanishes, too).

(c) Find matrix X from the following system of linear equations:

$$\begin{pmatrix} 2 & 1 \\ 3 & 2 \end{pmatrix} X \begin{pmatrix} -3 & 2 \\ 5 & -3 \end{pmatrix} = \begin{pmatrix} -2 & 4 \\ 3 & -1 \end{pmatrix}$$

(d) What is the minimum value of the function $u(x, y, z) = x + y^2/4x + z^2/y + 2/z$ in the positive orthant $x > 0, y > 0, z > 0$?

(e) Solve the differential equation $\frac{dy}{dx} = \frac{\sqrt{y+1}}{1+e^x}$.

3. Problem group III.

(a) For what values of α, β and γ is the Cobb-Douglas production function $F(K, L, H) = K^\alpha L^\beta H^\gamma$ concave?

(b) Given the matrix $A = \begin{pmatrix} 0 & 2 & 2 \\ 2 & 3 & -1 \\ 2 & -1 & 3 \end{pmatrix}$, evaluate its exponent, e^A.

(c) Find extrema of the function $z = z(x, y)$ implicitly defined as $x^2 + y^2 + z^2 - xz - yz + 2x + 2y + 2z - 2 = 0$.

(d) Classify the equilibria of the system of differential equations
$$\begin{cases} \dot{x} = xy + 4, \\ \dot{y} = x^2 + y^2 - 17, \end{cases}$$ and make a basic sketch of trajectories of this system in the $(x - y)$-plane.

(e) A policymaker desires to double in 10 periods of time the value of GDP (Y_t) produced in period t.

   i. An expert suggests that the evolution of GDP over time is given by the second-order difference equation $4Y_{t+2} - 4Y_{t+1} + Y_t = 0$, and currently, at $t = 0$, the value of GDP is Y_0. Is the objective of the policymaker feasible (provided the expert is right)?

   ii. Assume now that in each period the economy is exposed to a 'shock' $I_t = 2^t + t^2$, such that $4Y_{t+2} - 4Y_{t+1} + Y_t = I_t$. Is doubling of GDP feasible now? Can you find the period t when the value of Y_t will first exceed $2Y_0$?

(f) Solve the system of linear differential equations
$\dot{X}(t) = AX(t) + B$, where

$$X(t) = \begin{pmatrix} x_1(t) \\ x_2(t) \\ x_3(t) \end{pmatrix}, \quad A = \begin{pmatrix} 0 & 1 & 1 \\ 1 & 0 & 1 \\ 1 & 1 & 0 \end{pmatrix}, \quad B = \begin{pmatrix} 2 \\ 1 \\ 1 \end{pmatrix},$$

given that at $t = 0$, $X(0) = (1, -1, 0)^\mathsf{T}$.

Solutions to Sample Problem Set–IV

1. Problem group I.

   (a) At the point of tangency, $y_1(x) = y_2(x)$, and $y'_1(x) = y'_2(x)$. Thus $a = 1/2e$, and the point of tangency is $x = \sqrt{e}$, $y = 1/2$.

   (b) Apply linear approximation by first derivative at $x = 0$:
   $f(x) \approx f(0) + f'(0)x = a + x/2a$.

   (c) Two conditions should be satisfied: $x_0^2 = ax_0 + b$, and $2x_0 = a$; thus $a = 2x_0$ and $b = -x_0^2$.

   (d) $Det(A) = 15a + 15$. Therefore, if $a = -1$ then $rank(A) = 2$, otherwise $rank(A) = 3$.

   (e) Integrate by parts twice: $\int (x\ln(x))^2 dx = \frac{1}{3}x^3(\ln(x))^2 - \frac{2}{9}x^3\ln(x) + \frac{2}{27}x^3$. By Newton-Leibniz formula, $\int_1^e (x\ln(x))^2 dx = \frac{5}{27}e^3 - \frac{2}{27}$.

   (f) Minimum at $x = 0$, $f(0) = 0$, maximum at $x = 1$, $f(1) = 1$. Note that both at $x = 0$ and $x = 1$ the first derivative of f does not exist and, therefore, F.O.C are not applicable; however, F.O.C. detect the local maximum at $x = -1$, $f(-1) = 1/e^2$.

   (g) Evaluate $f'(x) = -\frac{8(x+1)^3}{(x-1)^5}$ and $f''(x) = \frac{(x+1)^2(16x+64)}{(x-1)^6}$. $f(x)$ attains minimum $f(x) = 0$ at $x = -1$; the inflection point $x = -4$, $f(-4) = \frac{81}{625}$; $f(x)$ is concave for $x < -4$ and convex for $x > -4$. As $x \to 1$, $f(x) \to \infty$ (vertical asymptote); as $x \to \pm\infty$, $f(x) \to 1$ (vertical asymptote).

   (h) $\Phi(\alpha, \beta, \gamma) = f(x, y, z) = f(\sqrt{\beta\gamma}, \sqrt{\alpha\gamma}, \sqrt{\alpha\beta})$

   $\Phi_\alpha = \frac{1}{2}\frac{\sqrt{\gamma}}{\sqrt{\alpha}}f_y + \frac{1}{2}\frac{\sqrt{\beta}}{\sqrt{\alpha}}f_z \Rightarrow \alpha\Phi_\alpha = \frac{1}{2}\sqrt{\alpha\gamma}f_y + \frac{1}{2}\sqrt{\alpha\beta}f_z = \frac{1}{2}yf_y + \frac{1}{2}zf_z$

   $\Phi_\beta = \frac{1}{2}\frac{\sqrt{\gamma}}{\sqrt{\beta}}f_x + \frac{1}{2}\frac{\sqrt{\alpha}}{\sqrt{\beta}}f_z \Rightarrow \beta\Phi_\beta = \frac{1}{2}\sqrt{\beta\gamma}f_x + \frac{1}{2}\sqrt{\alpha\beta}f_z = \frac{1}{2}xf_x + \frac{1}{2}zf_z$

   $\Phi_\gamma = \frac{1}{2}\frac{\sqrt{\beta}}{\sqrt{\gamma}}f_x + \frac{1}{2}\frac{\sqrt{\alpha}}{\sqrt{\gamma}}f_y \Rightarrow \gamma\Phi_\gamma = \frac{1}{2}\sqrt{\beta\gamma}f_x + \frac{1}{2}\sqrt{\alpha\gamma}f_y = \frac{1}{2}xf_x + \frac{1}{2}yf_y$

   Summing up, $\alpha\Phi_\alpha + \beta\Phi_\beta + \gamma\Phi_\gamma = xf_x + yf_y + zf_z$.

(i) $\lim\limits_{x\to 3_+} (x^2-5x+6)^{1/\ln(x-3)} = \lim\limits_{x\to 3_+} e^{\frac{\ln(x^2-5x+6)}{\ln(x-3)}} = \left|\text{L'Hôpital's rule}\right| =$

$\lim\limits_{x\to 3_+} e^{\frac{\frac{2x-5}{(x-2)(x-3)}}{\frac{1}{x-3}}} = \lim\limits_{x\to 3_+} e^{\frac{2x-5}{(x-2)}} = e.$

2. Problem group II.

  (a) By direct substitution, RHS=LHS.

   Just note that for the function $\phi = u(z)$ with $z = x + v(y)$, the chain rule implies that $\phi_x = u'$, $\phi_y = u'v'$, $\phi_{xx} = u''$, and $\phi_{xy} = u''v'$.

  (b) As x goes to infinity, either $f(x)$ is bounded or $f(x) \to \infty$. In the former case the ratio $f(x)/x$ obviously tends to zero; in the latter, L'Hôpital's rule yields the result.

  (c) A matrix equation $AXB = C$ has a solution $X = A^{-1}CB^{-1}$, provided A and B are non-singular. In our case, $X = \begin{pmatrix} 24 & 13 \\ -34 & -18 \end{pmatrix}$.

  (d) F.O.C. imply

   (a) $1 - \frac{y^2}{4x^2} = 0 \;\Rightarrow\; x = y/2$ (recall $x > 0, y > 0$);

   (b) $\frac{y}{2x} - \frac{z^2}{y^2} = 0 \;\Rightarrow\; z = y$;

   (c) $\frac{2z}{y} - \frac{2}{z} = 0 \;\Rightarrow\; z = 1$.

   Therefore, the only stationary pont is $(0.5, 1, 1)$. Checking for S.O.C. verifies that this stationary point is indeed minimum (that is, $d^2u(0.5, 1, 1) > 0$), and $u_{min} = u(0.5, 1, 1) = 4$.

  (e) Separating variables, we obtain $\dfrac{dy}{\sqrt{y}+1} = \dfrac{dx}{1+e^x}$

$$\int \frac{dy}{\sqrt{y}+1} = \begin{vmatrix} y = t^2 \\ dy = 2tdt \end{vmatrix} = \int \frac{2tdt}{t+1}$$

$$= 2 - 2\int \frac{dt}{t+1} = 2 - 2\ln(1+\sqrt{y})$$

$$\int \frac{dx}{1+e^x} = \begin{vmatrix} e^x = z \\ e^x dx = dz \\ dx = \frac{dz}{z} \end{vmatrix} = \int \frac{dz}{z(z+1)}$$

$$= \int \frac{dz}{z} - \frac{dz}{z+1} = \ln\frac{e^x}{1+e^x}$$

Finally, the solution is given by the implicit function

$2 - 2\ln(1 + \sqrt{y}) = \ln\frac{e^x}{1+e^x} + C.$

3. Problem group III.

(a) Construct the Hessian matrix:

$$H = K^\alpha L^\beta H^\gamma \begin{pmatrix} \frac{\alpha(\alpha-1)}{K^2} & \frac{\alpha\beta}{KL} & \frac{\alpha\gamma}{KH} \\ \frac{\alpha\beta}{KL} & \frac{\beta(\beta-1)}{L^2} & \frac{\beta\gamma}{LH} \\ \frac{\alpha\gamma}{KH} & \frac{\beta\gamma}{LH} & \frac{\gamma(\gamma-1)}{H^2} \end{pmatrix}$$

F is concave $\Leftrightarrow H$ is NSD $\Leftrightarrow -H$ is PSD $\Leftrightarrow$ all principal minors of $-H$ are non-negative. Direct computations show that the latter is equivalent to the conditions $0 \le \alpha \le 1$, $0 \le \beta \le 1, 0 \le \gamma \le 1, \alpha + \beta + \gamma \le 1$.

(For instance, let us compute the leading principal minors of $-H$:

$\Delta_1 = K^{\alpha-2}L^\beta H^\gamma \alpha(1-\alpha) \ge 0,$

$\Delta_2 = K^{\alpha-2}L^{\beta-2}H^\gamma \alpha\beta(1-\alpha-\beta) \ge 0,$

$\Delta_3 = K^{\alpha-2}L^{\beta-2}H^{\gamma-2}\alpha\beta\gamma(1-\alpha-\beta-\gamma) \ge 0.)$

(b) Eigenvalues of A are $\lambda_1 = -2$, $\lambda_2 = \lambda_3 = 4$. Corresponding orthogonal eigenvectors are

$$V_1 = \begin{pmatrix} -2/\sqrt{6} \\ 1/\sqrt{6} \\ 1/\sqrt{6} \end{pmatrix},$$

$$V_2 = \begin{pmatrix} 1/\sqrt{5} \\ 2/\sqrt{5} \\ 0 \end{pmatrix},$$

$$V_3 = \begin{pmatrix} 2/\sqrt{30} \\ -1/\sqrt{30} \\ 5/\sqrt{30} \end{pmatrix}$$

and the diagonalization matrix $U = (V_1, V_2, V_3)$. Therefore, $e^A = U' diag(e^{-2}, e^4, e^4)U$ (here $diag(...)$ denotes a diagonal matrix).

(c) The Implicit Function Theorem yields the F.O.C.:

$$\frac{\partial z}{\partial x} = \frac{2x - z + 2}{2z - x - y + 2} = 0,$$

$$\frac{\partial z}{\partial y} = \frac{2y - z + 2}{2z - x - y + 2} = 0.$$

Thus F.O.C. imply $y = x$, and $z = 2x + 2$. Substituting these values in the original explicit equation and solving for x we find two stationary points of $z(x, y)$: $x = y = -3 + \sqrt{6}$ and $x = y = -3 - \sqrt{6}$. We can check directly that $z_{min} = -(4 + 2\sqrt{6})$ at $x = y = -3 - \sqrt{6}$ and $z_{max} = 2\sqrt{6} - 4$ at $x = y = -3 + \sqrt{6}$.

(d) By setting $\dot{x} = 0$ and $\dot{y} = 0$ we find the equilibrium points of the system. That is,

$$\begin{cases} x^*y^* + 4 = 0 \\ x^{*2} + y^{*2} - 17 = 0 \end{cases}$$

(take square roots of the second $\Rightarrow$ line with added/subtracted first line multiplied by 2) $\begin{cases} x^* + y^* = \pm 3 \\ x^* - y^* = \pm 5 \end{cases}$

and there are four equilibrium points: $(-1, 4)$, $(1, -4)$, $(4, -1)$, $(-4, 1)$.

The matrix of the linearized system is $\begin{pmatrix} y^* & x^* \\ 2x^* & y^* \end{pmatrix}$. The characteristic equation associated with this matrix is $\lambda^2 - 3y^*\lambda + 2(y^{*2} - x^{*2}) = 0$, and its roots are

$$\lambda_{1,2} = \frac{3y^* \pm \sqrt{y^{*2} + 8x^{*2}}}{2}.$$

Solving for λ at different equilibrium points of (x^*, y^*) we find that

- at $(-1, 4)$ the eigenvalues are real and positive;
- at $(1, -4)$ the eigenvalues are real and negative;
- at $(4, -1)$ and $(-4, 1)$ the eigenvalues are real and of opposite signs.

Therefore, $(-1, 4)$ is an unstable node, $(1, -4)$ is an stable node, $(4, -1)$ and $(-4, 1)$ are saddle nodes.

(e)   i. Solve the difference equation $4Y_{t+2} - 4Y_{t+1} + Y_t = 0$. The roots of the characteristic equation $\lambda^2 - 4\lambda + 1 = 0$ are $\lambda_{1,2} = \frac{1}{2}$. Therefore, $Y_t = C_1 \left(\frac{1}{2}\right)^t + C_2 t \left(\frac{1}{2}\right)^t$, or, recalling that at $t = 0$, $Y_t = Y_0$, we evaluate $C_1 = Y_0$ to get $Y_t = Y_0(0.5)^t + C_2 t(0.5)^t$. Since both of the eigenvalues are less than one in absolute value, then Y_t is monotonically declining for $t > 1$ and the problem of doubling GDP in 10 years is not meaningful. (Technically, we can find such C_2 that $Y_{10} = 2Y_0$ but it just means that at $t = 1$ the output jumps to a very high level and from $t = 2$ onward will gradually decline.)

ii. The particular solution to the non-homogeneous equation $4Y_{t+2} - 4Y_{t+1} + Y_t = 2^t + t^2$ is searched in the form $at^2 + (bt^2 + ct + d)$. By direct substitution, we find $a = \frac{1}{9}$, $b = 1$, $c = -8$, $d = 20$. Since the particular solution increases as t increases, the doubling of GDP is feasible. However, the number of periods depends on C_2.

(f) Eigenvalues of A are $\lambda_1 = \lambda_2 = -1$, $\lambda_3 = 2$. There are two independent eigenvectors corresponding to $\lambda = -1$: $v_1 = (1, -1, 0)^{\mathsf{T}}$ and $v_2 = (1, 0, -1)^{\mathsf{T}}$. The eigenvector corresponding to $\lambda_3 = 2$ is $v_3 = (1, 1, 1)^{\mathsf{T}}$. The particular solution to the non-homogeneous equation is a constant vector $X_{\mathrm{p}} = -A^{-1}B = (0, -1, -1)^{\mathsf{T}}$. Thus, the general solution to the non-homogeneous system is

$$X(t) = C_1 \begin{pmatrix} 1 \\ -1 \\ 0 \end{pmatrix} e^t + C_2 \begin{pmatrix} 1 \\ 0 \\ -1 \end{pmatrix} e^{-t} + C_3 \begin{pmatrix} 1 \\ 1 \\ 1 \end{pmatrix} e^{2t} +$$

$$+ \begin{pmatrix} 0 \\ -1 \\ -1 \end{pmatrix}$$

From the initial conditions $X(0) = (1, -1, 0)^{\mathsf{T}}$ we find $C_1 = C_3 = \frac{2}{3}$ and $C_2 = -\frac{1}{3}$.

Sample Problem Set–V

1. For what values of the real parameter a does the system

$$\begin{cases} x+y+z &= a \\ ax-y+z &= -1 \\ x+y-z &= a \end{cases}$$

have a unique solution?

Does the system have a solution for $a = 1$? If yes, find this solution using the inverse matrix method.

2. Find all eigenvalues of the matrix

$$A = \begin{pmatrix} 2 & 2 & -2 \\ 2 & 5 & -4 \\ -2 & -4 & 5 \end{pmatrix}$$

Is it possible to diagonalize A (i.e. find matrix U such that $U'U = I$ and $U'AU$ is diagonal)? If yes, do so.

3. Given a polynomial function $f(x) = ax^4 + bx^3 + cx^2$, what restrictions (if any) should be imposed on the parameters a, b, and c, if we need this function to be convex for $x \geq 0$?

4. Find extrema of the function $x^2 + y^2$, subject to $ax^2 + by^2 = 1$, and classify them (here a and b are positive parameters). Illustrate your solution graphically.

5. The consumer's and producer's surplus in equilibrium is defined as the area in the positive quadrant, bounded by the demand and supply curves. If you are told that the demand function for a certain commodity as a function of price is linear, $Q^d = 11 - P$, and the industry supply of this commodity is subject to decreasing returns, $Q^s = 2\sqrt{P} + 3$, what is the value of the surplus?

6. Sketch the graph of the function $\quad y(x) = \dfrac{x-1}{(x-2)^2}$. Be as specific as you can.

Exercises

7. Prove or disprove:

    (a) If a matrix A has its inverse A^{-1}, then all eigenvalues of A are non-zero.

    (b) If x^* is a local maximum of $f(x)$ then $f''(x^*) < 0$.

    (c) If $f(x) = -f(-x)$ for all x then $\int_{-a}^{a} f(x)dx = 0$ for all a.

8. Find maximum and minimum of the function $f(x,y) = (x^2 + y^2)\ln(x^2 + y^2)$ in the closed domain $x^2 + y^2 \le 0.81$.

9. Let $\phi(\cdot)$ be an arbitrary differentiable function. Function z is implicitly defined as $\dfrac{z}{x} = \phi\left(\dfrac{y}{x}\right)$. Show that $z = x\dfrac{\partial z}{\partial x} + y\dfrac{\partial z}{\partial y}$.

10. Consider a cost function $f(x)$ and its average cost function $A(x) = \dfrac{f(x)}{x}$. Show that the necessary condition to minimize average cost is to equalize marginal cost and average cost. What restriction should you impose on $f(x)$ to make the necessary condition also a sufficient one?

11. Use the Lagrange multiplier method to write the first-order conditions for the maximum of the function $f(x,y) = \sqrt{x} + \sqrt{y}$, subject to $ax + y = 1$, where a is a real parameter.

    Are the first order conditions also sufficient?

    Could you show (without solving the first-order conditions explicitly) what happens with optimal x and y if a increases by a small number?

Solutions to Sample Problem Set–V

1.

$$
\det A = \det \begin{pmatrix} 1 & 1 & 1 \\ a & -1 & 1 \\ 1 & 1 & -1 \end{pmatrix} = \det \begin{pmatrix} 1 & 1 & 1 \\ a+1 & 0 & 0 \\ 1 & 1 & -1 \end{pmatrix}
$$

$$
= -(a+1)\det \begin{pmatrix} 1 & 1 \\ 1 & -1 \end{pmatrix} = 2(a+1).
$$

Therefore, $\det A = 0 \Leftrightarrow a = -1$.

If $a = 1$, $A = \begin{pmatrix} 1 & 1 & 1 \\ 1 & -1 & 1 \\ 1 & 1 & -1 \end{pmatrix}$ and $A^{-1} = \begin{pmatrix} 0 & 0.5 & 0.5 \\ 0.5 & -0.5 & 0 \\ 0.5 & 0 & -0.5 \end{pmatrix}$.

Thus $X = A^{-1}B = \begin{pmatrix} 0 & 0.5 & 0.5 \\ 0.5 & -0.5 & 0 \\ 0.5 & 0 & -0.5 \end{pmatrix} \begin{pmatrix} 1 \\ -1 \\ 1 \end{pmatrix} = \begin{pmatrix} 0 \\ 1 \\ 0 \end{pmatrix}$.

2. Calculating $\det(A - \lambda I)$, we get

$$\det \begin{pmatrix} 2-\lambda & 2 & -2 \\ 2 & 5-\lambda & -4 \\ -2 & -4 & 5-\lambda \end{pmatrix} = \det \begin{pmatrix} 2-\lambda & 2 & -2 \\ 0 & 1-\lambda & 1-\lambda \\ -2 & -4 & 5-\lambda \end{pmatrix} =$$

$$= (1-\lambda)\det \begin{pmatrix} 2-\lambda & 2 & -2 \\ 0 & 1 & 1 \\ -2 & -4 & 5-\lambda \end{pmatrix} = (1-\lambda)(\lambda^2 - 11\lambda + 10) =$$

$(1-\lambda)(\lambda - 1)(\lambda - 10)$.

Eigenvalues of A are $\lambda_1 = 10$, $\lambda_2 = \lambda_3 = 1$.

A is a real symmetric matrix, thus it should be diagonalizable, and U can be found as a normalized matrix formed by orthogonal eigenvectors.

The eigenvector corresponding to $\lambda = 10$ is $V_1 = \begin{pmatrix} 1 \\ 2 \\ -2 \end{pmatrix}$. $\lambda = 1$

is a repeated eigenvalue of order two, and we can find two linearly independent and orthogonal eigenvectors corresponding to this

repeated eigenvalue: $V_2 = \begin{pmatrix} 0 \\ 1 \\ 1 \end{pmatrix}$ and $V_3 = \begin{pmatrix} 4 \\ -1 \\ 1 \end{pmatrix}$.

Thus $U = \begin{pmatrix} 1/3 & 0 & 4/3\sqrt{2} \\ 2/3 & 1/\sqrt{2} & -1/3\sqrt{2} \\ -2/3 & 1/\sqrt{2} & 1/3\sqrt{2} \end{pmatrix}$.

You can check yourself that $U'AU = \begin{pmatrix} 10 & 0 & 0 \\ 0 & 1 & 0 \\ 0 & 0 & 1 \end{pmatrix}$.

3. $f(x) = ax^4 + bx^3 + cx^2$ is convex $\Leftrightarrow f''(x) \geq 0$, i.e. $12ax^2 + 6bx + 2c \geq 0$.

Suppose $a \neq 0$. Then the quadratic polynomial $12ax^2 + 6bx + 2c$ is non-negative $\forall x \geq 0$ if (i) $a > 0$ and (ii) either roots of this polynomial exist and both roots are positive, or the discriminant $D = (6b)^2 - 4 \cdot 12a \cdot 2c \leq 0$. The latter implies that $c \geq 0$ and

$$b \geq -\sqrt{\frac{8}{3}ac}.$$

If $a = 0$ then $6bx + 2c$ should be non-negative $\forall x \geq 0$, thus $b \geq 0$ and $c \geq 0$.

Summing up, the convexity of $f(x)$ requires $a \geq 0$, $c \geq 0$, and

$$b \geq -\sqrt{\frac{8}{3}ac}.$$

4. Without loss of generality let us assume $a < b$ (The case $a = b$ is meaningless because the objective function coincides with the constraint). Contours of the objective function are circles centered in the origin. The constraint is ellipse. To solve the problem, you may try the standard Lagrangian approach, but it is easier to see the solution graphically: Just plot contours of the objective function and the constraint.

Maximum is attained at $\left(\pm \dfrac{1}{\sqrt{a}}, 0 \right)$ and minimum is reached at $\left(0, \pm \dfrac{1}{\sqrt{b}} \right)$.

5. Q^d and Q^s schedules intersect at equilibrium price $P = 4$.

$$\text{Surplus} = \int_0^4 Q^d(P)dP - \int_0^4 Q^s(P)dP$$

$$= \left(11P - \frac{P^2}{2} \right) \Big|_0^4 - \left(\frac{4}{3}P^{3/2} - 3P \right) \Big|_0^4 = \frac{40}{3}.$$

6. $y(x) = (x-1)(x-2)^{-2}$, $y'(x) = -x(x-2)^{-3}$, $y''(x) = 2(x-1)(x-2)^{-4}$.

$y(x) = 0$ at $x = 1$.

$y'(x^*) = 0$ at $x^* = 0$; $y''(x^*) > 0 \Rightarrow x^* = 0$ is a local minimum; $y(0) = -1/4$.

$y''(-1) = 0 \Rightarrow x = -1$ is an inflection point; $y(-1) = -2/9$.

If $x > -1$ then $y''(x) > 0 \Rightarrow$ convexity; if $x < -1$ then $y''(x) < 0 \Rightarrow$ concavity.

As $x \to 2$, $y(x) \to \infty$; as $x \to \pm\infty$, $y(x) \to 0$.

7. (a) TRUE. A^{-1} exists $\Rightarrow \det A = \lambda_1 \cdot \ldots \cdot \lambda_n \neq 0 \Rightarrow \lambda_i \neq 0 \ \forall i = 1, 2, \ldots, n$.

   (b) FALSE. Counterexample: $f(x) = -x^4$ has maximum at $x^* = 0$, but $f''(0) = 0$.

   (c) TRUE. $\int_{-a}^{0} f(x)dx = -\int_{0}^{a} f(x)dx$.

8. Let $z = x^2 + y^2$. So, we have to find the minimum and maximum of $f(z) = z \ln z$ over a closed interval $z \in [0, 0.81]$.

F.O.C.: $f'(z) = \ln z + 1 = 0 \Rightarrow z^* = e^{-1} = 1/e$.

S.O.C.: $f''(z^*) = 1/z^* > 0 \Rightarrow z^*$ is minimum.

$$\lim_{z \to 0} f(z) = \lim_{z \to 0} z \ln z = \lim_{z \to 0} \frac{\ln z}{1/z} \overset{\text{L'Hopital's rule}}{=} \frac{1/z}{-1/z^2} = 0.$$

Since $f(0) = 0$, $f(0.81) < 0$, $f(z^*) = -z^* < f(0.81)$, we conclude that in $[0,0.81]$ $f(z)$ reaches its maximum at $z = 0$ (i.e. at $x = y = 0$); $f(z)$ is minimized over $[0,0.81]$ at $z^* = 1/e$ (i.e. at all points (x, y) such that $x^2 + y^2 = 1/e$).

9. Note that $z = x\phi(y/x)$.

$z_x = \phi(y/x) - \phi'(y/x) \cdot y/x;$ $z_y = \phi'(y/x)$.

Thus $xz_x + yz_y = z\phi(y/x) = z$.

10. F.O.C.: $A'(x) = 0 = \dfrac{f'(x)}{x} - \dfrac{f(x)}{x^2} = \dfrac{1}{x}(f'(x) - A(x))$.

Therefore, at optimal x^*, $f'(x^*) = A(x^*)$, i.e. we have just derived the standard economic principle of optimization: $MC = AC$.

Sufficient conditions for minimum: at x^*, $A''(x^*) > 0$.

$A''(x^*) = \dfrac{f''(x^*)}{x^*} - \dfrac{2f'(x^*)}{x^{*2}} + \dfrac{2f(x^*)}{x^{*3}} = \dfrac{f''(x^*)}{x^*}.$ (Here we take advantage of the F.O.C. at x^*)

Thus $A''(x^*) > 0 \Leftrightarrow f''(x^*) > 0$, i.e. $f(x)$ is convex at x^*.

11. The Lagrangian function is $L = \sqrt{x} + \sqrt{y} + \lambda(1 - ax - y)$. L is concave $\Rightarrow$ the F.O.C. are also sufficient.

F.O.C.: $\begin{cases} \frac{1}{2\sqrt{x}} - a\lambda &= 0, \\ \frac{1}{2\sqrt{y}} - \lambda &= 0, \\ 1 - ax - y &= 0. \end{cases}$ ($\leftarrow$note that this condition implies $\lambda > 0$)

Jacobian matrix:

$$J = \begin{pmatrix} -x^{-3/2}/4 & 0 & -a \\ 0 & -y^{-3/2}/4 & -1 \\ -a & -1 & 0 \end{pmatrix},$$

$$\det J = x^{-3/2}/4 + a^2 y^{-3/2}/4 > 0,$$

$$J^{-1} = \frac{1}{\det J} \begin{pmatrix} -1 & a & -ay^{-3/2}/4 \\ a & a^2 & -x^{-3/2}/4 \\ -ay^{-3/2}/4 & -x^{-3/2}/4 & (xy)^{-3/2}/16 \end{pmatrix}.$$

Finally, the implicit function theorem yields:

$$\begin{pmatrix} \frac{\partial x}{\partial a} \\ \frac{\partial y}{\partial a} \\ \frac{\partial \lambda}{\partial a} \end{pmatrix} = -J^{-1} \cdot \begin{pmatrix} -\lambda \\ 0 \\ -1 \end{pmatrix}.$$

$\frac{\partial x}{\partial a} = -(\lambda + ay^{-3/2}/4) < 0$, and the sign of $\frac{\partial y}{\partial a}$ is ambiguous.

Sample Problem Set – VI

1. Prove or disprove:

    (a) Given two square matrices A and B, $(A - B)^2 = A^2 + B^2$.

    (b) Let $u(x, y) = 0.5 \ln(x^2 + y^2)$. Then $u_{xx} + u_{yy} = 0$.

    (c) If $f(x)$ and $g(x)$ are two convex functions on R^n, then for any real positive numbers a and b, $af(x) + bg(x)$ is a convex function.

2. Show that two vectors (a, b) and (c, d) are linearly independent if and only if $ad - bc \neq 0$.

3. Consider a function $f(x) = \ln(x + 1)$. Show that $\lim_{x \to 0} \frac{f(x)}{x} = 1$. Further, without using a calculator, approximate $f(0.01)$.

4. Find the minimum and the maximum values of the function $f(x) = (x - 3)\sqrt{x}$ defined on $x \in [0, 4]$ and plot its graph.

5. Given a constant elasticity of substitution production function $F(K, L) = A(aK^\rho + bL^\rho)^{1/\rho}$, find the degree of its homogeneity.

6. Evaluate $\displaystyle\int_0^1 \frac{dx}{\sqrt{1 - x}}$

7. For a dynamic supply-demand model

$$Q_d = a + bP(t) + c\dot{P}(t),$$

$$Q_s = \alpha + \beta P(t),$$

find a rationale for the demand equation and solve the system for $P(t)$ in the equilibrium case $Q_d = Q_s$.

8. Suppose the dynamic behavior of two economic variables x and y is given by the autonomous system

$$\dot{x}(t) = f(x, y),$$

$$\dot{y}(t) = g(x, y).$$

What could be the appropriate phase diagram? (consider all possible cases and explain).

9. Solve the Metzler equation of inventory cycles

$$y_t = (2 + a)by_{t-1} - (1 + a)by_{t-2} + N$$

where $b \in (0, 1)$, $a > 0$, and N is a constant.

10. Find the optimal path of variables x and u in the optimal control problem

$$\max \int_0^1 (u - u^2)dt$$

subject to the constraint $\dot{x} = u$ and the terminal conditions $x(0) = 4$, $x(1) = 2$.

Solutions to Sample Problem Set – VI

1. (a) False; (b) True; (c) True.

2. By definition we have two vectors $x = (a, b)$ & $y = (c, d)$ linearly independent when $\alpha x + \beta y = \mathbf{0}$ holds if and only if $\alpha = 0$ & $\beta = 0$.

   Let us rewrite the condition $\alpha x + \beta y = \mathbf{0}$: $\alpha(a, b) + \beta(c, d) = (0, 0)$, that is,

$$\begin{cases} a \cdot \alpha + c \cdot \beta = 0 \\ b \cdot \alpha + d \cdot \beta = 0 \end{cases} \Rightarrow \begin{pmatrix} a & c \\ b & d \end{pmatrix} \cdot \begin{pmatrix} \alpha \\ \beta \end{pmatrix} = \mathbf{0}.$$

In order to satisfy the if and only if condition ($\alpha = 0$ & $\beta = 0$) we need to have $\begin{pmatrix} a & c \\ b & d \end{pmatrix}$ be nonsingular, that is, $ad - bc \neq 0$.

3. Taking the limit,

$$\lim_{x \to 0} \frac{f(x)}{x} = \lim_{x \to 0} \frac{\ln(x+1)}{x}$$
$$|\text{by L'Hôpital's Rule}|$$
$$= \lim_{x \to 0} \frac{\frac{1}{x+1}}{1} = \lim_{x \to 0} \frac{1}{x+1} = 1.$$

The derived result implies that for small x we can approximate $\ln(x + 1)$ by x. In particular, $f(0.01) = \ln(0.01 + 1) \approx 0.01$.

4. Rewriting, $f(x) = x^{\frac{3}{2}} - 3x^{\frac{1}{2}}$. It follows that $f'(x) = \frac{3}{2}x^{\frac{1}{2}} - 3x^{-\frac{1}{2}}$ and $f''(x) = \frac{3}{4}x^{-\frac{1}{2}} + \frac{3}{4}x^{-\frac{3}{2}}$. Equalizing the first derivative to zero we find a stationary point $x = 1$. The second derivative of f at this point is positive, therefore $f_{\min}(1) = -2$. Taking into account the values of the function at boundary points ($f(0) = 0$ and $f(4) = 2$) we conclude that on $[0,4]$ $f_{\min}(1) = -2$ and $f_{\max}(4) = 2$. Moreover, the function is strictly convex on the given interval $[0,4]$.

5. Homogeneous of degree one.

6. This improper integral is converging and its value equals 2.

7. Rationale for the demand equation could be as follows: If price is changing over time then the demanded quantity depends not only on the current price level but also on an intertemporal price change $\dot{P}$, i.e. the term $c\dot{P}$ reflects consumers' anticipations. The equilibrium condition $Q_s = Q_d$ yields $c\dot{P} + (b - \beta)P = \alpha - a$. The latter has a solution $P(t) = P_0 e^{-\frac{b-\beta}{c}t} + \frac{\alpha-a}{b-\beta}$, where the constant term P_0 is determined by the initial conditions.

8. First, assume that f and g are smooth enough to secure existence and uniqueness of a solution. If $f \neq 0$ then $dy/dx = g(x,y)/f(x,y)$. Local behavior in the neighborhood of stationary point(s) is defined by signs and values of the eigenvalues of the Jacobian matrix evaluated at the stationary point.

9. Let $\lambda_{1,2}$ be the roots of the characteristic equation $\lambda^2 - b(2+a)\lambda + b(1+a) = 0$, so that $\lambda_{1,2} = (b(2+a) \pm \sqrt{b^2(2+a)^2 - 4b(1+a)})/2$. If $b(2+a)^2 > 4(1+a)$ then $y_t = y_p + c_1\lambda_1^t + c_2\lambda_2^t$. If $b(2+a)^2 = 4(1+a)$ then $y_t = y_p + (c_1 + c_2 t)\lambda^t$. If $b(2+a)^2 < 4(1+a)$ then $y_t = y_p + \rho^t(c_1 \cos(\theta t) + c_+ c_2 \sin(\theta t))$, given complex roots of the characteristic equation. The particular solution $y_p = N/(1 - b(2+a) + b(1+a))$.

10. The Hamiltonian $H = u - u^2 + \lambda u$.

   F.O.C: $\dot{\lambda} = H_x = 0$ and $0 = H_u = 1 + \lambda - 2u$.

   The first equation means that λ is constant, and the second equation implies $u = (1+\lambda)/2$; thus from the equation of motion $\dot{x} = u$ it follows that $x(t) = (1+\lambda)t/2 + c$. From the terminal condition

$x(0) = 4$ and $x(1) = 2$ we find $\lambda = -5$ and $c = 4$. Therefore the optimal solution is $u_{opt} = -2$, $x_{opt} = -2t + 4$.

5.6   Unsolved Problems

5.6.1   More Problems

1. Find the maximum and minimum values of the function

$$f(x, y, z) = \sqrt{x^2 + y^2 + z^2}\, e^{-(x^2+y^2+z^2)}.$$

   (HINT: Introduce the new variable $u = x^2 + y^2 + z^2$; note that $u \geq 0$.)

2. a) In the method of least squares in regression theory the curve $y = a + bx$ is fit to the data (x_i, y_i), $i = 1, 2, \ldots, n$ by minimizing the sum of squared errors

$$S(a, b) = \sum_{i=1}^{n}(y_i - (a + bx_i))^2$$

   by the choice of two parameters a (the intercept) and b (the slope). Determine the necessary condition for minimizing $S(a, b)$ by the choice of a and b, solve for a and b and show that the sufficient conditions are met.

   b) The continuous analog of the discrete case of the method of least squares can be set up as follows:

   Given a continuously differentiable function $f(x)$, find a, b which solve the problem

$$\min_{a,b} \int_0^1 (f(x) - (a + bx))^2 dx.$$

   Again, find the optimal a, b and show that the sufficient conditions are met.

   c) Compare your results.

3. This result is known in economic analysis as *Le Chatelier Principle*. But rise to the challenge and figure out the answer yourself!

Consider the problem

$$\max_{x_1, x_2} F(x_1, x_2) = f(x_1, x_2) - w_1 x_1 - w_2 x_2,$$

where the Hessian matrix of the second order partial derivatives $\left(\frac{\partial^2 f}{\partial x_i \partial x_j} \right)$, $i, j = 1, 2$, is assumed negative definite and w_1, w_2 are given positive parameters.

a) What are the first-order conditions for a maximum?
b) Show that x_1 can be solved as a function of w_1, w_2 and $\frac{\partial x_1}{\partial w_1} < 0$.
c) Suppose that to the problem is added the linear constraint $x_2 = b$, where b is a given non-zero parameter. Find the new equilibrium and show that:

$$\left(\frac{\partial x_1}{\partial w_1} \right)_{\text{without added constraint}} \leq \left(\frac{\partial x_1}{\partial w_1} \right)_{\text{with added constraint}} < 0.$$

4. Consider the following minimization problem:
   min $(x - u)^2 + (y - v)^2$ s.t. $xy = 1$, $u + 2v = 1$.
   a) Find the solution.
   b) How can you interpret this problem geometrically? Illustrate your results graphically.
   c) What is the solution to this minimization problem if the second constraint is replaced with the constraint $u + 2v = 3$? In this case, do you really need to use the Lagrange multiplier method?

   (HINT: Given two points (x_1, y_1) and (x_2, y_2) in the Euclidean space, what is the meaning of $\sqrt{(x_1 - x_2)^2 + (y_1 - y_2)^2}$?)

5. Solve the integral equation

$$\int_0^x (x - t) y(t) dt = \int_0^x y(t) dt + 2x.$$

   (HINT: Take the derivatives of both parts with respect to x and solve the differential equation.)

6. a) Show that Chebyshev's differential equation

$$(1 - x^2)y'' - xy' + n^2 y = 0$$

(where n is constant and $|x| < 1$) can be reduced to the linear differential equation with constant coefficients

$$\frac{d^2 y}{dt^2} + n^2 y = 0$$

by substituting $x = \cos t$.

b) Prove that the linear differential equation of the second order

$$a(x)y''(x) + b(x)y'(x) + c(x)y(x) = 0$$

where $a(x)$, $b(x)$, $c(x)$ are continuous functions and $a(x) \neq 0$, can be reduced to the equation

$$\frac{d}{dx}\left(p(x)\frac{dy}{dx}\right) + q(x)y(x) = 0.$$

(HINT: In other words, you need to prove the existence of a function $\mu(x)$ such that $p(x) = \mu(x)a(x)$, $q(x) = \mu(x)c(x)$ and $\frac{d}{dx}(\mu(x)a(x)) = \mu(x)b(x)$.)

7. Find the equilibria of the system of differential equations

$$\dot{x} = \sin x, \quad \dot{y} = \sin y,$$

and classify them.

Draw the phase portrait of the system.

8. Given the matrix

$$A = \begin{pmatrix} 2 & 1 & 1 \\ 0 & 3 & \alpha \\ 2 & 0 & 1 \end{pmatrix},$$

a) find all values of α, for which this matrix has three distinct real eigenvalues.

b) let $\alpha = 1$. If I tell you that one eigenvalue $\lambda_1 = 2$, find two other eigenvalues of A and the eigenvector, corresponding to λ_1.

9. You are given the function $f(x) = xe^{1-x} + (1-x)e^x$.

   a) Find $\lim_{x \to -\infty} f(x)$, $\lim_{x \to +\infty} f(x)$

   b) Write down the first order necessary condition for maximum.

   c) Using this function, prove that there exists a real number c in the interval $(0, 1)$, such that $1 - 2c = \ln c - \ln(1 - c)$.

10. Find local extrema of the function

$$u(x, y, z) = 2x^2 - xy + 2xz - y + y^3 + z^2$$

and classify them.

11. Consider a problem of extremizing the objective function $U(x, y) = x^2 + y^2$ subject to the constraint $ax^2 + by^2 \in [\underline{M}, \overline{M}]$, where $a, b, \underline{M}$, and $\overline{M}$ are positive parameters, $\underline{M} < \overline{M}$.

   Give a geometric interpretation of the problem and its solution(s), and solve it analytically to find the maxima and minima of $U(x, y)$. Prove that the problem indeed has maximum (minimum).

12. Let $f_1(x_1, \ldots, x_n), \ldots, f_m(x_1, \ldots, x_n)$ be convex functions. Prove that the function

$$F(x_1, \ldots, x_n) = \sum_{i=1}^{m} \alpha_i f_i(x_1, \ldots, x_n)$$

is convex, if $\alpha_i \geq 0$, $i = 1, 2, \ldots, m$.

Is the statement true for quasi-convex functions?

13. Find the absolute minimum and maximum of the function

$$z(x, y) = x^2 + y^2 - 2x - 4y + 1$$

in the domain, defined as $x \geq 0$, $y \geq 0$, $5x + 3y \leq 15$ (a triangle with vertices in $(0,0)$, $(3,0)$ and $(0,5)$).

(HINT: Find local extrema, then check the function on the boundary and compare).

14. Consider the following optimization problem:

    extremize $f(x,y) = \frac{1}{2}x^2 + e^y$ subject to $x + y = a$, where a is a real parameter.

    Write down the first order necessary conditions.

    Check whether the extremum point is minimum or maximum.

    What happens to the optimal values of x and y if a increases by a very small number?

    (HINT: Don't waste your time trying to solve the first-order conditions.)

15. Solve the differential equation $y''(x) - 6y'(x) + 9y = xe^{ax}$, where a is a real parameter. Consider all possible cases!

16. a) Solve the system of linear differential equations $\dot{x} = Ax$, where

$$A = \begin{pmatrix} 2 & -1 & 1 \\ 1 & 2 & -1 \\ 1 & -1 & 2 \end{pmatrix}, \quad x = \begin{pmatrix} x_1(t) \\ x_2(t) \\ x_3(t) \end{pmatrix}.$$

    b) Solve the system of differential equations $\dot{x} = Ax + b$ where A is as in part a) and $b = (1, 2, 3)'$.

    c) What is the solution to the problem in part b) if you know that

$$x_1(0) = x_2(0) = x_3(0) = 0?$$

17. Solve the difference equation $3y_{k+2} + 2y_{k+1} - y_k = 2^k + k + (1/3)^k$.

18. Find the equilibria of the system of non-linear differential equations, classify them, and draw the phase diagram:

$$\begin{aligned} \dot{x} &= \ln(1 - y + y^2), \\ \dot{y} &= 3 - \sqrt{x^2 + 8y}. \end{aligned}$$

5.6.2 Sample Tests
Sample Test – I

1. Easy problems:

   a) Check whether the function $f(x,y) = 4x^2 - 2xy + 3y^2$, defined for all $(x, y) \in \mathbf{R}^2$, is concave, convex, strictly concave, strictly convex, or neither. At which point(s) does the function reach its maximum (minimum)?

   b) Evaluate the indefinite integral $\int x^3 e^{-x^2} dx$.

   c) You are given the matrix A and the real number x, $x \in \mathbf{R}$:

   $$A = \begin{pmatrix} 0 & 0 & 2 & 4 \\ 1 & 3 & 0 & 0 \\ 0 & 0 & 1 & 3 \\ 3 & x & 0 & 0 \end{pmatrix}.$$

   For which values of x is the matrix non-singular?

   d) The function $f(x,y) = \sqrt{xy}$ is to be maximized under the constraint $x + y = 5$. Find the optimal x and y.

2. Using the inverse matrix, solve the system of linear equations

   $$\begin{cases} x + 2y + 3z &= 1, \\ 3x + y + 2z &= 2, \\ 2x + 3y + z &= 3. \end{cases}$$

3. Find the eigenvalues and the corresponding eigenvectors of the following matrix:

   $$\begin{pmatrix} 2 & 2 \\ 2 & -1 \end{pmatrix}.$$

4. Sketch the graph of the function $y(x) = \dfrac{2 - x}{x^2 + x - 2}$. Be as specific as you can.

5. Given the function $\dfrac{1 - e^{-ax}}{1 + e^{ax}}$, $a > 1$

a) Write down the F.O.C. of an extremum point.

b) Determine whether an extremum point is maximum or minimum.

c) Find $\dfrac{dx}{da}$ and determine the sign of $\dfrac{dx}{da}$.

6. The function $f(x, y) = ax + y$ is to be maximized under the constraints

$x^2 + ay^2 \leq 1$, $x \geq 0$, $y \geq 0$, where a is a positive real parameter.

a) Use the Lagrange multiplier method to write down the F.O.C. for the maximum.

b) Taking as given that this problem has a solution, find how the optimal values of x and y change if a increases by a very small number.

7. a) Solve the differential equation $y''(x) + 2y'(x) - 3y(x) = 0$.

b) Solve the differential equation $y''(x) + 2y'(x) - 3y(x) = x - 1$.

8. Solve the difference equation $x_t - 5x_{t-1} = 3$.

9. Consider the optimal control problem

$$\max_{u(t)} I(u(t)) = \int_0^{+\infty} (x^2(t) - u^2(t))e^{-rt}dt,$$

subject to

$$\frac{dx}{dt} = au(t) - bx(t),$$

where a, b and r are positive parameters.

Set up the Hamiltonian function associated with this problem and write down the first-order conditions.

Sample Test – II

1. Easy problems:

   a) For what values of p and q is the function
   $$f(x,y) = \frac{1}{x^p} + \frac{1}{y^q}$$
   defined in the region $x > 0,\ y > 0$
   a) convex,
   b) concave?

   b) Find the area between the curves $y = x^2$ and $y = \sqrt{x}$.

   c) Apply the characteristic root test to check if the quadratic form $Q = -2x^2 - 2y^2 - z^2 + 4xy$ is positive (negative) definite.

   d) A consumer is known to have a Cobb-Douglas utility of the form $u(x,y) = x^\alpha y^{1-\alpha}$, where the parameter α is unknown. However, it is known that when faced with the utility maximization problem
   $$\max u(x,y) \text{ subject to } x + y = 3,$$
   the consumer chooses $x = 1, y = 2$. Find the value of α.

2. Find the unknown matrix X from the equation
   $$X \begin{pmatrix} 1 & 1 & -1 \\ 2 & 1 & 0 \\ 1 & -1 & 1 \end{pmatrix} = \begin{pmatrix} 1 & -1 & 3 \\ 4 & 3 & 2 \\ 1 & -2 & 5 \end{pmatrix}.$$

3. Find local extrema of the function $z = (x^2 + y^2)e^{-(x^2+y^2)}$ and classify them.

4. A curve in the $(x - y)$ plane is given parametrically as $x = a(t - \sin t)$, $y = a(1 - \cos t)$, a is a real parameter, $t \in [0, 2\pi]$.

   Draw the graph of this curve and find the equation of the tangent line at $t = \pi/2$.

5. In the method of least squares in regression theory the curve $y = a + bx$ is fit to the data (x_i, y_i), $i = 1, 2, \ldots, n$ by minimizing the sum of squared errors
   $$S(a,b) = \sum_{i=1}^{n} (y_i - (a + bx_i))^2$$

by the choice of two parameters, a (the intercept) and b (the slope).

Determine the necessary condition for minimizing $S(a, b)$ by the choice of a and b.

Solve for a and b and show that the sufficient conditions are met.

6. Use the Lagrange multiplier method to write the first-order conditions for the maximum of the function

$$f(x, y) = \ln x + \ln y - x - y, \text{ subject to } x + cy = 2,$$

where c is a real parameter, $c > 1$.

If c is such that these conditions describe the maximum, show what happens to x and y if c increases by a very small number.

7.   a) Solve the differential equation $y''(x) - 2y'(x) + y(x) = 0$.

   b) Solve the differential equation $y''(x) - 2y'(x) + y(x) = xe^x$.

8. Solve the difference equation $x_{t+1} - x_t = b_t, \quad t = 0, 1, 2, \ldots$.

9. Write down the Kuhn-Tucker conditions for the following problem:

$$\max_{x_1, x_2} 3x_1 x_2 - x_2^3,$$

$$\text{subject to } 2x_1 + 5x_2 \geq 20, \quad x_1 - 2x_2 = 5, \quad x_1 \geq 0, x_2 \geq 0.$$

Sample Test – III

1. Solve any three of the given four problems:

   (a) Find the critical value(s) of the function $u = xy^2 z^3(a - x - 2y - 3z)$ and classify them (a is a positive parameter).

   (b) Solve the difference equation $y_{k+2} + 2y_{k+1} + y_k = 3^k + 3k + 10$.

   (c) Solve the differential equation $2xy' + y^2 = 1$.

(d) In economics, the two most common ways of averaging values $x_1, x_2, \ldots, x_n$ are to compute the arithmetic average $\bar{x} = \frac{1}{n}(x_1 + \ldots + x_n)$, and the geometric average $\hat{x} = \sqrt[n]{x_1 \ldots x_n}$. Show that $\hat{x} \le \bar{x}$.

2. Prove or disprove any four out of five statements below:

(a) For any non-singular $[n \times n]$ matrix A and any real number λ, $\lambda \ne 0$,

$$\det(A^{-1} - \lambda I) = (-\lambda)^n \cdot \det(A^{-1}) \cdot \det\left(A - \frac{1}{\lambda}I\right).$$

(b) In the limit $x^n e^{-ax} \to 0$ for any positive a and n as $x \to \infty$.

(c) $\int_o^x e^{x^2} dx \sim \frac{1}{2x} e^{x^2}$ as $x \to \infty$.

(d) A concave function is not necessarily quasi-concave.

(e) Given a differential equation $y' = f(x, y)$ where the function $f(x, y)$ is continuous, for any point (x_0, y_0) in the $(X - Y)$ plane you can always find a unique solution $y = y(x)$ to this differential equation such that $y(x_0) = y_0$.

Try to be specific by justifying in each case your answer.

3. For the following system of linear differential equations

$$\frac{dx}{dt} = 4x - y - z,$$

$$\frac{dy}{dt} = x + 2y - z,$$

$$\frac{dz}{dt} = x - y + 2z$$

(a) Find all eigenvalues and eigenvectors of the matrix of the system.

(b) Find the general form of the solution.

(c) Could you find a solution that at $t = 0$ passes through the origin?

(d) Does the system have an equilibrium? If so, what can you say about the stability of the equilibrium?

4. Solve any of the two static optimization problems:

(a) Consider two-commodity utility maximization problem

$$\max_{x,y} U(x, y) = xy^2$$

$$\text{s.t. } x + 2y = M$$

where income level M is assumed to be strictly positive.

   i. Write down the first-order conditions (F.O.C.) for this problem.
   ii. Use the implicit function theorem to derive dx/dM, dy/dM, $d\lambda/dM$ at the point of optimality (here λ is the Lagrange multiplier (shadow price)).

(b) For the nonlinear optimization problem

$$\max_{x,y}(x + \ln y)$$

$$\text{s.t. } x + y \leq 4, \ x + 2y \leq 6, \ x \geq 0, \ y \geq 0$$

   i. Check whether this problem has a unique optimal solution.
   ii. Write down the Kuhn-Tucker conditions and find the optimal solution numerically.

5. Solve the following dynamic optimization problem:

$$\max \int_0^1 (2x - u^2)dt$$

$$\text{s.t. } \dot{x} = u, \quad x(0) = 1, \quad x(1) \text{ free}, \quad x \geq u$$

MATHEMATICS FOR ECONOMISTS MADE SIMPLE

Viatcheslav V. Vinogradov

Charles University in Prague
Karolinum Press
Ovocny trh 3–5, 116 36 Prague 1, Czech Republic
Prague 2010

Vice-rector-editor Professor PhDr. Mojmir Horyna
Editor Sergey Slobodyan
Cover Katerina Rezacova
Print by Karolinum Press
First edition

ISBN 978-80-246-1657-5